LIMNOLOGY
Current Perspective

LIMNOLOGY
Current Perspective

Editor
Dr. Vishwas B. Sakhare
Head,
Post Graduate Department of Zoology
Yogeshwari Mahavidyalaya,
Ambajogai – 431 517
Maharashtra

2011
DAYA PUBLISHING HOUSE®
Delhi - 110 035

© 2011 VISHWAS BALASAHEB SAKHARE (1974–   )
ISBN 9789351240921

Proceedings of National Conference on Current Perspectives in Limnology 25th–26th December 2009.

Published by	:	**Daya Publishing House®**
		A Division of
		Astral International Pvt. Ltd.
		– ISO 9001:2008 Certified Company –
		4760-61/23, Ansari Road, Darya Ganj,
		New Delhi - 110 002
		Phone: 23245578, 23244987
		Fax: (011) 23260116
		E-mail: dayabooks@vsnl.com
		website: www.dayabooks.com
Laser Typesetting	:	**Classic Computer Services**
		Delhi - 110 035
Printed at	:	**Chawla Offset Printers**
		Delhi - 110 052

PRINTED IN INDIA

Dedicated to My Guru

Late Dr. M.S. Kodarkar

Secretary
Indian Association of Aquatic Biologists, Hyderabad
For his constant support and encouragement in all academic endeavours

Preface

Freshwater is one of our most precious, yet undervalued resources. Although water may seem like a renewable resources, in fact only 2.5 per cent of the water on earth is suitable for drinking. Furthermore, only 0.01 per cent of the total amount of drinkable water is readily accessible. In other words, the world's supply of freshwater is indeed limited. Inland waters are a potential source for a variety of food organisms. Basic limnological knowledge is essential not only for sound management of our aquatic resources, but also, more importantly, for the development of aquaculture industry. Freshwater, critical for human survival is facing alround degradation all over the world. India is not an exception to this global phenomenon. Deterioration of water quality, loss of biodiversity and fast depletion of water resources are the main challenges which need urgent attention. Not only step taking is essential but also there is a early need to evolve guidelines and strategies for conservation and sustainable development of the precious ecosystems.

Considering the plethora of problems, Department of Zoology, Yogeshwari Mahavidyalaya, Ambajogai in collaboration with Indian Association of Aquatic Biologists, Hyderabad organized a National Conference on Current Perspectives in Limnology, during 25th–26th December 2009, at Ambajogai (Maharashtra).The conference had attracted participants from many parts of the country and also from abroad. It was organized with the financial assistance from University Grants Commission, Western Regional Office, Pune. The conference was organized into 7 sessions *viz.*, aquatic biodiversity and conservation, fisheries and aquaculture, aquatic pollution, estuarine ecosystems, stygobiology, eco-restoration and habitat protection management. Forty five papers were presented and 130 participants took part in the deliberations. The participants were from Indian Council of Agricultural Research, Marine Products Development Authority, Zoological Survey of India, Non-Governmental Organizations, Universities and Colleges.

This volume is based on the proceedings of the conference. It is hoped that the contents of this volume would provide a new orientation and impetus to freshwater fisheries development in India and offer better inputs for scientific management and utilization of the resources on a long term sustainable basis, contributing to food security, economic viability of operations, ecological and environmental safety.

The present book is compendium of recent original research in the field of limnology. It is an assemblage of up-to-date information of rapid advances and developments taking place in the field of limnolgy. The book is a unique compilation of 40 chapters which discusses exhaustive studies on planktonology, physico-chemical environment, fisheries, toxicology, physiology etc. With its application oriented and interdisciplinary approach, the book would be immensely useful to students, teachers, researchers, scientists, policymakers, environmental lawyers and others interested in limnology.

I am able to release this publication within a very short period from the date of conference. In many cases, the research papers could not be sent back to the authors for incorporating the modifications suggested by the experts and such corrections had to be made by me. Therefore, I am equally responsible for any mistakes that might have crept in, despite my efforts to avoid them. Any suggestions on the contents and presentation of this document will be welcome.

I am grateful to distinguished scientists who had sent messages and good wishes for the conference. Iam thankful to a number of limnologists from various parts of the country for discussing and finalizing the recommendations of the conference. Thanks are also due to the eminent scientists who had delivered the inaugural address, the key note address, and the valedictory address at the conference.

I wish to acknowledge with the gratitude the splendid contributions of Prof. M. B. Mule, Dr. G. D. Khedkar and Dr. R.J. Chavan of Dr. Babasaheb Ambedkar Marathwada University(Aurangabad), Dr. Indranil Ghosh of West Bengal University of Animal and Fishery Sciences (Kolkata), Dr. C.M. Bharmbe of Vidnyan Mahavidyalaya (Malkapur), Dr. P.K. Joshi, Dr. Shivajirao Pawar and Dr. Md. Babar of Dnyanopasak College (Parbhani), Dr. M. S. Kodarkar of Indian Association of Aquatic Biologists (Hyderabad), Dr. Vishwas Shembekar of Rajarshi Shau College (Latur), who skillfully chaired the different sessions of the conference.

Last but not least, Iam thankful to Dr. S. P. Chavan of Dnyanopasak College (Parbhani), Dr. J. M. Gaikwad and Dr. Sunil Ahirrao of Shri Shivaji College (Parbhani), Dr. R. P. Mali of Yeshwant College (Nanded), Dr. Milind Girkar and Dr. V. B. Sutar of College of Fisheries (Udgir), Dr. V. Ravi and P. Murugesan of CAS in Marine Biology, Annamalai University (Parangipettai), Dr. M. G. Babare of A. S. C. College (Naldurg), Dr. M. N. Kadam of R. P. College (Osmanabad) and Dr. R. J. Chaudhari of Marathwada Agricultural University (Parbhani) for giving me encouragement and valuable suggestions.

I wish to express my sincere gratitude to Dr. S. T. Khursale (President), Yogeshwari Education Society, Ambajogai, Adv. V. K. Chousalkar (Secretary), Adv. Shivajirao Karhad (Vice President), Shri V. P. Jaju (Vice President), Shri P. Y. Dixit (Treasurer), Prof. S. D. Moharir (Joint Secretary) of the College governing body for their interest, inspiration and guidance in the organization of the conference.

I feel it's my duty to express heartfelt gratitude towards our Principal, Dr. P.S. Prayag. I shall never foreget his valuable inspiring guidance, constant encouragement, keen interest and constructive criticism for organization of national conference.

My departmental colleagues and research scholars deserve much appreciation and thanks for their wholehearted cooperation. I am deeply indebted to all the organizations and firms who have helped us raise the funds through advertisements.

Dr. Vishwas B. Sakhare

Contents

List of Contributors

Ahirrao, S.D.
Fisheries Research Laboratory, Shri Shivaji College, Parbhani – 431 401

Ambore, N.E.
Post Graduate Department of Zoology, Yeshwant Mahavidyalaya, Nanded – 431 604

Babar, Md.
Department of Geology, Dnyanopasak College, Parbhani – 431 401

Babare, M.G.
A.S.C. College, Naldurg

Babare, R.S.
PG Department of Zoology and Fishery Science, DSM'S College of ACS, Parbhani – 431 401

Bhagwan, H.K.
Post Graduate Department of Zoology, S.M.D. Mohekar Mahavidyalaya, Kallam – 413 507,

Bharambe, C.M.
Department of Zoology, Vidnyan Mahavidyalaya, Malkapur – 443 101

Bhosale, P.R.
Department of Zoology, Dr. Babasaheb Ambedkar Marathwada University, Aurangabad – 431 004

Buktar, P.P.
Department of Zoology, Yogeshwari Mahavidyalaya, Ambajogai – 431 517

Chaudhari, R.J.
Department of Zoology, Dr. Babasaheb Ambedkar Marathwada University, Aurangabad – 431 004

Chavan, R.J.
Department of Zoology, Dr. Babasaheb Ambedkar Marathwada University, Aurangabad – 431 004

Chavan, S.P.
PG Department of zoology and Fishery Science, DSM'S College of A.C.S., Parbhani – 431 401

Deshmukh, M.P.
PG Department of zoology and Fishery Science, DSM'S College of A.C.S., Parbhani – 431 401

Deshmukh, R.S.
Department of Botany, B. Raghunath College, Parbhani – 431 401

Divde, A.A.
Department of Environmental Science, Yeshwant College, Nanded – 431 604

Gaikwad, A.M.
Department of Zoology, Dr. Babasaheb Ambedkar Marathwada University, Aurangabad – 431 004

Ghosh, Indranil
Department of Aquaculture, West Bengal University of Animal Sciences and Fisheries, Kolkata, West Bengal.,

Hande, G.J.
Department of Zoology, The Institute of Science, Mumbai

Ingole, S.B.
Shri. Siddheshwar Mahavidhyalaya, Majalgaon, District Beed

Jadhav, H.K.
A.S.C. College, Naldurg, Dist. Osmanabad

Jadhav, R.R.
Zoology Research Center, A.S.C. College, Naldurg, Dist. Osmanbad

Jagtap, A.R.
Post Graduate Department of Zoology, Vivek Vardhini Day College, Jambagh, Hyderabad – 95

Jagtap, V.A.
S.G.R.G. Shinde Mahavidyalaya, Paranda, District Osmanabad

Jaiswal, N.R.
Research and P.G. Department of Zoology, Yeshwant Mahavidyalaya, Nanded – 431 603

Jawale, R.S.
Department of Zoology, Dr. Babasaheb Ambedkar Marathwada University, Aurangabad – 431 004

Jetithor, S.G.
Post Graduate Department of Zoology, Yogeshwari Mahavidyalaya, Ambajogai – 431 517

Jindal, Meenakshi
Department of Zoology and Aquaculture, CCS Haryana Agricultural University, Hisar – 125 004

Joshi, P.K.
Department of Zoology, D.S.M.College, Parbhani – 431 401

Kadam, G.A.
Shri. Siddheshwar Mahavidhyalaya, Majalgaon, Dist. Beed

Kadam, M.S.
Post Graduate Department of Zoology, Yeshwant Mahavidyalaya, Nanded – 431 603

Kadam, S.U.
Department of Zoology and Fishery Science, Dnyanopasak College, Parbhani – 431 401

Kamble, S.M.
Post Graduate Department of Zoology, S.M.D. Mohekar Mahavidyalaya, Kallam – 413 507

Kashid, N.G.
Department of Botany, Vasant Mahavidyalaya, Kaij, Dist. Beed

Khaire, B.S.
Department of Zoology, Anandrao Dhonde Alias Babaji College, Kada – 414 202

Khobragade, Kshama
S.B.E.S. College of Science, Aurangabad – 431 001

Khole, A.M.
Department of Zoology, B. Raghunath College, Parbhani – 431 401

Kodarkar, M.S.
Member, Sci-Com, ILEC, Japan, Secretary, Indian Association of Aquatic Biologists (IAAB), P.O. Box – 517, Putlibowli Post Office, Jambagh, Hyderabad – 500 095

Kolekar, M.A.
Department of Animal Biotechnology, College of Veterinary and Animal Sciences, Parbhani – 431 402

Kothole, S.D.
P.G Department of Zoology, Vivek Vardhini Day College, Jambagh, Hyderabad – 95

Kulkarni, G.K.
Department of Zoology, Dr. Babasaheb Ambedkar Marathwada University, Aurangabad – 431 004

Kumar, Birendra
Department of Zoology, B.N. Mandal University, Madhepura – 854 105,

Lokhande, M.V.
Department of Zoology, Indira Gandhi (Sr) College, CIDCO, New Nanded – 431 603

Mali, R.P.
Department of Zoology, Yeshwant Mahavidyalaya, Nanded – 431 602

Mallikarjuna, Ashashree Hassan
Department of Zoology, Sahyadri Science College (Autonomous), Shivamogga – 577 201

Mane, U.H.
Department of Zoology, Dr. Babasaheb Ambedkar Marathwada University, Aurangabad – 431 004

Menkudale, D.V.
Adarsh Mahavidyalaya, Omerga, Dist. Osmanbad

Mohekar, A.D.
Department of Zoology, S.M.D.M. College, Kallam – 413 507

Mohite, V.T.
Thakur College of Science and Commerce, Kandivali (E), Mumbai – 400 101

Morey, C.D.
Department of Zoology, Vidnyan Mahavidyalaya, Malkapur – 443 101

Murugesan, P.
Centre of Advanced Study in Marine Biology, Annamalai University, Parangipettai – 608 502

Naik, S.R.
Yeshwant Mahavidhyalaya, Nanded – 431 602

Niture, S.D.
Department of Zoology, J.E.S College, Jalna – 431 203

Parra, Luciano
Universidad Católica de Temuco, Facultad de Recursos Naturales, Escuela de Ciencias Ambientales, Casilla 15 – D, Temuco, Chile

Parveen, S.R.
Department of Environmental Science, Yeshwant College, Nanded – 431 604

Pathan, T.S.
Department of Zoology, Dr. Babasaheb Ambedkar Martahwada University, Aurangabad – 431 004

Patil, B.V.
Department of Zoology, Vidnyan Mahavidyalaya, Malkapur – 443 101

Patil, D.W.
Department of Aquaculture, College of Fisheries, Shirgaon, Ratnagiri – 415 612

Patil, N.B.
Department of Zoology, Vidnyan Mahavidyalaya, Malkapur – 443 101

Pawar, C.V.
Department of Zoology, S.C.S. Colloge, Omerga, Dist. Osmanbad

Pawar, J.H.
Department of Zoology, Vidnyan Mahavidyalaya, Malkapur – 443 101

Pawar, V.B.
D.S.M. College of Arts, Commerce and Science, Parbhani – 431 401

Pedge, S.S.
Department of Zoology, Shri Shivaji College, Parbhani – 431 401

Pulle, J.S.
Head, Department of Chemistry, SGBS College, Purna (Jn), Dist. Parbhani

Rahate, N.R.
Department of Zoology, Vidnyan Mahavidyalaya, Malkapur – 443 101

Rakh, R.R.
Department of Microbiology, GBS College, Purna (Jn), Dist. Parbhani

Rathod, D.S.
Department of Zoology and Fishery Science, Rajurshi Shahu College, Latur – 413 512

Raut, K.S.
Department of Zoology, L.L. Deshmukh Mahila College, Parli (V)

Ravi, V.
Centre of Advanced Study in Marine Biology, Annamalai University, Parangipettai – 608 502

Ríos, Patricio De los
Universidad Católica de Temuco, Facultad de Recursos Naturales, Escuela de Ciencias Ambientales, Casilla 15 – D, Temuco, Chile

Sabry, S.W.
Department of Environmental Science, Yeshwant College, Nanded – 431 604

Sakhare, V.B.
Post Graduate Department of Zoology, Yogeshwari Mahavidyalaya, Ambajogai – 431 517

Salok, D.P.M.
Department of Zoology, Vidnyan Mahavidyalaya, Malkapur – 443 101

Sawant, P.L.
B.S.S. Arts and Science College, Makani, Dist. Osmanabad

Sayed, J.A.
Department of Environmental Science, Yeshwant College, Nanded – 431 604

Shaikh, Afsar
Post Graduate Department of Zoology, Vivek Vardhini Day College, Jambagh, Hyderabad – 95

Shaikh, Y.A.
Department of Zoology, Dr. Babasaheb Ambedkar Marathwada University, Aurangabad – 431 004

Sharma, Kavita
Department of Zoology and Aquaculture, C.C.S. Haryana Agricultural University, Hisar – 125 004

Shelke, S.R.
P.G. Department of Zoology and Fishery Science, D.S.M.'S College of A.C.S., Parbhani – 431 401

Shembekar, V.S.
Department of Zoology and Fishery Science, Rajurshi Shahu College, Latur – 413 512

Shinde, S.E.
Department of Zoology, Dr. Babasaheb Ambedkar Marathwada University, Aurangabad – 431 004

Singh, H.
Department of Aquaculture, College of Fisheries, Shirgaon, Ratnagiri – 415 612

Sirsath, D.B.
Head, Department of Zoology, Kai Shankarrao Gutte College, Dharmapuri, District Beed

Sonawane, D.L.
Department of Zoology, Dr. Babasaheb Ambedkar Marathwada University, Aurangabad – 431 004

Sunnap, N.V.
P.G. Department of Zoology, Yeshwant Mahavidyalaya, Nanded – 431 603

Suryawanshi, G.D.
Post Graduate Department of Zoology, Yogesgwari Mahavidyalaya, Ambajogai – 431 517

Tabassum, Alfia
Department of Zoology, Vidnyan Mahavidyalaya, Malkapur – 443 101

Thakur, Abhay
Department of Zoology, The Institute of Science, Mumbai

Thorat, D.H.
Department of Zoology, Yogeshwari Mahavidyalaya, Ambajogai – 431 517

Vega, Marcela
Universidad Católica de Temuco, Facultad de Recursos Naturales, Escuela de Ciencias Ambientales, Casilla 15 – D, Temuco, Chile

Viadya, R.B.
Department of Zoology, The Institute of Science, Mumbai

Waghmare, V.N.
P.G. Department of Zoology, Yeshwant College, Nanded – 431 601

Zambre, S.P.
Department of Zoology, Dr. Babasaheb Ambedkar Marathwada University, Aurangabad – 431 004

Chapter 1

Application of Limnological Studies for Conservation of Urban Lakes in India: A Case Study of Lake Hussainsagar, Hyderabad, Andhra Pradesh

☆ *M.S. Kodarkar*

Introduction

Limnology, the science of lakes, is emerging as one of the most important subjects as the lentic ecosystems hold more than 96 per cent freshwater used by man. Further, lake based primary and secondary productivity supports aquaculture and fishery contributing substantially to food security and fight against protein deficiency and malnutrition. Brakatullah University, Bhopal, Madhya Pradesh, is the first University in India to introduce Limnology as an independent discipline.

Scientific information and knowledge should form the basis for policy formulation and decision making, however, often such knowledge remains buried in research papers and theses and never finds an application. Tus there is great need of application of limnological knowledge and information in decision and policy making processes to ensure sustainability. How laborious this process can be realized from case study of lake Washington by Dr. Edmondson. It took 16 years and a lot of scientific evidence to prove that phosphates released by basin detergent industry that is responsible for algal blooms in the lake. Of course, the industry was shifted down stream and blooms disappeared saving people dependent on the lake from health and environmental hazards.

Lake Hussainsagar is an ecological, social, recreational and cultural land mark on the map of historical city of Hyderabad. Unfortunately, in the last 60 years the lake is subjected to un-precedented environmental degradation from sewage and industrial waste. A number of actions based on technological approaches were implemented on the lake without much effect on the key factor *i.e.* improvement of lake water quality.

On this background, actions based on ecosystem approach were proposed and are being partly implemented. The limnology based actions are showing promising results. However, ecology based solutions need longer gestion period and sustained action. It is advocated that an integrated approach based on technology and eco-technology could be effective in mitigation of pollution impacts and restoration of goods and services of a lake ecosystem.

Hussainsagar, the picturesque lake situated between the twin cities of Hyderabad and Secunderabad, is an ecological and cultural landmark on the map of state of Andhra Pradesh, India (Figure 1.1). It was constructed in 1562 mainly to store drinking water brought from the river Musi, a tributary of Krishna, one of the major rivers of South India. The lake represents one of the thousands of impoundments on Deccan plateau in peninsular India, developed for storage of surface water run off in this semi-arid region with an average of 800 mm annual rainfall (Table 1.1).

Table 1.1: Physico-graphic Features of Lake Hussainsagar

Year of construction	1562	Average depth	5.2 m
Basin area	240 km²	Depth variable	1 to 12 m
Direct Catchment area	67 km²	Storage volume (spill)	28.6 X 10⁶ m³
Shoreline length	14 km	Maximum operating level	514.93 m
Maximum water area	5.7 km²	Normal operating level	513.43 m
Capacity	27.1 million m³	Road bund level	5.18.16 m

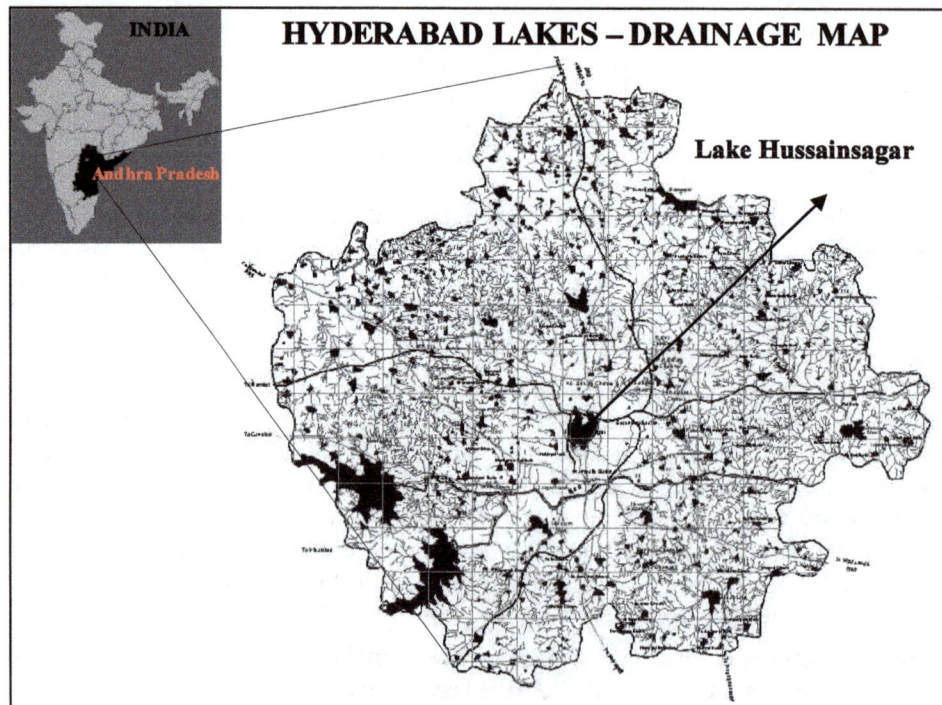

Figure 1.1: Lake Hussainsagar: Drainage Basin of Hyderabad and Satellite Imagery of the Lake

Located 15°N 78°E, the lake is 510 mt. above the mean sea level (MSL) the lake basin is bounded on west by Banjara hills. The 275 Km² watershed of Hussainsagar is divided into four sub basins *viz.* Kukatpally, Dullapally, Bowanpally and Yusufguda. The highest peak in the catchment is at 642 m north and lowest contour near tank bund at 500 meters, the effective north south drop being 142 m covering a distance of 17 km.

The lake hydrology is sustained by four feeding channels (nullahs); Kukatpally (70 mld), Picket (5.7), Banjara (6) and Balkapur (13.3 mld). Of the four in-lets, Kukatpally with total catchment of 168 Km² passes through two major industrial areas *viz.* Kukatpally and Balanagar and is the main feeding channel of the lake (Table 1.2).

Table 1.2: Dry Weather Flows in to Lake Hussainsagar

Channel	Flow in Mld			Remarks
	Sewage	Industrial Effluents	Total	
Picket	05.7	–	05.7	Interception and diversion after pumping Proposed STP (30 mld capacity)
Kukatpally	55.0	15.0	70.0	Interception and diversion (I and D)
Banjara	06.0	–	06.0	Interception and diversion (I and D)
Balkapur	13.3	–	13.3	STP (20 mld capacity)
Total	80.0	15.0	95.0	50 mld treatment by 2 STPs

In the last few years intensive efforts have been made to restore the lake environment including management of lake basin and water quality. Further, protection of lake Hussainsagar in terms of improved water quality, control of pollution and beautification of its environment has assumed greater significance in view of its socio-cultural and tourism potential. In this context Asia's biggest monolithic Buddha statue standing majestically on the rock of Gibraltar in the center of the lake, has emerged as the major attraction like the Statue of Liberty in USA. The extensive development of lake environment with the help of National and International funding has bought Hyderabad on the tourism map of the world and the lake has emerged as the landmark representing socio-cultural ethos of the historical city and the region.

Issues and Problems

Reclamation and Encroachments

With phenomenal urbanization, once peripheral lake finds itself in the densely populated zone of the mega-city and in the last 50 years there has been drastic reduction in its morphometry due to encroachments and large scale reclamation for developmental activities around the lake ecosystem (Table 1.3). What stands today as the lake area is hardly 2/3rd of its original in the middle of last centuary.

In view of its ecological, economical and recreational importance conservation of lake Hussainsagar is high on the agenda of the State Government and to achieve the same, a special Buddha Purnima Project Authority (BPPA) was established in 2000 to look after the lake and its environment covering special development area of 902 hectors.

Table 1.3: Developments Around the Lake

Name	Area in Hectors (Acres)	Characteristics
NTR Garden	13.736 hectors (34 acres)	Greenery, Party zone, Recreation
NTR Memorial	0.81 ha. (2 acres)	Greenery and memorial of former Chief Minister Dr N.T.Rama Rao
Lumbini Park	2.025 ha (5 acres)	Floral clock, Fountain, Toy train, Recreation, Jetty for boating on the lake, Laser show
Sanjeevaiah Park	36.45 ha (90 acres)	Lung space, Sprawling garden, Palm garden, Recreational centre
P.V. Memorial	1.1745 ha (2.9 acres)	Memorial of former Prime Minister of India, P.V.Narsimha Rao
Necklace road	1.1458 ha (3.6 acres)	Garlanding road on rear side of the lake, recreational zone
People's plaza	1.4175 ha (3.5 acres)	Promenades and areas for exhibitions
Wetland eco-conservation zone	5.0615 ha (12.5 acres)	Great floral diversity, '*in situ*' conservation, greenery

Pollution from Sewage and Industrial Effluents

One of the major impacts of urbanization and industrialization on the lake was in the form of poor water quality due to pollution. Since beginning of the last centaury when lake basin was undergoing rapid change very little attention was paid to this vital issue. Thus, failure of lake basin management in terms of proper sewerage system and industrial waste disposal infrastructure, was responsible for rapid degradation of lake environment.

Two sub-basins, Kukatpally and Dullapally, of the lake are highly industrialized zones. The Kukatpally sub-basin has three industrial areas viz, Kukatpally, Balanagar and Sanathnagar while in Dullapally one sub-basin, Jeedimetla, are developed as industrial hubs under a planned programme of industrialization. The range of products manufactured by 300 odd industrial units include chemical reagents, organics, pharmaceuticals, drugs, bio-chemicals, synthetic chemicals, detergents, aircraft batteries, distillation products, alloys and rubber products and effluents generated bring in a cocktail of toxic waste in to the lake. Though interception, diversion and treatment of industrial effluents was undertaken as a part of management intervention, the volume of waste generated exceeds treatment capacity of CETP in the Jeedimetla area and toxic effluents continue entering the lake through Kukatpally stream, the main feeding channel.

Siltation from Natural and Cultural Factors

Soil erosion due to construction activities in the catchment generates a lot of silt which along with the surface runoff ultimately ends up into the lake. Similarly, traditional festivals like Ganesh and Durga Puja conclude by immersion of massive idols of the deities in the lake. These events apart from addition of tons of silt also pollute the lake by floral offerings, paints, pigments, wooden and iron frames that go into making of idols, year after year.

Eutrophication

Lake Hussainsagar has been a subject of extensive research for the last 50 years and earlier reports on its limnology clearly indicate that up to 1950 it was virtually pollution free. Subsequent degradation of the ecosystem is directly linked to nutrient loading from domestic sewage, the magnitude of which could be assessed from a study that has estimated daily influx of 1041 kg of Phosphates and 1204 kg Nitrates in to the lake (Kodarkar *et al.*, 1991; Reddy *et al.*, 2002). Of the enormous amounts of nutrients entering the lake most are trapped in the sediment, while a fraction enters the food chain and webs sustaining eutrophic state of the water body (Zafar,1959,1966; Associated Industrial Consultant Pvt. Ltd., 1993).

Some of the perceptible manifestations of eutrophication are (a) Algal blooms and wild growth of aquatic macrophytes like water hyacinth, (b) Breeding of vectors, (c) Foul smells and (d) Fish kills (Kodarkar *et al.*, 1991). The Pollution has also an adverse effect on the biodiversity of the lake and consequent disruptions in bio-geo-chemical cycling of organic load on the lake.

Groundwater Pollution

Due to inadequate treatment and disposal facilities, groundwater pollution from seepage of effluents let out in open drains is a common phenomenon (Kodarkar *et al.*, 1991). For example, groundwater along seepage zone of Kukatpally stream, the main feeding channel of lake Hussainsagar, is heavily polluted with all physico-chemical parameters exceeding the WHO and IS standards (Table 1.4).

Table 1.4: Groundwater Quality in the Kukatpally Industrial Area in the Catchment

Parameter	WHO	IS	Average	Range
pH	7.5	8.5	8.06	7.3–9.0
TDS	500	500	827	435–1531
Alkalinity	–	–	313	200–475
Hardness	150	300	249	160–440
Sulphates	400	250	618	242–985
Nitrates	45	20	57	40–90
Nitrites	–	–	2.3	0.12–5.8
Phosphates	0.1	0.1	0.48	0.11–0.68
Chlorides	250	25	358	10–760

All values except pH in mg/L.

Breeding of Vectors

The nutrient-rich lake supports wild growth of aquatic weeds like water hyacinth and, in turn, provide ideal breeding sites for vectors (mosquitoes and snails) of diseases like Malaria, Dengue fever and Filariasis. Lake pollution thus has a very heavy cost in terms of expenditure on public health.

Loss to Lake Dependent Communities

Fishermen, washer-men and small dairy farmers are traditional lake dependent socio-economically backward communities and degradation of the lake has direct effect on their livelihood. Since studies in 1980s that indicated contaminated nature of fish, fishing activities in the lake are banned and once flourishing fishermen community has almost disappeared. Historically, the lake is a great community asset, particularly for floating population with no access to water and its pollution has deprived people at large ecological, economical and recreational benefits bestowed by a healthy lake ecosystem.

Strategies and Actions

In the South Asian context, lake management is relatively a young subject and unique by virtue of its highly inter-disciplinary and cross sectoral nature. Since 1990 after pollution of Hussainsagar reached an alarming proportion, culminating in to massive fish kills of 1993, otherwise complacent state government was forced in to action. Credit for this turn around also should go to civil society movements and Judicial interventions in response to Public Interest Litigation (PIL).

Apart from earlier actions for conservation of the lake, presently Hyderabad Urban Development Authority (HUDA) with support from Japan Bank of International Cooperation (JBIC) has initiated an ambitious project titled–Hussainsagar lake and the catchment area improvement initiative in 2006. The 3700 US$ (370 crores rupees) project has following important components:

Interception and Downstream Diversion of Sewage and Industrial Effluents

At six locations along the main in-flow channels I and D weirs are proposed to prevent pollution of the lake from sewage and industrial effluents.

Establishment of a Proposed and Upgradation of Existing Sewage Treatment Plants (STP)

To sustain hydrology of the lake in 1998 an STP of 20 mld capacity was commissioned, treated water from which is let in to the lake to sustain its hydrology (Table 1.5).

Table 1.5: The Water Quality Before and After Treatment by the STP

Parameter	Influents	Effluents	Per cent Reduction	Limits
pH	7.15	7.51	–	7–8
Suspended Solids	252	20	92	30
BOD	250	8	96.8	20
COD	540	46	91	250
Phosphates	5.9	2.5	58	5
TKN	48	5	90	100

Further, a second STP of 30 mld capacity is planned at the entry point of Picket stream. With this total 50 mld of treated water will be let in to the lake for sustaining its hydrology (Table 1.3). This measure is also expected to improve the lake water quality to the level of SW-II; suitable for bathing, contact water sports and commercial fishing (Table 1.6).

Table 1.6: Central Pollution Control Board (CPCB) Water Quality Standards/Norms

Parameter	Surface Water Quality of Hussainsagar Lake	Water use SW-II (Bathing)	CPCB Revised Criteria		
			A Excellent	B Desirable	C Acceptable
pH	7.6	6.5–8.5	–	–	–
BOD	30–48	<3	<2	<3	<6
DO	0.99	>4	>90 per cent	>80 per cent	>60 per cent
Turbidity	High	<30	–	–	–
Coliform count	>1600	<100	<20	<200	<2000

Lake Shore Line Improvement

1.2 km stretch will be developed by beautification of lake front through development of paved walk ways, railings, fountains and pavilions.

Dredging of 1,000,000m³ Silt

Massive removal of sediment in the 500 m radius from 4 nalla confluence is proposed to reduce nutrient and toxic waste load on the lake.

Solid Waste Management

A comprehensive programme of collection and disposal of solid waste from catchment localities and lake environment will be implemented with the help of community participation to rid the lake of solid waste pollution.

Environmental Awareness and Community Participation

The lake environ has been promoted as people's plaza for socio-cultural events. Annual lake festival and establishment of Lake Interpretation Centre (LIC) for spreading awareness about benefits from the lake and it's role in the over all water economy of the city. A lake information centre will be established to propagate the environmental awareness among visitors and communities around the lake.

Ecosystem Approach

Ecosystem approach underlines the idea of facilitating the nature to sustain structural and functional integrity of natural ecosystems. Too much anthropogenic pressure beyond carrying capacity of natural ecosystems is identified as the basic cause for their constant and sometimes irreversible degradation. Internationally, Convention on Biological Diversity (CBD) was perhaps the first effort to integrate ecology with management of natural and man-made ecosystems on the earth. Article 2 of the Convention, while elaborating on ecosystem approach has defined "Ecosystem" as a dynamic complex of plant, animal and micro-organism communities and their non-living environment interacting as a functional unit. The ecosystem approach is a strategy for the integrated management of land, water and living resources that promotes conservation and sustainable use in an equitable way. It is based on the application of appropriate scientific methodologies focused on levels of biological organization which encompass the essential processes, functions and interactions among organisms and their environment. It recognizes that humans, with their cultural diversity, are an integral component of ecosystems.

The ecosystem approach underscores the importance of basic principles on which ecosystems function and emphasizes the socio-economic dimensions of nature management when implementing the CBD. It further advocates that human life, activities and well-being must be included as basic factors in the wider geographical application of the ecosystem approach. Biodiversity has to be integrated into the economy of the relevant communities, and the various values of biodiversity should be captured and realized at the local level to give the right incentives to those that are nearest to guard it.

At the heart and soul of ecosystem approach is the biodiversity which is the life insurance of life itself. The intra-specific diversity is the insurance for the species survival in difficult times, the inter-specific diversity is the guarantee for ecosystem functioning and services, and the variation of functional ecosystems is the life insurance for sustainable development.

Ecosystem Approach and Conservation of Hussainsagar Lake

A review of conservation efforts on Hussainsagar undertaken so far with National and International funding clearly bring out the fact that most of the work done is of civil engineering nature and did not comprehensively address two basic issues *viz*. Lake basin and water quality managements.

The basin of Hussainsagar lake is highly urbanized and heavily industrialized. Further, lack of comprehensive management of sewage and toxic industrial waste are the basic reasons for environmental degradation of the lake ecosystem. There is also a historical perspective to this situation; after independence of India, industrialization was considered as the major activity to generate employment and work for millions. Secondly since developments were mainly concentrated in the urban areas, population from impoverished countryside migrated in large numbers to urban centers in search of better livelihood. These developments led to large-scale development of urban centers and when such developments happened in the catchment of rivers and lakes like Hussainsagar, the water bodies were the victims of pollution and environmental degradation.

Actions Proposed Based on Ecosystem Approach

Lake Basin Management

A lake is reflection of its catchment and the developments in the latter has a direct bearing on the status of the former. Such developments in the catchment include urbanization, industrialization, waste generation and topographical alterations affecting the in-lets channels draining the catchment. Lack of proper lake basin management invariably leads to sewage, industrial effluents and solid waste finally ending up in to the water body. Thus eutrophication, toxification of water and siltation are direct results, in turn, leading to reclamation and loss of precious water body. In the last 50 years hundreds of lakes are lost all over the country.

1. *Sewage management:* Decentralization of sewage management and introduction of septic tank system could reduce sewage related problems of eutrophication. Further, wherever possible emphasis should be on recycle and reuse of sewage as a resource.

2. The catchment topography needs to be protected so that rain water flows in to the lake to sustain its hydrology. Laying of proper sewage network is necessary and separation of sewage lines from storm water drains needs priority.

3. For reducing pressure on water supply system rain water harvesting, recycle and reuse needs to be given priority.

4. Industries should be encouraged to adapt zero discharge policies through effective use of water in its processes

5. Common Effluent Treatment Plant (CETP) to treat effluents through public-private partnership should be encouraged to minimize effluent load on the lake.

6. Sewage Treatment through STP needs to be complemented through introduction of cost effective eco-technologies for restoration of the lake water quality.

Water Quality

Restoration of water to the quality to SW-II level so as to use the lake for its intended purpose is at the heart of conservation of Hussainsagar. This needs actions to reduce nutrients load and toxic materials through remedial measures both at catchment and water body levels. Following are suggested actions for improvement of water quality in the lake.

Maintenance of Bioconservation Zone

A well de-marked bio-conservation zone around the lake extending between 100 to 1000 meters in width depending on topography, will act as an effective barrier to moderate the negative impacts from developmental activities in the lake environment. In this eco-sensitive zone only plantation and other eco-friendly activities should be allowed.

Treatment of Sewage in the Course of its Flow in to the Lake by Green Bridge Filtration System

It can be integrated as a part of nalla (stream) improvement proposed in the JBIC funded project. As effective as STP, the green bridge technology is not only cost effective and eco-friendly but also effectively treat entering sewage to the SW-II grade quality as recommended by the Ministry of Environment and Forests (MOEF), Government of India (GOI) in its revised guidelines for National Lake Conservation Programme (NLCP) (Figure 1.2).

Construction of Green Bridge Filtration System (GBFS)

The system consists of an easy four step filtration process involving a green Bridge and linearly arranged three stages of the fine gravel (size 10mm to 40mm), the course gravel (size 40mm to 60mm) and the large sized gravel. The Green Bridge filter consists of layers of Coir over which a bed of floating aquatic weeds (*e.g.* water hyacinth) are compactly woven to form a bridge. All the floatable and suspended solids are trapped in this biological bridge and the turbidity of water is reduced. The aquatic weeds are very efficient at nutrient stripping. The over-flowing sewage passes over three sections of gravels during which microbes and algae remove the nutrients. Finally, the filtered water is sent to a pond where the organic matter will settle down by the process of sedimentation and clear water will flow into the lake.

Micro Habitat and Feeding Ground Along Shore Line

Shore line along margins of a lake generates feeding habitats and help in reducing nutrients due to development of benthic food chain/web. In the case of lake Hussainsagar such habitats can be created along shore line; particularly the zone from where watermen community is relocated down stream. Such sand-gravel shore line provides habitats for development of micro and macro-benthos and create feeding ground for resident and migratory birds visiting the lake.

Phytoremediation

Macrophytes in the lake are important components of biotic community and should not be viewed as a menace. These species basically absorb nutrients and toxicants from the water thus improving its quality. If allowed to grow in controlled manner and regularly harvested and composted, there is a great possibility of utilizing this natural resource for improvement of lake water quality. Allowing towable flotillas of weeds like water hyacinth and other floating species will not only enhance beauty but also rid the lake of excess of nutrients and toxic materials.

Introduction of Composite Fish Culture

Fishes constitute very important biotic community in a lake. They not only harvest nutrients and live biomass but also provide much needed proteins. Hussainsagar lake once use to harbour 27 species of fishes (Babu Rao and Siva Reddy, 1984).

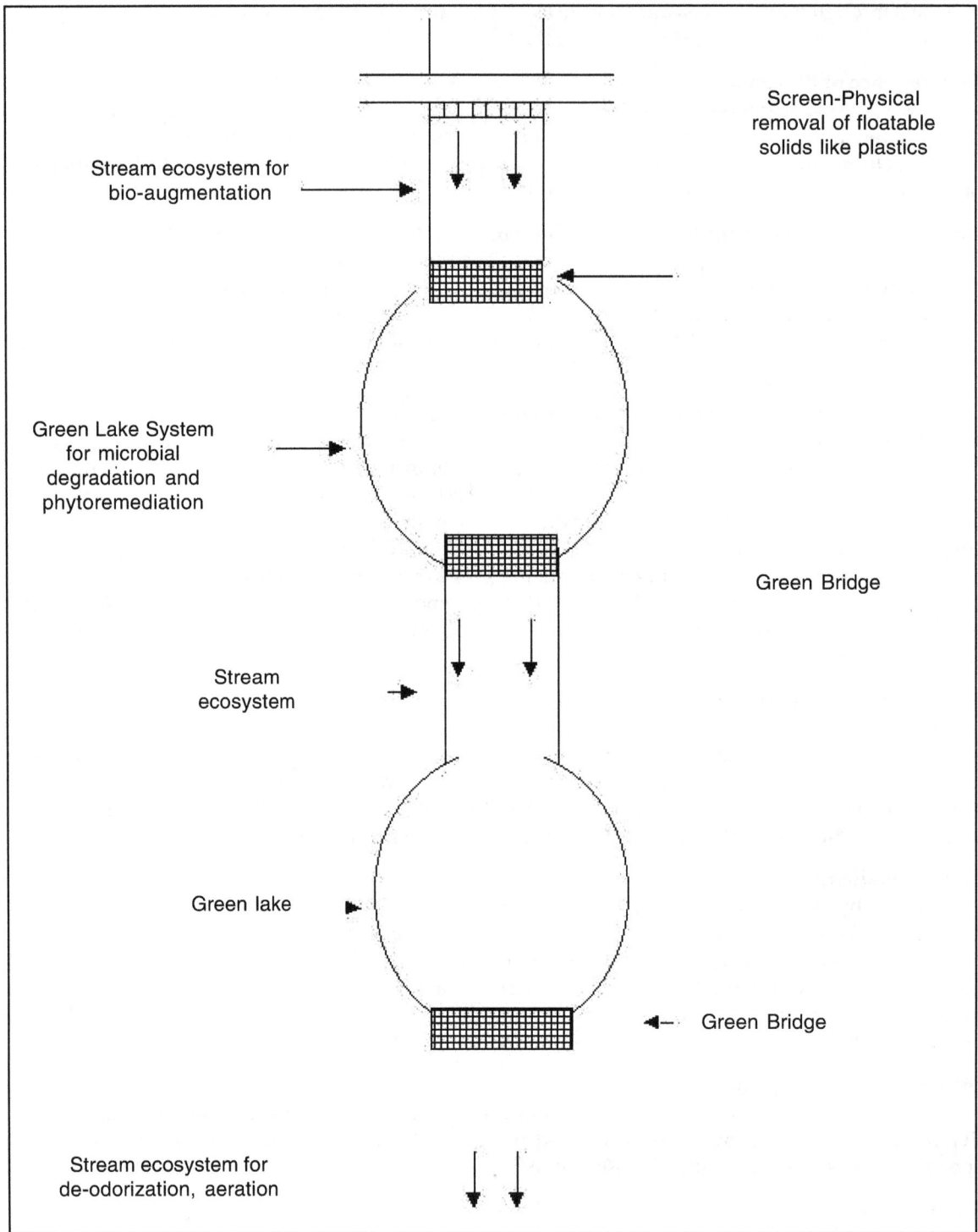

Screen-Physical removal of floatable solids like plastics

Stream ecosystem for bio-augmentation

Green Lake System for microbial degradation and phytoremediation

Green Bridge

Stream ecosystem

Green lake

Green Bridge

Stream ecosystem for de-odorization, aeration

Figure 1.2

Table 1.7: Fish Diversity of Lake Hussainsagar, Hyderabad, India
(Babu Rao and Siva Reddy, 1984)

Species	*Species*
Channa punctatus (Bloch)	*Channa marulius* (Ham)
Channa striata (Bloch)	*Channa orientalis* (Schneider)
Hetereopneustes fossilis (Bloch)	*Clarias batrachus* (Lin)
Mystus vittatus (Bloch)	*Mystus bleekeri* (Day)
Wallago attu (Schneider)	*Chela bacaila* (Ham)
Amblypharyngodon mola (Ham)	*Rasbora daniconius* (Ham)
Puntius sophore (Ham)	*Puntius ticto* (Ham)
Puntius chola (Ham)	*Puntius amphibian* (Valencinennes)
Puntius sarana (Ham)	*Puntius filamentosus* (Valenciennes)
Labeo boggut (Sykes)	*Gerra mullya* (Sykes)
Lepidocephalus (Lepidocephallchthys) *guntaa* (Ham)	*Gambusia affinis patruelis* (Baird and Girard)
Poecilla reticulata (Peters)	*Etroplus maculates* (Bloch)
Glossogobius giuris (Ham)	*Osphronemus goramy lecepode*
Notopterus notopterus (Pallas)	

However, recent survey has reported loss of sensitive species and existence of only few species of hardy cat fishes belonging to the genera *Channa, Clarias, Mystus,* and *Heteropneustes.*

Once the lake use to support flourishing fishermen community but with on-set of pollution and reports of contaminated fish, the fishery was banned. There is a need to reassess the situation and introduction of large scale composite fish culture. To begin with some controlled cage culture can be introduced to begin with and based on the results fisheries should be introduced on large scale.

Aeration

Low levels of Dissolved Oxygen (DO) often lead to anaerobic conditions, interruption of normal bio-geochemical cycling of energy and matter and reduction of biodiversity. Aeration in such situation can improve DO levels and thereby reduction of BOD of water.

Establishment of Lake Conservation Society (Hussainsagar Sarovar Samvardhini) to Ensure Peoples Participation

It is envisioned as a common platform to bring together diverse stake holders on a common platform to develop common action plan, sectoral conflict resolution and a mechanism to create environmental awareness.

Lessons Learned

As per one prediction, India sub-continent will be a water stressed region by 2017 with decline of per capita availability of water from 2,200 m^3 in 2000 to 1600 m^3 in 2017. By 2025 water will be the limiting factor for development and sustainability in all the sectors of national activities. On this background there is an urgent need to conserve and manage available freshwater resources in the form mainly of rivers, lakes and reservoirs.

Fortunately with reference to conservation and management of lakes in the last one decade a number of initiatives with the help of National, International and private funding are being undertaken. A look at the actions implemented under these programmes brings out a common pattern in which funds are mainly utilized for beatification of lake environment which includes a number of civil engineering actions like strengthening of bund, laying of ring roads, developments of gardens, pathways, promenades and other recreational facilitates. The two key aspects as out lined by International Lake Environment Committee (ILEC) Foundation, Japan, *viz.* lake basin management and water quality are not adequately addressed. Further, limno–logically these actions transform of a lake in tank with loss of self-sustaining capacity of the ecosystem. Unlike a lake, tanks need annual/ periodical maintenance and finances are seldom allocated for such activity. Thus, key to successful lake management lies on approaching conservation based on an integrated ecosystem approach.

Traditionally in India water sector is dominated by the people with engineering and technological backgrounds who often have very less knowledge about the ecological functions that sustain structural and functional integrity of ecosystems like lakes, reservoirs and wetlands. Further, in conservation and management actions cost intensive technological solutions are often over emphasized. This imbalance at decision making and execution levels comes in the way of integrating alternate approaches. The difficulty with ecological solutions is, they are cost effective and benefits generated are invisible in a system where big (budgeted work) is beautiful and politically attractive. On this backdrop, an integrated ecosystem approach advocated by CBD has a potential to generate a synergy between technological solutions and solutions based on ecological principles. An emphasis on such an integrated approach by National Government and International funding agencies can certainly make a difference.

References

Anitha, G., Chandrasekhar, S.V.A. and Kodarkar, M.S., 2005. Hydrography in relation to benthic macro-invertebrates in Mir Alam lake, Hyderabad, Andhra Pradesh. *Rec. Zool. Surv., India,* Occ. Paper No. 235: 1–148.

Associated Industrial Consultants Pvt. Ltd., 1993. Hussainsagar lake, Hyderabad: The profile of a highly eutrophic lake, Vol. 7(10).

Babu Rao, M. and Reddy, Y. Siva, 1984. Fish fauna of Hussainsagar. *Jantu,* 2: 1–6.

Hyderabad Urban Development Authority (HUDA), 2005. *National Seminar on Management of Urban Lakes,* 1–2nd December. Proceedings, pp. 28.

ILEC, 2005. Managing lakes and their basins for sustainable use: A report for lake basin managers and stakeholders. International Lake Environment Foundation. Kusatsu, Japan.

Kodarkar, M.S., 1995. Conservation of lakes, case study of five lakes in and around Hyderabad, Andhra Pradesh, India. Publ. 3. Indian Association of Aquatic Biologists (IAAB), Hyderabad. p. 85.

Kodarkar, M.S. and Joshi, Sandeep, 2006. Conservation and management of lakes in urban environment: bioremediation, a new frontier in the control of Eutrophication of urban lakes. In: *Proceedings of 11th World Lake Conference,* Nairobi, Kenya, pp. 79–82.

Kodarkar, M.S., Joshi, Sandeep and Bhatnagar, Manu, 2005. Conservation and management of lakes in urban environment: bioremediation, a new frontier in the control of Eutrophication of urban lakes. In: *Proceedings of 11th World Lake Conference,* Nairobi, Kenya, 31st Oct to 4th Nov., Abstract, p. 24.

Kodarkar, M.S., Ranade, Vidyanand, Joshi, Sandeep, Supate, A.R., Yeole, Viswas and Vaidya, Shanti, 2009. Integrated Lake Basin Management (ILBM)–A case study of Yeshwantsagar (Ujjani), Maharashtra, India. In: *World Lake Conference*, 1–5th November, Wuhan, China, Abstract, pp. 105.

Reddy, R.C., Kelkar, P.S., Rao, R.R. and Pande, S.P., 2002. Eutrophic status of Hussainsagar lake in Hyderabad. *Insitutte of Engineers (IE) Journal EN*, 83: 14–19.

World Bank, 2005. Lessons for Managing Lake Basins for Sustainable Use. Environment Department Report No. 32877. World Bank, Washington, DC.

World Lake Vision Committee, 2003. World lake vision. Kasatsu, Japan. International Lake Environment Committee and International Environment Technology Centre (IETC), United Nations Environment Programme (UNEP).

Zafar, A.R., 1959. Taxonomy of lakes. *Hydrobiologia*, 13(3): 187–299.

Chapter 2

Water Quality of Tridhara: A Holy Place in Parbhani District, Maharashtra

☆ *S.D. Ahirrao and S.S. Pedge*

Introduction

Surface water, groundwater, local streams are extensively used for purposes such as domestic, industrial and agricultural use. Among that the domestic use is also important to perform various activities. In India most of the civilization, urbanization, and industrialization has been developed science ancient times. Similarly the most of religious places and temples are also developed along the bank of river. Indirectly or directly for the various types of pollution's are taking origin from the above places. On that base line of pollution Tridhara is one of the important religious spot of Parbhani town. During the new moon and full moon the traditional religious peoples are taking holy bath in this water of Tridhara for physically and for mentally to get a peace of mind. Due to over crowding and several religious activities the water is becoming polluted to greater extent and they may affect the health problems. Therefore the present study is under take to evaluate the pollution level of water of that holy places.

Materials and Methods

The present study sample was collected by monthly (2008) on the occasion of fully moon (FM) and new moon (NM) as well as on local festivals. The samples were collected from the surface into the clean plastic container of two-liter capacity for analysis. In the present study following parameter of Tridhara water have been observed. Such as dissolved oxygen, free carbondioxide (Free CO_2), acidity, total hardness, calcium hardness, magnesium hardness, chlorides, alkalinity, salinity, pH, turbidity, Electrical conductivity, Residual chloride, BOD etc. were analyzed and shown in Table 2.1 and compared with various values. The dissolved oxygen was determined by using Winkler's titration method as directed Golterman (1969). The turbidity is measured with the help of turbidity meter and reported in nephelometric turbidity units (NTU). All remaining parameters were analyzed with the help of standards methods described by APHA (1989), Trivedi *et al.* (1987), Michale (1986) IAAB (1989) Maiti (2001).

Table 2.1: Physico-chemical Parameters of Tridhara Water (2008)
(Average values)

Sl.No.	Parameters	Standard Values	Tridhara Water
1.	Dissolved oxygen (mg/l)	6.0	3.75
2.	Free CO_2 (mg/l)	No. Standard	9.6
3.	Acidity (mg/l)	No. Standard	5.8
4.	Total Hardness (mg/l)	300	195.0
5.	Calcium hardness (mg/l)	200	164.0
6.	Magnesium hardness (mg/l)	100	31.0
7.	Chlorides (mg/l)	200	86.94
8.	Alkalinity (mg/l)	200	110.6
9	Turbidity (NTU)	5	85.0
10.	EC (μmhos/cm)	No. Standard	1.88
11.	Residual chlorine (mg/l)	0.2	0.0
12	BOD (mg/l)	2–3	5.3

Result and Discussion

pH

Most of the natural water falls within the range of 4 to 9. The mostly waters are slightly basic became of the present of carbonate and bicarbonate. The pH is having more importance in environmental engineering such as water supplies, water softening, disinfecting and corrosion control, low pH causes corrosion high pH causes tastes soapy feel in the present investigation pH were recorded as acidic of *i.e.* 6.1. This is beyond the level for drinking and bathing purposes.

Alkalinity

The alkalinity of water is a measure of its capacity to neutralize acid. Alkalinity is a measure of the water ability to absorbed H+ without significant pH change. In natural water most of the alkalinity is caused due to CO_2. The free CO_2 dissolves in water to form carbonic acid H_2CO_3 in large quantities, alkalinity imparts bitter taste of water. In the present investigation it is found with in the range of standards value *i.e.* 110.6 mg/l.

Acidity

It is a quantitative capacity to react with strong base to a designated pH. In the present study the acidity is observed as 5.8 mg/l may due to pollution load but no standard value are established for drinking and other use of water as per records.

Biochemical Oxygen Demand

Biochemical oxygen demand (BOD) is defined as the amount of oxygen required by bacteria in decomposing organic material in samples under aerobic condition. It is also used to assess the self, purification capacity of receiving water bodies in the present observation it is recorded as 5.3 mg/l. that is not useful for domestic use as per standard guidelines.

Total Hardness

The hardness can be defined as concentration of multivalent metallic cations in solution. It varies from place to place. It reflects the nature of geological formulation with which it has been in contact. it seems as basis of routine control of softening process. In the present investigation it is records as 195.0 mg/l which is within the range of permissible limits.

Calcium and Magnesium Hardness

Calcium and magnesium are common constituent of natural water and important contributor to the hardness of water in the present investigation both calcium hardness (164.0 mg/l) and magnesium hardness (31.0 mg/l) are found with in the range of permissible limits.

Chlorides

Chlorides directly indicate the presence of various types of salts. The excess of chlorides makes the water quality inferior and bitter salty test. In the present investigation on the chloride values were observed as 86.94 mg/l which is a within the range of limit.

Turbidity

The turbidity can be defined as interference to the passage of light by suspended particles in water. It is caused by wide variety of suspended and colloidal materials. Run-off from barren area during rain is the most contributor of turbidity. The organic material reaching water bodies seems as food for bacteria, resulting growth of bacteria and the microorganism feed upon bacteria. In the present investigation the turbidity has been observed as 85.0 NTU, which is above the range of limits.

Residual Chlorine

Chlorine is primarily added tot he water *i.e.* chlorination for destroying the harmful microorganism in water. In the present investigation the residual chlorine is not found indicate the absence of chlorination of water by the concerned department.

Electrical Conductivity

It is a numerical expression of the ability of water to carry on electrical current. The electrolytes in solution dissociate with position (cations) and negative (anions) ions and impart electrical conductivity. In the present study the EC were recorded as 1.88 µmhos/cm. and there is on any standard is given for use of water.

Free Carbon Dioxide

The free CO_2 in water accumulates due to microbial activity and respiration of organism. Its excess imparts the acidity to the water because of the formation of carbonic acid in the present study it is observed as 9.6 mg/l. There is no standard for free CO_2.

The immersion of Ganesh idols also impact on environment studied by Skthival Veena *et al.* (2005) and Pejaver *et al.* (2001) has observed the disposal of nirmalayas (religious refuse) as a social problems.

The value of observed parameter in the study which are found more than the permissible limits are may contributes the load of various types of pollution and waste disposed in the water bodies. The nirmalya, oils, fruits, plant material, bathing, use of detergents, anthropogenic activity of religious peoples religious waste may be cause of pollution in the water. They may causes various pathogenic and non pathogenic contagious and water born diseases. Therefore, peoples are advised to do not use of water for drinking and bathing purpose without proper purification and various type of primary and other treatment to this water.

Acknowledgement

The author is grateful to principal of the College, Head and staff of Zoology Department, Shri Shivaji College, Parbhani (MS) for providing laboratory facilities in connection with this work and for encouragement.

References

APHA, 1989. *Standard Method for Examination of Water and Wastewater including both Sediments and Slugs*, 19th edn.

Golterman, H.L., 1969. *Method of Chemical Analysis of Freshwater*. Blackwell Scientific Publication, Oxford Edinburg, p. 210.

IAAB, 1998. *Method of Water Analysis*. Indian Association of Aquatic Biologist, Hyderabad.

Michael, P., 1986. *Ecological Methods for Field and Laboratory Investigations*. Tata McGraw-Hill, New Delhi.

Pejaver, M., Somani, V. and Quadros, G., 2001. Disposal of nirmalya (religious refuse) a social problem. *Envn. Eco. Cons.*, 19(2): 375–382.

Trivedi, R.K., Goel, P.K. and Trisal, C.K., 1987. *Practical Method in Ecological and Environmental Sciences*. Enviro Media, Karad.

Veena, Sakthivel, Parelkar, G. J. and Shingadia, H.U., 2005. Environmental impact of Ganesh idol immersion at Juhu and Mahim beaches along the Mumbai coast, Maharashtra. *J. Aqua. Biol.*, 20(2): 105–109.

Chapter 3

Studies on Physico-chemical Parameters of Wan Reservoir in Beed District, Maharashtra

☆ *P.P. Buktar and V.B. Sakhare*

Introduction

Life and water are said to be two faces of the same coin because all forms of life upon the earth depends upon water for there mere existence water covers about three fourth of earth surface.

The specific range of physico-chemical parameters decides water quality. The changes in the physico-chemical parameters of the water of an ecosystem tend to change the living condition especially on the number and diversity of the biota of that ecosystem. Intensive fish culture practices in reservoirs are gaining importance to meet fish requirement of the country. The physico-chemical analysis is the prime consideration to assess the quality of water for its best usage especially for drinking and fishing.

The Wan reservoir is a medium sized reservoir constructed across river Wan in year 1963.The reservoir is bounded by latitudes 18°53'N and longitudes 76°27'E. The water of reservoir is used for irrigation. It also supplies drinking water to Parli Vaijnath town and Vaijnath co-operative sugar factory. The area around reservoir comprises forest-covered hills. The reservoir is having catchment area of 379 sq.kms.The reservoir consists of two canals through which water goes for agricultural lands of the villages like Nagapur, Pangri, Deshmukh takli, Talegaon, Limbota, Kowthali, Sangam, Mandekhel, Gopalpur, Bhilegaon, Tandoli, Parchondi, Selu, Sirsala, Kanadi and Pimpalgon. The salient morphometric features of Wan reservoir are depicted in Table 3.1. The normal annual rainfall over the reservoir area varies from about 535 mm. The air is generally dry from February to May, the relative humidity during the afternoons being 20 per cent. During the southeast monsoon season, the humidity is as high as 80 per cent in the mornings and 60 per cent in the afternoons. The climate of the reservoir area is semiarid subtropical, characterized by a hot summer. Except for the monsoon period the weather is generally dry. The summer outsets in the area generally in March and continues up to May. This is followed by the southwest monsoon from June till September. However, there are periods of dry spells even within this monsoon period. The winter season stretches from December to the end of February.

In the study area, winds are generally moderate with appreciable increase in force during southwest monsoon season. During the monsoon season the wind is mainly from west and southwest directions. During the rest of the year, wind blows from north and east directions. The maximum-recorded wind speed during the month of May and August is 11.4 to 14.2 Km/hour with annual average wind speed being of 8.6 km/hour.

Considering the importance of water quality from fisheries point of view, the present investigation aims at to assess the water quality of wan reservoir at Nagapur near Parli Vaijnath town. The physico-chemical parameters such as air temp, water temp, transparency, pH, total dissolved solids (TDS), dissolved oxygen (DO), free CO_2, phenolphthalein alkalinity, total alkalinity, chlorides, total hardness were analyzed for a period from December 2008 to May 2009.

Table 3.1: The Salient Features of Wan Reservoir

1.	Catchment area	379.92 km²
2.	Year of construction	1963
3.	Annual rainfall at reservoir site	535 mm
4.	Water spread area	347 hectares
5.	Gross storage capacity	25.181 McM
4.	Full tank level	454.87 m.
6.	High flood level	456.40 m.
7.	Height of dam	19.81 m
8.	Length of canal	
	☆ Right canal	12.03 m.
	☆ Left canal	9.65 m.
9.	Area under irrigation	3602 hectares
		(right canal)
		3965 hectares
		(Left canal)
10.	Length of west weir	617.50 m.
11.	Irrigation potential	7567 hectares

Material and Methods

Representative samples were collected covering three sampling stations in the first week of every month. The analysis of water was done at the spot and few parameters in the laboratory, to do so water samples were collected in one liter plastic bottle. The collected samples were properly transferred to the laboratory and analysis was carried out immediately. The specification of instruments used are CD 600 TDS meter and pocket sized pH meter. The remaining parameters were analyzed by standard methods (Trivedy and Goel,1986; Kodarkar *et al.*, 2006).

Results and Discussion

Experimental results of physico-chemical parameters are shown in Table 3.2. Absence of objectionable colours or odour makes water good from various aspects.

Air and Water Temperature:

Air temperature was found to be in the range of 28 °C to 31 °C. It was minimum in month of December (28°C) and maximum was recorded in May (31 °C),where as water temp. was found to be in the range between 22.2 to 25.5 °C being minimum in month of December and maximum in Month of May.

Table 3.2: Physcio-chemical Profile of Wan Reservoir in Maharashtra

Physico-Chemical Parameter	Dec 2008			Jan 2009			Feb 2009		
	Site I	Site II	Site III	Site I	Site II	Site III	Site I	Site II	Site III
Temp. Air (°C)	28.00	28.00	28.00	31.00	31.00	31.00	36.00	37.00	37.00
Temp. Water (°C)	24.00	24.00	24.00	22.00	22.00	22.00	25.00	25.00	25.00
pH	8.20	8.30	8.30	8.20	8.30	8.20	8.10	8.30	8.00
Transparency (cm)	67.06	54.86	24.38	67.05	54.86	27.43	64.00	51.81	21.33
TDS (mg/lit)	160.00	170.00	170.00	160.00	160.00	160.00	160.00	180.00	180.00
DO (mg/lit)	2.27	2.20	1.80	2.53	2.42	2.47	1.20	1.40	1.20
PA (mg/lit)	–	–	–	–	–	–	–	–	–
TA (mg/lit)	285.00	290.00	310.00	287.00	300.00	263.00	282.00	253.00	242.00
Free CO_2 (mg/lit)	12.32	18.58	21.22	24.29	21.26	27.26	26.40	33.00	26.40
Chlorides (mg/lit)	87.00	86.00	84.00	119.45	118.20	112.18	45.40	42.60	38.50
Total Hardness (mg/lit)	48.00	46.00	56.00	59.89	56.60	58.60	46.00	44.00	54.00

Physico-Chemical Parameter	Mar 2009			Apr 2009			May 2009		
	Site I	Site II	Site III	Site I	Site II	Site III	Site I	Site II	Site III
Temp. Air (°C)	30.00	31.00	31.00	28.00	28.00	29.00	31.00	31.00	31.00
Temp. Water (°C)	23.00	23.00	23.00	24.00	24.00	25.00	23.00	23.00	24.00
pH	8.10	8.10	8.30	8.20	8.30	8.20	8.20	8.20	8.30
Transparency (cm)	60.96	45.72	21.33	33.83	42.67	21.33	54.86	42.67	18.28
TDS (mg/lit)	160.00	180.00	180.00	180.00	180.00	170.00	170.00	170.00	160.00
DO (mg/lit)	1.60	1.28	1.80	1.15	1.32	1.41	1.20	1.40	1.45
PA (mg/lit)	–	–	–	–	–	–	–	–	–
TA (mg/lit)	270.00	276.00	308.00	265.00	288.00	290.00	243.00	268.00	286.00
Free CO_2 (mg/lit)	24.20	23.00	23.60	24.30	28.20	24.40	23.40	26.50	28.40
Chlorides (mg/lit)	42.20	43.10	39.20	42.20	44.60	45.60	44.60	42.20	39.10
Total Hardness (mg/lit)	43.00	42.00	45.00	44.00	56.00	54.00	44.00	42.00	40.00

Both air and water temp. were recorded maximum during the summer season. This may be attributed to high solar radiation from sunlight and clear sky whereas the rain fall during wet season and cloudy skies brought down the air and water temperature to maximum.

pH

pH is the value expressed as the negative logarithm of the hydrogen ion concentration. pH can be an indicator of changing water quality due to land use changes. The pH value ranged from 8.0 to 8.4 with a mean of 8.0. The desirable pH of drinking water is 7.0–8.5 (W.H.O., 1984) and maximum permissible limit is 6.5 samples 9.2 (ICMR, 1975). The results show that water from wan reservoir was within permissible limits.

Transparency

The extent to which light can penetrate depends on transparency of standing water column. Water transparency is dependent on turbidity which is directly proportional to the amount (density) of suspended matter. Thus transparency and turbidity play an important role in energy dynamics of an aquatic ecosystem.In present study the highest value of transparency was recorded in the month of May (61cm) and lowest in December (21cm).

Total Dissolved Solids

Total dissolved solids may be chemically organic or inorganic dissolved solids in reservoir water vary both qualitatively and quantitatively with the season and it influences the chemical density of the environment and composition of the biotic community. In present investigation TDS was recorded in the range of 160 to 180 mg/lit. The maximum value 180 was recorded in the Month of May. It may be due to evaporative loss of water and consequent incurable in concentration of salt present in water.

Dissolved Oxygen

The wet lands receives oxygen mainly diffusion from the atmosphere at the surface and by photosynthesis of the chlorophyll bearing organism in it. Dissolved oxygen concentration is one of the most important parameter to indicate water quality and its relation to other aquatic flora and fauna.

In present investigations dissolved oxygen was found in the range of 1.32 to 6.2 mg/lit. Its maximum in the month of December (6.2 mg/l) lowest in May. The decrease in dissolved oxygen value of the Month of May could be attributed to the fact that higher water temperature causes decrease in the solubility of water.

Free CO_2

In an aquatic ecosystem sources of CO_2 are community respiration and decomposition while it is consumed in the photosynthesis. The free CO_2 was varied from Nil to 4.3 mg/lit during the present investigation.

Total Alkalinity

Total alkalinity is the quantitative capacity to neutralize and acidic solution. The total alkalinity analyzed in present investigation lies between 243 to 310 mg/lit but the phenolphthalein alkalinity was found to be zero.

Chlorides

Chloride is an indicator of contamination of water with animal and human waste. The chloride content was varied from 38.5 to 119 mg/lit. The higher value of chloride found in the month of May indicates the effect of breakdown of organic matter and reduced volume of water.

Total Hardness

Hardness is a measure of concentration of metallic ions *i.e.* Ca^{2+} and Mg^{2+} it is the indicator hydrogeology and aesthetic quality of water. It is used to classify waters as hard and soft. In present investigation the hardness was ranged from 40 to 59 mg/lit. The maximum value (59mg/lit) recorded in December and minimum value (40 mg/lit) in month of May.

Conclusion

Water quality analysis of wan reservoir during December 2008 to May 2009 revealed that water is suitable for use of domestic and drinking purpose. The analysis of water also revealed that the reservoir has good potential for fishery.

References

Kodarkar, M.S., Diwan, A.D., Misra, S.M., Dhanapati, M.V.S.S.S., Murugan, N., Thirumalai, G., Altaff, K. and Sakhare, V.B., 2006. Methodology for water analysis: Physico-chemical, biological and microbiological. *Indian Association of Aquatic Biologists*, Hyderabad, pp. 106.

Trivedy, R.K. and Goel, P.K., 1986. *Chemical and Biological Methods for Water Pollution Studies*. Env. Publications, Karad.

WHO, 1984. *Guidelines for Drinking Water*. CBS Publishers and Distributors, New Delhi, 2: 264.

Chapter 4

Studies on Seasonal Variation in Physico-chemical Characteristics of Sina-Kolegaon Reservoir in Osmanabad District, Maharashtra

☆ *V.A. Jagtap, H.K. Bhagwan and S.M. Kamble*

Introduction

The freshwater bodies of India include large number of rivers, ponds, dams, reservoirs and lakes. The quality of water and the living organisms were affecting directly due to excessive uses of industrial chemicals pesticides and fertilizers. The agricultural practices are responsible for the deterioration of water quality. The quality of any water resources is measured in form of its physico-chemical parameters decides its quality of water. The water quality is mainly depends upon its nature. Fertilizer, pesticides, animal wastes and sediments cause agricultural water pollution. The chemicals are used in fertilizers and pesticides contaminate water physico-chemical analysis is the prime consideration to assess the quality of water for its best utilization like drinking, irrigation, fisheries, industrial purpose and helpful in the understanding the complex processes interaction between the climatic and biological processes in the water. Several workers carried out studies on freshwater bodies (Sreenivasan, 1963 and 1964; Datta Munshi, 1975; Ganapati and Pathak, 1978; Khan and Zutshi, 1979 and 1980; Rao and Rama Sarma, 1991 and 1992; Kodarkar *et al.*, 2006).

Material and Methods

The water samples were collected in the first week of each sampling stations (S1, S2 and S3) of Sina-Kolegaon Prakalp for physico-chemical analysis during investigation from January 2008 to

December 2008. The samples were collected at regular intervals from selected sampling stations (S1, S2 and S3) in to plastic can of 5 liter between 10.00 to 11.00 am. The atmospheric and water temperature, pH and turbidity were estimated on the spot or sampling stations (S1, S2 and S3) while other parameters were estimated such as dissolved oxygen (DO), free carbon dioxide (CO_2), biochemical oxygen demand (BOD), chemical oxygen demand (COD), chloride, total alkalinity, calcium hardness, magnesium hardness, total solids, total dissolved solids (TDS), total suspended solids (TSS), etc. in the laboratory employing methods described by Welch (1948 and 1952), Trivedi and Goel (1984), APHA (1995), Kodarkar *et al.* (1998).

Result and Discussion

The physico-chemical parameters are very important for the aquatic ecosystem and for the biochemical and physiological processes in aquatic organisms such as temperature, pH, dissolved oxygen (DO), free carbon dioxide (CO_2), turbidity, chlorinity, total alkalinity, calcium hardness, magnesium hardness, total solids (TS), total dissolved solids (TDS), total suspended solids (TSS), biochemical oxygen demand (BOD), chemical oxygen demand (COD), etc. The seasonal variation in physico-chemical parameters of Sina-Kolegaon Prakalp obtained data has been summarized in Tables 4.1–4.3.

In present investigation, the range of atmospheric temperature was 26.0°C to 38.0°C at all sampling stations (S1, S2 and S3). The higher values was recorded in summer months April and May, moderate in monsoon and little slower in winter seasons in all sampling stations (S1, S2 and S3) of Sina-Kolegaon Prakalp during research period from January 2008 to December 2008, while water temperature also recorded higher during summer months April and May and ranged from 21.0 to 31.0°C and lowest in winter and moderate in monsoon seasons in all sampling stations (S1, S2 and S3). Increase in water temperature leads to decrease in dissolved oxygen concentrations. There is increase in temperature of water bodies during the March, April and May due to decreases in water level and exposed to maximum solar radiation while low temperature in winter season due to minimum solar radiation. Similar results were observed from some workers by Ramani (1990), Rice (1938) and temperature fluctuation also observed by Swarnalata and Narsing Rao (1991) from Sarovernagar Lake, Hyderabad Pawar and Pulle (2005), Bhagde (2005), Pawar and Mane (2006), Mane and Pawar (2007), Mishra *et al.* (2007), Kadam *et al.* (2007), Salve and Hiware (2007), Pejaver and Gurav (2008) and Gonjari and Patil (2008).

The pH of water ranged between 7.0 to 7.9 indicating alkaline nature of the reservoir water. The pH was maximum ranged from 7.6 to 7.9 in summer season while minimum ranged from 7.0 to 7.5 during monsoon and winter seasons. These findings are also supported by in Patra River; Saxena (1992), Rajalakshmi and Sreelatha (2005), Pawar and Pulle (2005), Bhagde (2006), Pawar *et al.* (2006), Tiwari and Chauhan (2006), Rajalakshmi and Krishnamoorthy (2007) and Jayabhaye *et al.* (2008).

The turbidity of water bodies were recorded ranged from 227 to 274 mg/lit. The minimum range was recorded 227 mg/lit turbidity at station S2 December 2008 in winter season and monsoon while maximum turbidity was recorded in summer season ranged from 230 to 274 mg/lit in all sampling stations (S1, S2 and S3) similar recorded was supported to earlier findings by Trivedi (1984), Anusha Pawar (2006), Rajashekhar *et al.* (2007) Rajalakshmi and Krishnamoorthy (2007) and Delphine Rose *et al.* (2008).

The chlorides are generally present in natural waters. The chlorinity of water was recorded during present investigation ranged from 20.0 to 25.5 mg/lit in all sampling stations (S1, S2 and S3).

Table 4.1: Physico-chemical Analysis of Surface Water of Sina Kolegaon Prakalp, Paranda Station S1, from Jan 2008 to Dec 2008

Sl.No.	Parameters	Jan	Feb	Mar	Apr	May	Jun	Jul	Aug	Sep	Oct	Nov	Dec
1.	Atmospheric Tem.	27	28	29	38	38	35	30	31	29	29	27	26
2.	Water Temp.	24	26	24.2	30	31	31	28	29	26	25	24	22
3.	pH	7.1	7.0	7.7	7.6	7.8	7.4	7.2	7.5	7.3	7.4	7.5	7.4
4.	Turbidity	229	230	240	270	257	258	250	255	240	238	231	230
5.	Chloride	22.9	23.1	24.5	23.5	24.0	25.0	20.0	21.0	22.0	21.0	22.0	22.5
6.	Alkalinity	131	126	130	141	109	113	102	119	145	137	145	139
7.	Calcium Hardness	73.12	68.34	70.00	62.00	62.00	65	67.00	71.00	70.00	82.00	85.00	91.00
8.	Magnesium Hardness	12.65	8.70	10.00	7.07	10.49	10.00	9.51	10.98	9.76	8.78	8.54	8.05
9.	Dissolved Oxygen	7.9	6.8	7.1	7.5	7.18	7.25	7.7	7.9	7.7	7.4	8.0	7.5
10.	Carbon Dioxide	0.1	0.1	0.1	0.2	0.2	0.3	0.3	0.1	Nil	Nil	Nil	Nil
11.	Total Solids	370	375	200	450	480	450	496	565	506	338	260	200
12.	Total Dissolved Solids	257	264	283	427	344	304	344	394	374	211	163	105
13.	Total Suspended Solids	113	111	117	123	136	146	152	172	132	127	97	95
14.	Biochemical Oxygen Demand (BOD)	3.01	2.75	2.39	2.22	2.52	2.49	5.5	4.82	4.39	4.10	3.82	4.10
15.	Chemical Oxygen Demand (COD)	5.24	5.75	6.28	8.17	8.68	7.78	6.72	6.44	7.4	6.20	6.65	5.15
16.	Average Rainfall	Nil	Nil	Nil	Nil	21mm	92mm	189mm	161mm	296mm	116mm	Nil	Nil

Table 4.2: Physico-chemical Analysis of Surface Water of Sina Kolegaon Prakalp, Paranda Station S2 from Jan 2008 to Dec 2008

Sl.No.	Parameters	Jan	Feb	Mar	Apr	May	Jun	Jul	Aug	Sep	Oct	Nov	Dec
1.	Atmospheric Temp.	27	28.2	29.2	38	38	35	30	31	29	28	26	26
2.	Water Temp.	24.1	26.0	27.2	30	31.0	30.0	28.0	24.0	26.0	25.0	24.5	22.0
3.	pH	7.6	7.5	7.7	7.6	7.8	7.4	7.2	7.5	7.5	7.6	7.6	7.5
4.	Turbidity	231	230	242	271	255	256	251	255	241	237	230	227
5.	Chloride	23.0	23.2	24.6	23.6	24.1	25.0	20.1	21.0	22.0	20.0	22.0	22.7
6.	Total Alkalinity	141	138	130	130	113	111	122	128	139	131	133	145
7.	Calcium Hardness	73.30	64.00	67.0	62.00	62.00	65.00	67.00	70.00	71.0	69.00	82.00	92.00
8.	Magnesium Hardness	13.64	10.73	9.02	10.49	7.56	8.78	4.14	10.98	9.27	10.73	8.78	7.56
9.	Dissolved Oxygen	7.6	6.7	7.1	7.0	7.1	7.3	7.5.	7.9	7.7	7.3	7.7	7.4
10.	Carbon Dioxide	0.1	02.	0.1	0.2	0.2	0.3	0.3	0.1	Nil	Nil	Nil	Nil
11.	Total Solids	345	364	410	470	401	440	446	540	520	290	300	216
12.	Total Dissolved Solids	201	321	317	339	312	389	372	376	372	196	190	119
13.	Total Suspended Solids	144	143	193	131	189	151	174	150	128	146	110	097
14.	Biochemical Oxygen Demand (BOD)	3.46	2.18	3.67.	2.60	2.20	4.67	4.60	5.17	4.18	4.28	3.60	4.12
15.	Chemical Oxygen Demand (COD)	5.26	5.95	6.34	8.49	8.89	7.99	7.33	6.28	7.18	6.34	5.89	5.56
16.	Average Rainfall	Nil	Nil	Nil	Nil	21mm	92mm	189mm	161mm	296mm	116mm	Nil	Nil

Table 4.3: Physico-chemical Analysis of Surface Water of Sina Kolegaon Prakalp, Paranda Station S3, from Jan 2008 to Dec 2008

Sl.No.	Parameters	Jan	Feb	Mar	Apr	May	Jun	Jul	Aug	Sep	Oct	Nov	Dec
1.	Atmospheric Tem.	27.1	28.5	29.2	38	38.0	35	30	31	29	29.1	27.0	27
2.	Water Temp.	23	26	27	30.1	31	31	28	29	26	25	24.5	21
3.	pH	7.22	7.6	7.9	7.7	7.8	7.3	7.4	7.5	7.5	7.5	7.5	7.4
4.	Turbidity	241	232	245	274	258	257	252	253	245	238	235	233
5.	Chloride	23.1	23.2	24.7	23.4	24.5	25.5	20.1	20.2	22.0	20.1	22.1	22.3
6.	Alkalinity	121	114	109	119	104	102	107	116	123	129	132	137
7.	Calcium Hardness	73.30	67.00	72.00	58.00	59.00	62.00	67.00	67.00	64.00	70.00	70.00	89.00
8.	Magnesium Hardness	10.90	10.24	14.15	7.7	7.07	6.83	7.80	9.91	9.02	8.78	10.98	7.07
9.	Dissolved Oxygen	7.0	7.0	7.1	7.0	7.1	7.3	7.5.	7.9	7.7	7.3	8.6	7.5
10.	Carbon Dioxide	0.1	0.2	0.1	0.2	0.2	0.3	0.2	0.1	Nil	Nil	Nil	Nil
11.	Total Solids	271	409	450	460	459	430	476	570	560	360	316	250
12.	Total Dissolved Solids	271	286	333	327	339	372	342	444	416	229	204	154
13.	Total Suspended Solids	100	123	217	233	120	158	134	190	152	97	112	96
14.	Biochemical Oxygen Demand (BOD)	3.82	3.35	3.98	3.45	2.35	2.72	4.89	5.17	4.67	4.29	3.90	4.13
15.	Chemical Oxygen Demand (COD)	5.68	5.28	6.12	8.80	8.22	7.64	7.34	6.12	7.10	6.29	5.56	5.25
16.	Average Rainfall	Nil	Nil	Nil	Nil	21mm	92mm	189mm	161mm	296mm	116mm	Nil	Nil

The minimum chlorinity value was recorded in monsoon season. Moderate in winter and maximum in summer season and ranged chlorinity value was record from 23.0 to 25.5 mg/lit from February to June in sampling stations (S1, S2 and S3) in Sina-Kolegaon Prakalp during investigation from January 2008 to December 2008.

The total alkalinity values were shows marked seasonal variation. The total alkalinity values was ranged from 102 to 145 mg/lit at all sampling stations (S1, S2 and S3) of Sina-Kolegaon Prakalp. The maximum value was noted at station S1 in September and November 2008 in winter season while the total alkalinity was ranged minimum 102 mg/lit in monsoon and moderate during summer seasons. Similar observations are also made by Shastry *et al.* (1999), Singh (2000), Kulkarni Rajendra Rao (2002), Sakhre and Joshi (2003), Muley and Patil (2006), Rajashekhar *et al.* (2007), Kadam *et al.* (2007), Mane and Pawar (2007), Aher *et al.* (2007) and Delphine Rose *et al.* (2008).

The calcium hardness values was ranged from 58.0 to 92.0 mg/lit at all sampling stations (S1, S2 and S3) from Sina-Kolegaon Prakalp. The maximum calcium hardness was ranged from 71.0 to 92.0 mg/lit in winter months and high in month of December 2008 while minimum calcium hardness was ranged from 58.0 to 68.34 mg/lit and in April 2008 very less as compared to other months.

The magnesium hardness of water samples was ranged from 4.14 to 14.15 mg/lit at all sampling stations (S1, S2 and S3) of Sina-Kolegaon Prakalp. The minimum magnesium hardness was ranged from 4.14 while maximum value was recorded range from 7.07 to 14.15 mg/lit at all sampling stations (S1, S2 and S3).

The calcium and magnesium are an important elements influencing flora of ecosystem and play an important role in metabolism and growth. Some workers also recorded calcium and magnesium hardness by Somashekar (1984), Ansari and Prakash (2000), Pawar and Pulle (2005), Rajashekhar *et al.* (2007), Jayabhaye *et al.* (2008), Delphine Rose *et al.* (2008)and Manjappa *et al.* (2008).

The dissolved oxygen is one of the most important parameters in water quality assessment. The dissolved oxygen is some time referred to as measure of the pulse of an aquatic ecosystem and bringing out various biochemical changes and many ecologists discussed its effect on metabolic activities of organisms Hancock (1973), Mishra and Yadav (1978) and Adebisi (1981). The dissolved oxygen values ranged from 6.7 to 8.6 mg/lit. The maximum dissolved oxygen was ranged from 7.3 to 8.6 mg/lit in winter season and moderate in monsoon season while minimum dissolved oxygen was recorded ranged from 6.7 to 7.1 mg/lit in summer season. Similar observation were recorded by Singh and Trivedi (1979), Adebies (1981) discussed seasonal average fluctuations in dissolved oxygen. He reported its maxima in winter and minima in summer. Earlier workers also supported by Sakhare and Joshi (2003), Nasar Shaikh and Yeragi (2004), Pawar and Pulle (2005), Rai and Shrivastava (2005), Tamlurkar and Ambore (2006), Kadam *et al.* (2007), Srinivasarao *et al.* (2007), Rajashekhar *et al.* (2007), Gonjari and Patil (2008) and Kamble and Muley (2008).

Free carbon dioxide is added to aquatic ecosystem by directly being mixed from atmosphere. Respiratory activity of aquatic organisms and process of decomposition are important sources of carbon dioxide in water bodies. In present investigation free carbon dioxide (CO_2) values ranged from 0.0 to 0.3 mg/lit nil values recorded in winter season from October to December at all sampling stations while maximum value was ranged from 0.3 mg/lit in June and July in sampling stations (S1, S2 and S3) in Sina-Kolegaon Prakalp during January 2008 to December 2008. Similar results were observed from some workers by Arvind Kumar (1995), Kadam *et al.* (2007) and Gonjari and Patil (2008).

The total solids of water samples was ranged from 200 to 570 mg/lit and minimum 200 mg/lit and maximum 570 mg/lit at station S1 in March and December and station S3 in August 2008 respectively. Seasonal analysis revealed that during summer season total solids ranged from 200 to 480 mg/lit. Minimum 200 mg/lit at station S1 March and in May. The total solids was recorded maximum in summer and monsoon while minimum in winter season. Similar finding was also recorded by Gonzalves and Joshi (1946), Tripati and Pandey (1990), Sawarialatha and Narsing Rao (1980), Pawar and Pulle (2005) and Shubhash Chandra Meitei *et al.* (2004).

The total suspended solids range was recorded from 95 to 193 mg/lit maximum range of the total suspended solids was recorded from 144 to 193 mg/lit during summer and monsoon seasons while minimum range was recorded from 95 to 136 mg/lit in winter season in all sampling stations (S1, S2 and S3) of Sina-Kolegaon Prakalp. The results are similar from earlier worker carried out by Dutta (1978), Zutshi (1992) and Datta *et al.* (2001), Mohammad Musaddin and Fokmare (2002).

The total dissolved solids ranged from 96 to 444 mg/lit was recorded in all sampling stations (S1, S2 and S3). The total dissolved solids was maximum about 440 mg/lit at station S3 in August 2008 and minimum total dissolved solids was recorded at stations (S1, S2 and S3). Seasonal analysis reveals that total dissolved solids ranged from 243 to 444 mg/lit in monsoon and summer seasons and minimum in winter in the range from 96 to 229 mg/lit in all smapling stations (S1, S2 and S3) of Sina-Kolegaon Prakalp. The results are coinciding with earlier work carried out by Verma *et al.* (1978), Salodia (1996), Masood Ahmed and Krishnamurthy (1990), Mahajan (1996), Kadam *et al.* (2007), Salve and Hiware (2007), Rajalakshmi and Sreelatha (2005), Deshmukh and Ambore (2007), Rajashekhar *et al.* (2007) and Rajalakshmi and Krishnamoorthy (2007).

The biochemical oxygen demand of water samples was ranged from 2.18 to 5.5 mg/lit maximum biochemical oxygen demand (BOD) was recorded in monsoon and winter seasons while minimum range was recorded from 2.18 to 3.82 mg/lit in summer season at all sampling stations (S1, S2 and S3) of Sina-Kolegaon Prakalp during January 2008 and December 2008. Similar trands are also reported by Patki (2002) and Salve and Hiware (2007).

The chemical oxygen demand was recorded ranged from 5.15 to 8.89 mg/lit maximum COD was recorded range from 6.20 to 8.89 mg/lit in monsoon and summer while minimum COD was recorded ranged from 5.15 to 5.75 mg/lit in winter at all sampling stations (S1, S2 and S3) of Sina-Kolegaon Prakalp. Similar trends was also recorded by Thirumala *et al.* (2006), Salve and Hiware (2007), Chandanshive *et al.* (2008).

Acknowledgement

The authors are thankful to Dr. A. D. Mohekar, Principal, S.M.D. Mohekar Mahavidyalaya, Kallam for giving laboratory facilities and encouragement.

References

Adebisi, B.A., 1981. The physico-chemical hydrobiology of a tropical river upper-ugan river, Nigeria. *Hydrobiol.*, 79(2): 757–165.

Aher, S.K., Mane, U.H. and Pawar, B.A., 2007. A study on physico-chemical parameters of Kagdipura swamp in relation to Pisciculture, near Aurangabad (MS).

APHA, AWWA, WPCF, 1995. *Standard Methods for the Examination of Water and Wastewater,* 18th Edn., New York.

Bhagde, Rupendra V., 2005. Study of physico-chemical parameters of the Bhatye Estuary on Ratnagiri Coast of (MS). *J. Aqua. Biol.*, 20(2): 113–116.

Bhagde, Rupendra V., 2006. Hydrochemical study of sea water from the Mandvi Shore in Ratnagiri District of Maharashtra State. *J. Aqua. Biol.*, 21(1): 97–100.

Chandanshive, N.E., Pahade, P.M. and Kamble, S.M., 2008. Physico-chemical aspects of pollution in River Mula-Mutha at Pune, Maharashtra. *J. Aqua. Biol.*, 23(2): 51–55.

Delphine Rose, M.R., Jeyaseeli, A., Mery, A. Joice and Rani, J.A., 2008. Characteristics of groundwater quality of selected areas of Dindugal District, Tamil Nadu. *J. Aqua. Biol.*, 23(2): 40–43.

Deshmukh, J.V. and Ambore, N.E., 2006. Seasonal variation in physical aspects of pollution in Godavari River at Nanded (MS) India. *J. Aqua. Biol.*, 21(2): 93–96.

Gonjari, G.R. and Patil, R.B., 2008. Hydrobiological studies on Triputi Reservoir near Satara, Maharashtra. *J. Aqua. Biol.*, 23(2): 73–76.

Gonzalves, E.A. and Joshi, D.B., 1946. Freshwater algae near Bombay. J. The seasonal succession.

Jayabhaye, U.M., Pentewar, M.S. and Hiware, C.J., 2008. A study of physico-chemical parameters of a minor reservoir Sawana, Hingoli District of Maharashtra. *J. Aqua. Biol.*, 21(2): 56–60.

Kadam, M.S., Nanware, S.S. and Ambore, 2007. Physico-chemical parameters of Masoli Reservoir with respects to fish production. *J. Aqua. Biol.*, 22(1): 81–84.

Kadam, M.S., Pampatwar, D.V. and Mali, R.P., 2007. Seasonal variation in different physico-chemical characteristics in Masoli Reservoir of Parbhani District (MS). *J. Aqua. Biol.*, 22(1): 110–112.

Kamble, S.P. and Muley, D.V., 2008. Study on some physico-chemical poarameters of Kalbadevi estuary in Ratnagiri District of Maharashtra. *J. Aqua. Biol.*, 23(2): 61–66.

Kodarkar, M.S., Diwan, A.D., Murugan, N., Kulkarni, K.M. and Anuradha, R., 1998. *Methodology for Water Analysis*. Indian Association of Aquatic Biologists, IAAB Publication No. 2.

Kulkarni Rajender Rao, Rita N., Sharma, Mehtab, Burkari, 2002. Diurnal variations of physico-chemical aspects of pollution in Khushavati River at Quepem, Goa. *J. Aqua. Biol.*, 17(1): 27–28.

Kumar, Arvind, 1998. Some limnological aspects of the freshwater tropical wetland of santhal pergana (Bihar) India.

Madhuri, Pejavar and Gurav, Minakshi, 2008. Study of water quality of Jail and Kalwa Lake, Thane, Maharashtra. *J. Aqua. Biol.*, 23(2): 44–50.

Mane, A.M. and Pawar, S.K., 2007. Some physico-chemical properties of Manar River of Nanded District (MS). *J. Aqua. Biol.*, 22(2): 88–90.

Manjappa, S., Suresh, B. Aravinda, Puttaiah, H.B. and Thirumala, E.T., 2008. Studies on environmental status of Tungabhadra River near Harihar Karnataka (India). *J. Aqua. Biol.*, 23(2): 62–72.

Michael, R.G., 1969. Seasonal trends in physico-chemical factors and plankton of a freshwater fishpond and their role in fish culture. *Hydrobiol.*, 33: 144–159.

Mishra, G.P. and Yadav, A.K., 1978. A comparative study of physico-chemical characteristic of lake and river water in central India. *Hydrobiol.*, 59(3): 275–278.

Mishra, Vidya, Goldin, Quadros and Athalye, R.P., 2007. Hydrological study of Ulhas River estuary (MS).

Pawar, B.A. and Mane, U.H., 2006. Hydrography of a Sadatpur lake near Pravaranagar, Ahmadnagar District of Maharashtra. *J. Aqua. Biol.*, 21(1): 101–104.

Pawar, S.K. and Pulle, J.S., 2005. Study on physico-chemical parameters in Pethwadaj Dam, Nanded District of Maharashtra, India. *J. Aqua. Biol.*, 20(2): 123–128.

Rajashekhar, A.V.A., Lingaiah, M.S., Satyanarayana Rao and Ravi Shankar Piska, 2007. The studies on water quality parameters of a minor reservoir, Nadergul, Rangareddy District of Andhra Pradesh. *J. Aqua. Biol.*, 22(1): 118–122.

Rajalakshmi, S. and Krishnamoorthy, G., 2007. Hydrological variations in mangroves of Puducherry, India. *J. Aqua. Biol.*, 22(1): 77–80.

Srinivasarao, S., Khan, A.M. and Lova Rani, Y.V.S.S. and Raghuram, M.V., 2007. Variation of physical characteristics of Godavari River water at Nanded (Maharashtra) and Rajahumunotry (AP). *J.Aqua. Biol.* 22(2): 91–95.

Sakhare, V.B. and Joshi, P.K., 2003. Physico-chemical Limnology of Papnas, A minor wetland in Tuljapur Town, Maharashtra. *J. Aqua. Biol.*, 18(2): 93–95.

Salodia, P.K., 1996. *Freshwater Biology: An Ecological Approach*, pp. 64–68.

Salve, B.S. and Hiware, C.J., 2007. Studies on water quality of Wanparkalpa Reservoir, Nagapur Near Parali Vaijanath, Dist. Beed, Marathwada Region. *J. Aqua. Biol.*, 21(2): 113–117.

Shaikh, Nisar and Yeragi, S.G., 2004. Some physico-chemical aspects of Tansa River of Thane District of Maharashtra. *J. Aqua. Biol.*, 19(1): 99–102.

Singabal, S.Y.S., 1973. Diurnal variations of some physico-chemical factors in the zuary estuary of Goa. *Indian Journal of Marine Sciences*, 2: 90–93.

Singh, S. K. and Trivedi, R.K., 1979. The impact of sewage water on the quality of Ganga water. *Mendel.*, 6(10): 99–101.

Trivedy, R.K. and Goel, P.K., 1984. *Chemical and Biological Methods for Water Pollution Studies*. Enviro Media Pub., Karad (India), pp. 215.

Welch, P.S., 1952. *Limnology*. McGraw Hill Book company, New York. Toronto, London, pp. 538.

Chapter 5

Water Quality Analysis of Godavari River with Respect to Aquatic Pollution during the Tercenary Celebration of Guru-Ta-Gaddi at Nanded

☆ *A.A. Divde, J.A. Sayed, S.R. Parveen and S.W. Sabry*

Introduction

Nanded earlier known as 'Nandigram' is located to the southeastern part of Maharashtra state. It covers area of above 10,332 Sq. Kms. Nanded has a uniqueness of its own due to its historical, social and political importance. It is the second largest city in Marthawada after Aurangabad. Takhat Sachkhand Shri Hazur Abchalnagar Sahib is the main Gurudwara situated in Nanded, Maharashtra. It is one of the four High seats of Authority of the Sikhs. Guru Gobind Singh was the 10th and last Guru of the Sikhs who held his court and congregation here. The tercentenary celebration of Gur-ta-Gaddi to be celebrated at the Sachkhand Gurudwara at Nanded is of great relevance to the Sikh Community. This involves the remembrance of the 300th year of the consecration of Guru Granth Sahib as the last guru of Sikh community by Guru Gobind Singhji. The event will also commemorate the 300th anniversary of the Parlok Gaman of Guru Gobind Singhji. The event will be of 8 days duration and will be conducted in the month of October 2008.

The Godavari River originates at Trimbakeshwar in Nasik district of Maharashtra. The main tributaries of Godavari are Pravara, Purna, Penganga, and Wainganga. The domestic wastewater from various cities is discharged in the river Godavari without any treatment. Nanded city is one of them. Three sampling sites were selected along the course of river Godavari at Nanded city.

Water pollution can be defined as 'any adverse change in conditions or composites of the water so, it become less suitable for the purpose for which it would be stable in the natural state'. The solid content of sewage matter is much less than aqueous solution or water nitrogen in sewage is present either as ammonical nitrogen (free or saline ammonia) and originally bound nitrogen derived from proteinaceous matters. The ammonical nitrogen constitutes about 50–70 per cent of the total nitrogen. The domestic sewage contains large amount of urine, which consists of about 2.5 per cent urea, 1 per cent NaCl and other complex organic substances. NaCl is also derived from industrial wastes. The sewage also includes many inorganic substances such as nitrates and phosphates of detergents and Na^+, K^+, Ca^{2+}, Cl^-, HCO_3^- ions etc.

Present day sewage also contains appreciable amounts of synthetic detergents. In addition to surface active agents, they also contribute phosphates of sodium and other builders. The sewage also includes biodegradable faeces, animal wastes and certain household wastes in the form of organic compounds such as fats, carbohydrates, proteins etc. Latest reports have indicated that trace amounts of metals such as Cu, Cr, Zn, Mn, Pb and Ni are also present in domestic sewage.

Objectives

1. Status report of river Godavari in Nanded city during Guru-Ta-Gaddi Program.
2. Comparative study of water samples with different classes of water.
3. To determine potability of water.
4. To determine water quality analysis during Guru-Ta-Gaddi programme in Nanded city.

Materials and Methods

Collection and Preservation of Sample

Samples are collected for different physico-chemical and biological examination. Generally two liters of sample is sufficient for most of the physical and chemical examination. For certain special

Figure 5.1: Standard Method of Recording Temperature at the Sight

determinations however large volumes of samples are necessary same sample should not be used for chemical, physical and bacteriological examination.

Sample Container

For most of the purpose, the ordinary Stoppard Winchester quartz bottle of 2.5 liters capacity is enough. Generally glass containers are preferable to those made of polythene or other plastic materials polythene container are however used in some circumstances *e.g.* when the sample being examined extracts substances from the glass sample adhere to glass.

Table 5.1: Gur-ta-Gaddi Tercentenary Celebration 2008–Event Schedule

Sl.No	Event	Date
1.	Takhat Snan (Holy Bath)	27/10/2008
2.	Start of Samagam	29/10/2008
3.	Gur-ta-Gaddi–Shabad Kirtan	30/10/2008
4.	Parlok Gaman–Shabad Kirtan	03/11/2008
5.	Nagar Kirtan	04/11/2008

Sampling Equipment

Careful selection of sampling equipment is important if continuous or automatic sampling is carried out. For sample an expensive continuous sampler or an automatic sampling device is used. Variety of special sampling device is available to meet the specific requirement. Among there displacement sample depth and point sampler are of importance. When composite sampler are to be collected, it is better to use of wide mouthed bottle of capacity 200-300ml for the sub samples. They are particularly useful in transferring the content without leaving any settle able matter. New glass bottle should be treated either with diluted HCl or chromic acid and again rinsed with distilled water.

Table 5.2: Methods Used to Calculate the Parameters

Sl.No.	Parameter	Method
1.	Colour	Visual
2.	Temperature	Digital Thermometer
3.	pH	pH meter (Model No. DPH500) Global electronics.
4.	Dissolved Oxygen	Modified Azide Method
5.	Turbidity	Nephloturbidity meter (Model No. 132) Systronics
6.	Chlorides	Argentometric Method
7.	Hardness	EDTA Titrometric Method
8.	Oil and Grease	Partition–Gravematric Method
9.	Suspended Solids	Dried at $103°$ C To $105°$ C (Whatman Filter paper No. 42)
10.	Residual Chlorine	Iodometric Method–I
11.	Nitrite	NED Di Hydrochloride Method (Spectrophotometer Model No. 106, Systronics)
12.	Nitrate	PDA Method (Spectrophotometer Model No. 106, Systronics)
13.	COD	Open Reflux–Titrometric Method
14.	BOD	BOD_3 Days, at $27°$ C
15.	Total Coliforms	Multiple Tube MPN Test

Sampling Techniques

The collection of sample should be done very carefully as all interpretations are based on the analysis report. The sample water about 2 liters should be collected in a clean plastic bottle and should be sealed on the mount no floating material should enter the bottle. If possible sample should be collected below the surface of water. The choice of location, depth and frequency of sampling should be based on local conditions and upon the purpose of investigation.

Expression of Result

Analytical result are usually expressed in milligram per liter (mg/lit). The term part per million is equivalent to mg/lit. If the concentration generally less than 1 mg/lit, it is convenient to express the result in terms of micrograms per liter (µg/l).

Table 5.3: Permissible Limits of Water Quality

Sl.No.	Parameters	Standards for A-II class water
1.	Colour	Less than 300units (platinum cobalt units)
2.	Temperature	Not higher than ambient temperature by 5° C
3.	pH	6 To 9
4.	Dissolved Oxygen	Not less than 4 mg/L
5.	Turbidity	25 units JTU
6.	Chlorides	600 mg/L
7.	Hardness	500 mg/L
8.	COD	150 mg/L
9.	BOD	5 mg/L
10.	Nitrate–N	45 mg/L
11.	Nitrite–N	1 mg/L
12.	Oil and Grease	0.1 mg/L
13.	Suspended Solids	25 mg/L
14.	Fecal Coliforms	Nil
15.	Total Coliforms	5000/100ml
16.	Residual Chlorine	Nil

Results

The observation table clearly shows that most of the parameters under study were found to be within the permissible limits. The variations were observed in pH, Turbidity, B.O.D. and Total Coli forms which are shown in bold font style in the above observation table. The pH was found to significantly higher values at all the locations during the Guru-ta-Gaddi celebration. The observed values for pH were in between 8.9 to 9.3.

The Turbidity of potable water is expected to be less than 25JTU, but during the celebration period the turbidity of water was found to be increased above 25JTU. This makes the river water turbid. The observed values for the turbidity were in between 29-32. This was the expected observation prior to the tercentenary celebration but the values were not expected to be these high.

Table 5.4: Water Quality of River Godavari at Nanded

Sl.No.	Parameters*	23.10. 2008	24.10. 2008	25.10. 2008	26.10. 2008	27.10. 2008	28.10. 2008	29.10. 2008
1.	Colour	Dark pale yellow	Light pale yellow	Dark pale yellow	Dark yellow	Light yellow	Colorless	Colorless
2.	Temperature	27.9±0.5**	27.6±0.5	27.2±0.5	26.8±0.5	26.4±0.5	25.9±0.5	27±0.5
3.	pH	7.4±0.2	7.2±0.2	7.3±0.2	7.4±0.2	9.9±0.2	8.2±0.2	8.7±0.2
4.	D.O.	4.0±0.15	3.4±0.15	3.4±0.15	4.1±0.15	3.6±0.15	5.2±0.15	5.4±0.15
5.	Turbidity	ND	ND	ND	ND	30±0.2	23±0.1	30±0.1
6.	Chlorides	126±2	142±2	130±2	95±2	98±2	110±2	85±2
7.	Hardness	204±3	308±2	226±2	230±2	215±2	230±2	198±2
8.	C.O.D.	68±1.5	72±2	76±2	79±2	76±2	54±2	44±2
9.	B.O.D.	6.0±0.2	5.5±0.2	3.0±0.2	2.0±0.2	16.2±0.2	12.0±0.2	12.2±0.2
10.	Nitrates	0.280±0.005	0.268±0.005	0.372±0.005	0.368±0.005	0.302±0.005	0.282±0.005	0.290±0.005
11.	Nitrites	0.021±0.002	0.018±0.002	0.019±0.002	0.017±0.002	0.015±0.002	0.031±0.002	0.029±0.002
12.	Oil and Grease	0.04±0.003	0.05±0.003	0.06±0.003	0.07±0.003	0.08±0.003	0.09±0.003	0.10±0.003
13.	Suspended Solids	21±2	22±2	21±2	20±2	23±2	21±2	20±2
14.	Faecal Coliform	NIL	NIL	NIL	NIL	NIL	NIL	NIL
15.	Total Coliform	4500±150	5100±150	4800±150	4600±150	16000±150	15900±150	15750±150
16.	Residual Chlorine	NIL	NIL	NIL	NIL	NIL	NIL	NIL

Contd...

Table 5.4–Contd...

Sl.No.	Parameters*	30.10. 2008	31.10. 2008	1.10. 2008	2	3	4	5
1.	Colour	Light pale yellow	Light pale yellow	colorless	Pale yellow	Colorless	colorless	colorless
2.	Temperature	26.4±0.5	27.8±0.5	27.2±0.5	26.9±0.5	27.9±0.5	26.7±0.5	26.6±0.5
3.	pH	9.3±0.2	8.6±0.2	8.3±0.2	8.9±0.2	8.6±0.2	9.7±0.2	7.9±0.2
4.	D.O.	5.3±0.15	5.4±0.15	5.4±0.15	4.9±0.15	5.2±0.15	4.9±0.15	4.7±0.15
5.	Turbidity	18±0.1	28±0.1	33±0.1	29±0.1	21±0.1	ND	27±0.1
6.	Chlorides	86±2	90±2	95±2	82±2	78±2	86±2	82±2
7.	Hardness	210±2	198±2	212±2	236±2	240±2	220±2	170±2
8.	C.O.D.	36±2	34±2	36±2	32±2	32±2	36±2	32±2
9.	B.O.D.	13.0±0.2	13.0±0.2	11.2±0.2	12.0±0.2	13.0±0.2	14.0±0.2	13.0±0.2
10.	Nitrates	0.279±0.005	0.302±0.005	0.285±0.005	0.295±0.005	0.278±0.005	0.261±0.005	0.310±0.005
11.	Nitrites	0.032±0.002	0.028±0.002	0.030±0.002	0.033±0.002	0.032±0.002	0.031±0.002	0.17±0.002
12.	Oil and Grease	0.09±0.003	0.10±0.003	0.08±0.003	0.10±0.003	0.075±0.003	0.09±0.003	0.095±0.003
13.	Suspended Solids	18±2	24±2	23±2	25±2	23±2	21±2	19±2
14.	Faecal Coliform	NIL	NIL	NIL	NIL	NIL	NIL	NIL
15.	Total Coliform	15350±150	16000±150	16000±150	16000±150	16000±150	16000±150	16000±150
16.	Residual Chlorine	NIL	NIL	NIL	NIL	NIL	NIL	NIL

*Units are shown in Permissible limits tables.

**: The observed readings are mean readings of 4 different samples collected from, Old bridge, Bondar, Elichpur and Govardhan ghat

The most important parameter regarding the water quality is B.O.D. The B.O.D. value was found to be significantly higher during the tercentenary celebration. The permissible value for the B.O.D. should be less than or equal to 5mg/L. The during the study period the B.O.D. values were ranging from 13 to 15.

The total coliform was another parameter of concern. The values of total coliform were more than 16000. The permissible limit for this parameter is 5000/100mL

Conclusion

It was concluded that some of the parameters of water quality were showing the higher values. The following observations have been reported to authority with necessary suggestions for control and abatement.

The parameters pH, Turbidity, B.O.D. and Total Coli form showed higher values at the already mentioned four sampling sights (Old Bridge, Bondar, Elichpur and Govardhan ghat). These sights were the sewage water sampling stations where the outled was given without treatment, directly discharged in to the Godavari river. According to authorities about 5,00,000 people visited at Nanded to take the holy bath from 26th October to 4th November 2008. Due to activity of use of bathing soap and detergents by pilgrimage the water quality highly deteriorated. The authorities took immediate steps to solve the problems and ensure minimum required flow in the river bed on 30th October 2008 onwards. Also the doors of New Were, Wajegaon were opened to ensure flow resulting in to the natural purification process towards the downstream.

Chapter 6

Integrated Studies on Water Resource Management and Conservation in Karpara River Sub-basin in Parbhani District by Using Remote Sensing and GIS

☆ *Md. Babar*

Introduction

The population growth has been creating more and more stress on agriculture sector for increasing the food grain production, which consequently increased deforestation and demand for more water. The available surface water resources are inadequate to meet the entire water requirement for various purposes. So the demand for undergroundwater has increased over the years. Generally, groundwater is less prone to pollution in comparison to surface water. Hence, groundwater serves as an important source of water for various purposes in rural and urban areas. In recent years intensive use of satellite remote sensing has made it easier to define the spatial distribution of different groundwater prospect classes on the basis of geomorphology and other associated features (Agashe, 1990, Kulkarni, 1994; Adyalkar, 1996; Goswami *et al.*, 1996; Srinivasa Rao *et al.*, 1997; Bhan, 1998; Bhagwan, 1998; and Patil *et al.*, 1999). In many earlier studies (Karanath and Seshu Babu, 1978; Raju *et al.*, 1985; Satyanarayana, 1991; and Srinivasa Rao *et al.*, 1997) remote sensing techniques have been applied for groundwater prospecting.

Geographical Informational System (GIS) is a computer-assisted system designed to capture, store, edit, display and plot geographically referenced data. It has the capability to make quick and unbiased decision, which includes distance, directions, adjacency, relative locations and other spatial concepts. This new wave of information technology is relevant to organizations like Government, Corporation, Municipality, Industries, etc. which are involved with resources planning and management of facilities, designing and management of networks, monitoring, assessment, development and planning for the future, etc (Arya *et al.,* 2002, Bhushan, 2002, Gupta *et al.,* 2002, Jain and Reddy, 2002, Kumar *et al.,* 1996, Nayak, 2002, Sara Naaz and Venkateshwara Rao, 1998, and Vazir Mahmood and Durga Rao, 2002).

Study Area

The study area, Karpara river in Parbhani district is bounded by latitude of 19° 19^{I} N and 19° 39^{I} N and longitude 76° 30^{I} E and 76° 54^{I} E (Figure 6.1) and is included in the toposheet no. 56 A/10, A/14 and A/15. The Karpara river, a tributary stream of Purna river covers an area of 462.15 sq km in Parbhani District. The study area belongs to semiarid and subtropical climate characterized by hot summer and the normal annual rainfall of 909 mm. The satellite image used is given in Figure 6.2.

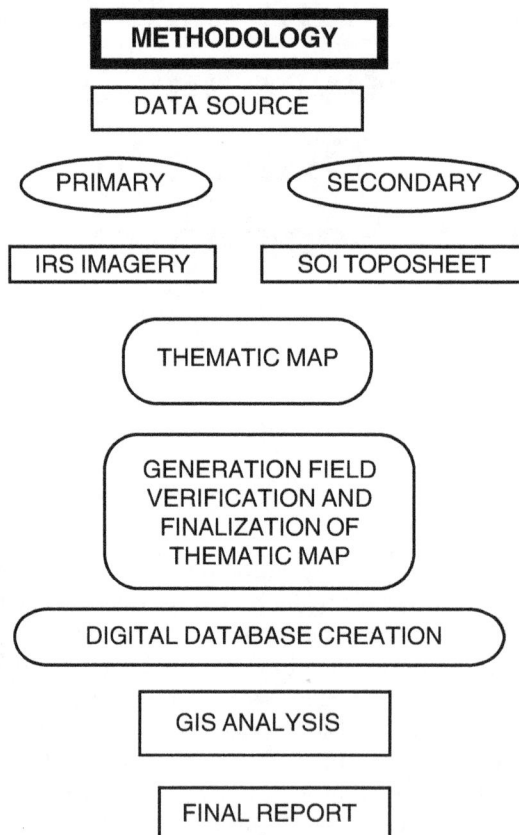

```
            ┌──────────────────────┐
            │     METHODOLOGY      │
            └──────────────────────┘
               ┌─────────────────┐
               │   DATA SOURCE   │
               └─────────────────┘
      ( PRIMARY )              ( SECONDARY )
   ┌──────────────┐         ┌───────────────┐
   │ IRS IMAGERY  │         │ SOI TOPOSHEET │
   └──────────────┘         └───────────────┘
               ( THEMATIC MAP )
            ┌────────────────────┐
            │  GENERATION FIELD  │
            │  VERIFICATION AND  │
            │  FINALIZATION OF   │
            │   THEMATIC MAP     │
            └────────────────────┘
        ( DIGITAL DATABASE CREATION )
               ┌───────────────┐
               │  GIS ANALYSIS │
               └───────────────┘
               ┌───────────────┐
               │  FINAL REPORT │
               └───────────────┘
```

Discussion and Conclusion

Remote sensing with its advantages of spatial, spectral and temporal availability of data covering large and inaccessible areas within short time has become a very handy tool in assessing, monitoring

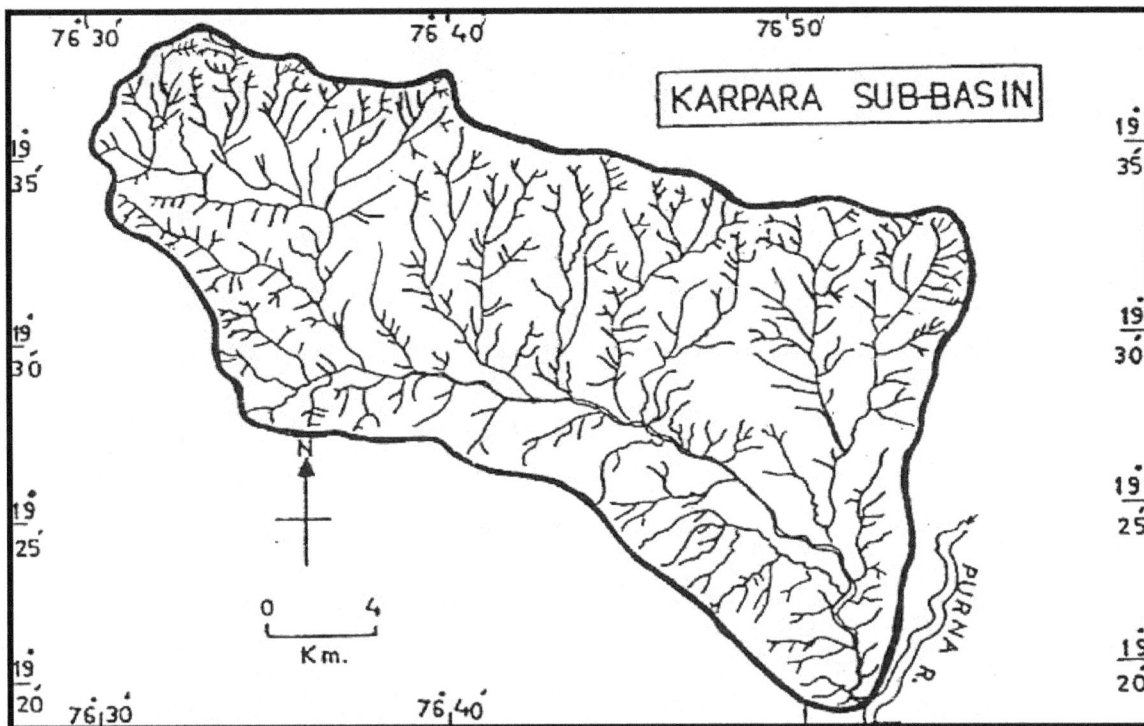

Figure 6.1: Drainage Map of Karpara Sub-basin

and conserving groundwater and surface water resources. Satellite data provides quick and useful baseline information on the parameters controlling the occurrence and movement of groundwater like geology, lithology/structural, geomorphology, soils, landuse/cover, lineaments etc. However, all the controlling parameters have rarely been studied together because of non-availability of data, integrating tools and modeling techniques. Hence a systematic study of these factors leads to better delineation of prospective zones in an area, which is then followed up on the ground through detailed hydrogeological and geophysical investigations. Visual interpretation has been the main tool for evaluation of groundwater and surface water prospective zones for over three decades (Karanath and Seshu Babu, 1978; Raju *et al.*, 1985; Satyanarayana, 1991; Srinivasa Rao *et al.*, 1997, Muley *et al.*, 2000 and Babar and Kaplay, 2003). It has also been found that remote sensing besides helping in targeting potential zones for water resources exploration provides inputs towards estimation of the total water resources in an area, the selection of appropriate sites for water conservation, artificial recharge and the depth of the weathering area.

By combining the remote sensing information with adequate field data, particularly well inventory and yield data, it is possible to arrive at prognostic models to predict the ranges of depth, the yield, the success rate and the types of wells suited to various terrains under different hydrogeological domains. Based on the status of groundwater development and groundwater irrigated areas (though remote sensing), artificial recharge structures such as percolation tanks, check dams and subsurface dykes can be recommended upstream of groundwater irrigated areas to recharge the wells in the downstream areas so as to augment groundwater resources.

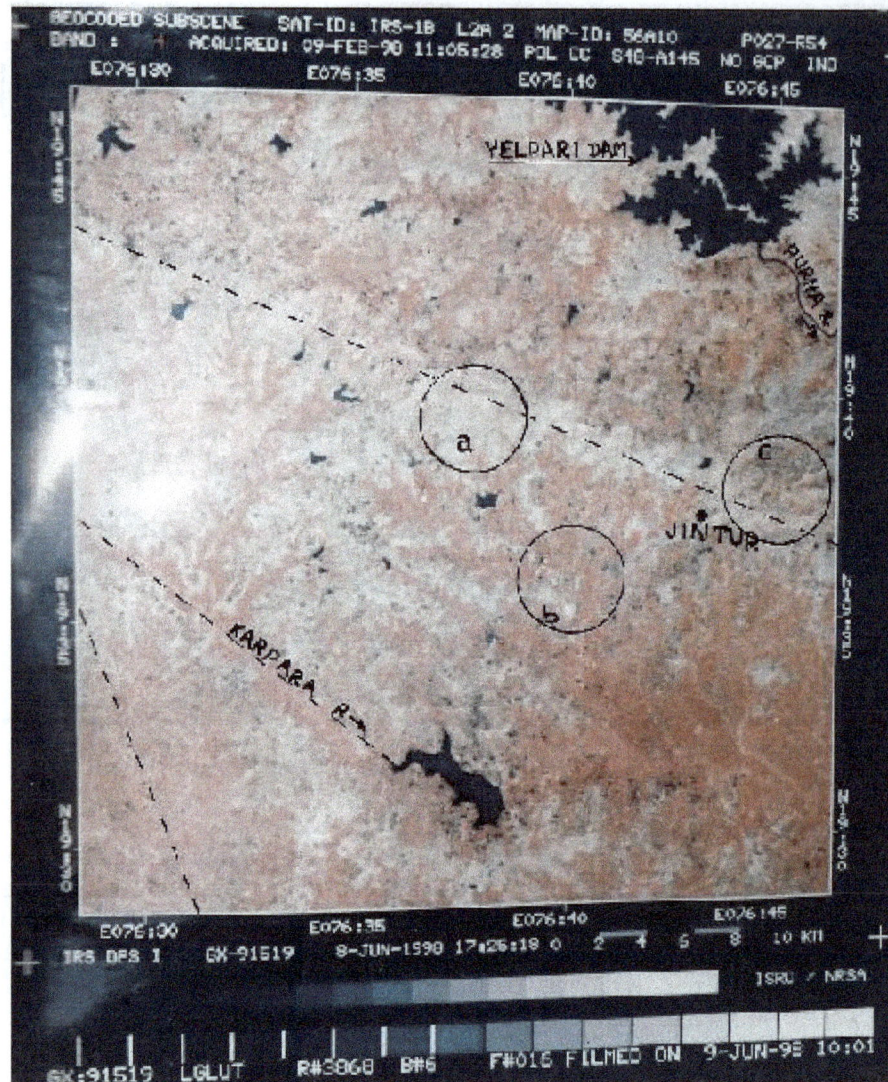

Figure 6.2: Satellite Imagery of the Area (IRS IB) On scale 1 : 50000

**(a) Dendritic with radial annular enclave drainage, (b) Sharp bend in stream,
(c) Radial drainage pattern - - - - - Lineaments**

Apart from visual interpretation, many researchers for deriving geological, structural and geomorphological details use digital techniques. The various thematic layers generated using remote sensing data like lithology/structural, geomorphology, landuse/cover, lineaments etc, can be integrated with slope, drainage density and other collateral data in a Geographic Information System (GIS) framework and analyzed using a model developed with logical conditions to derive at groundwater zones as well as artificial recharge sites. The GIS is useful for groundwater targeting, management and conservation of water resources that ensures optimum and judicious use of water and in identification of artificial recharge sites.

Geology

Geologically, the area is occupied by Deccan Basalts of late Cretaceous to early Eocene period. As the primary porosity is limited to gas cavities (vesicles), the groundwater in the area is therefore concentrated the zone of secondary porosity developed in the rocks due to fractures, joints and amount of weathering.

Hydrogeology

Surface Water

Surface water is a general term describing any water body which is found flowing or standing on the surface, such as streams, rivers, ponds, lakes and reservoirs. The quality and quantity of surface water depends on a combination of climate and geological factors. The recent pattern of rainfall, for example, is important in enclosed water bodies such as lakes and reservoirs where water is collected over a long period and stored. The water storage in streams and rivers is in a constant stage of movement and depend on the weather conditions. The reservoirs have played a significant role in the India's social and economic progress during the past five decades. Without the dams and reservoirs India would have been a thirsty, hungry dark land ravaged with floods and draughts every year. These reservoirs store precious rainwater to irrigate farmlands, generate electricity, supply drinking water and save land from floods and draughts.

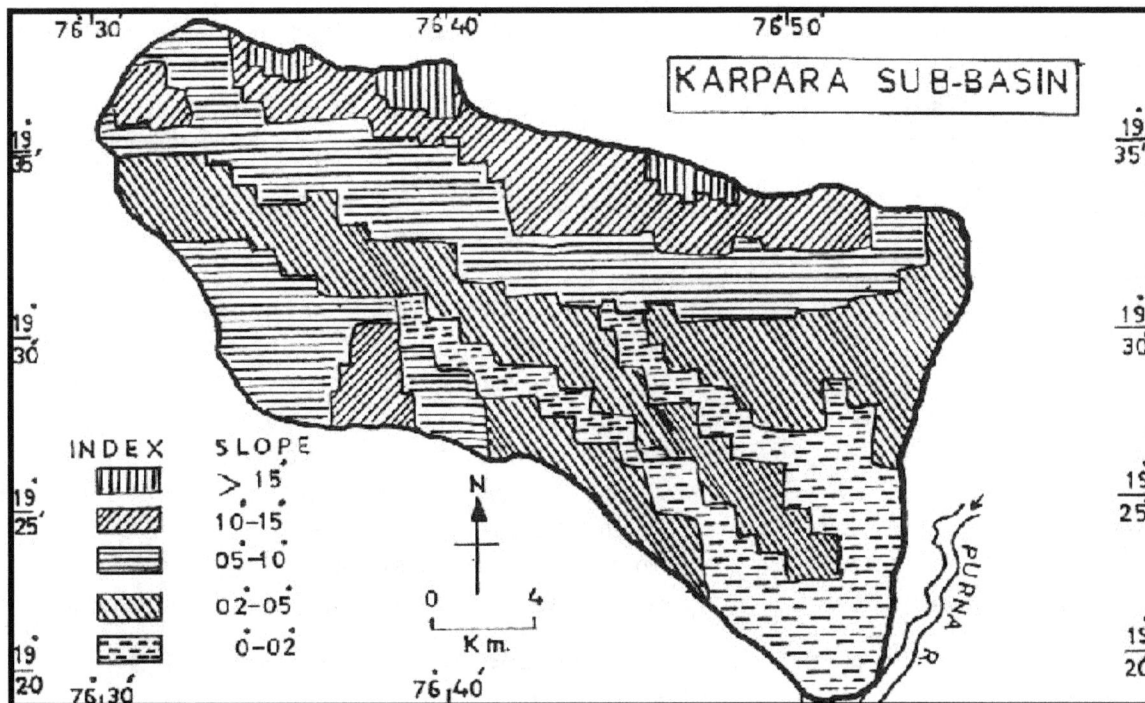

Figure 6.3: Slope Map of Karpara Sub-basin

The traditional management of reservoirs does not consider the sustainable use of these structures. Large water flows and measures to mitigate soil erosion are used to span the usable life of a reservoir. Due to the high cost of elimination of sediments, siltation has to be considered as a non-reversible process. A new philosophy for sustainable management of reservoirs has to focus on controlling the sediment accumulation. Water sustainability requires that the traditional concept of limited (< 100 years) reservoir usable life be replaced with a longer–term perspective on sustainable use. A balance between water quantity and quality available to human communities and preservation of biodiversity should be achieved, to protect the environment and conserve resources (Babar, 2005). The loss of water storage capacity caused by sediment accumulation makes reservoirs key non-sustainable components of the water supply systems. The estimated annual lost water storage capacity worldwide is about 1 per cent, however there is a great range of siltation rates.

Groundwater

In Deccan Basalt terrain groundwater occurs under phreatic conditions in the exposed lave flows and under semiconfined conditions in the flows at deeper level. Lithological constraints dictate that groundwater is present in the pore spaces of the vesicular basalt and in the jointed and fractured portions of massive parts of the flows. The primary porosity in the basalts is associated with the vesicles, which are the pore spaces developed due to the escape of volatile and gases when the lava erupts on the surface as a lava flow. This primary porosity in the basalt is naturally limited and related to the quantum of gasses/volatile in the eruptive phase, which resulted in the basalt flow. The groundwater in the study area therefore is restricted mostly to the zones of secondary porosity developed in these rocks due to fractures, joints and weathering.

Well inventory was conducted by observing the dug wells and lithologs of various bore wells. Well depth, depth to water below ground level (bgl), water level fluctuations and yield particulars are given in Table 6.1.

Table 6.1: Well Inventory and Yield Particulars of Wells in Karpara Sub-basin

Sl.No.	Geomorphic Unit	Observation Wells		Range of Depth (m)	Range of Depth to Water (m bg1)		Water Table Fluctuation (m)	Range of Yield (1 pm)	Av. Yield (1 pm)
		Type	Nos		Pre monsoon	Post-monsoon			
1.	Valley Fills	DW	7	8.0–9.5	1.5–5.0	0.5–3.4	1.0–2.6	80–210	150
		BW	4	25.0–45.0	2.7–7.2	1.1–4.0	1.1–3.2	220–395	320
2.	Pediplain	DW	8	9.2–12.4	2.6–5.6	1.1–3.2	1.5–2.4	95–175	140
		DBW	4	10.5–16.5	4.1–7.8	1.6–3.8	2.5–4.0	120–205	185
		BW	8	30.0–60.0	5.9–10.3	2.1–4.3	3.8–6.0	188–294	260
3.	Pediments	DW	7	9.5–14.1	5.3–10.5	2.3–4.1	3.0–6.4	40–81	50
		DBW	5	15.0–20.5	7.6–13.8	3.1–5.2	4.5–8.6	65–106	85
		BW	5	40.0–65.0	9.4–17.6	4.4–6.6	5.0–11.0	110–170	140
4.	Highly Dissected Plateau	DW	7	8.6–12.3	6.3–12.2	3.1–5.2	3.2–7.0	18–36	25
		BW	6	50.0–70.0	11.5–28.9	5.0–13.2	6.5–15.2	42–91	70
5.	Denudational Hills	BW	2	48.0–56.0	13.5–31.9	7.0–14.5	6.5–17.4	40–82	65

DW: Dug well; DBW: Dug-cum bore well; BW: Bore well.

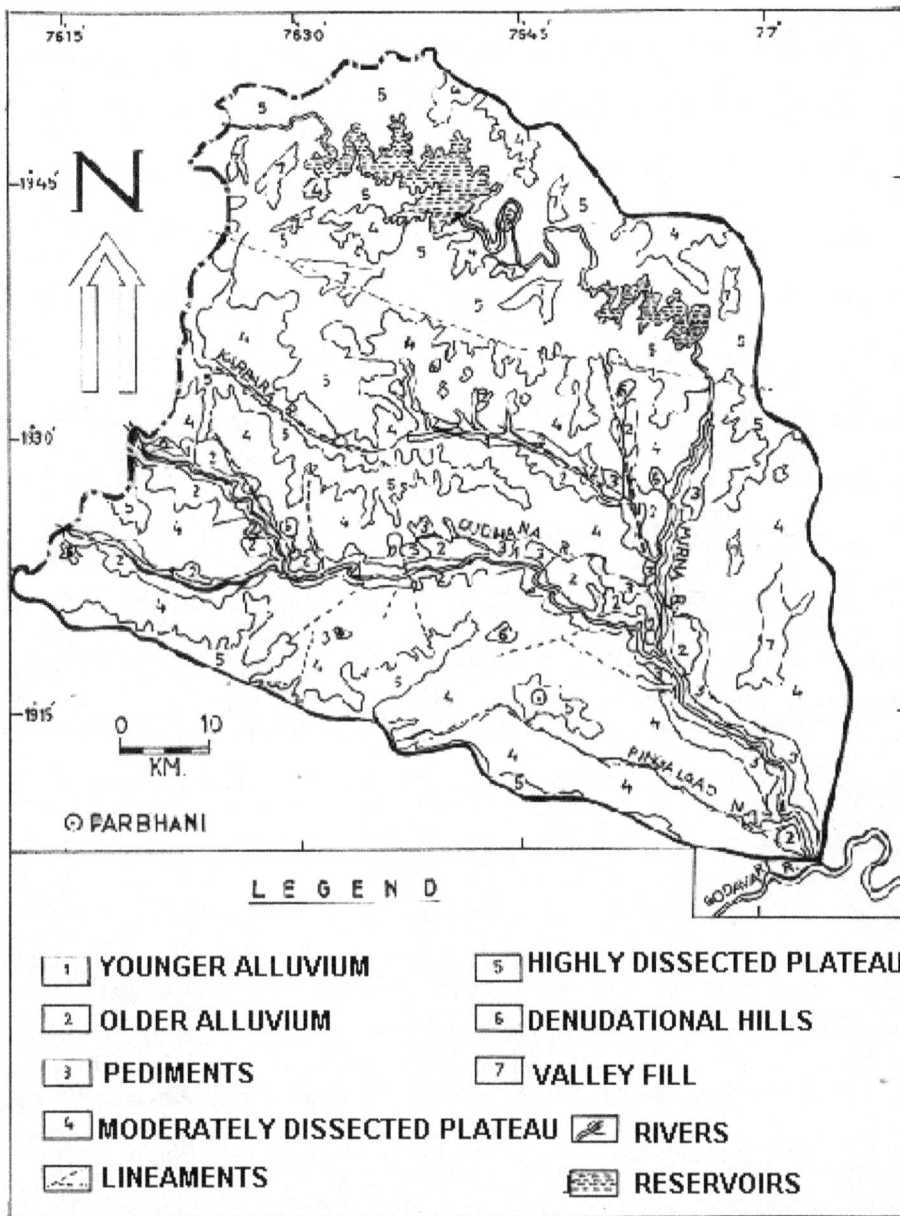

Figure 6.4: Geomorphological Map of the Part of Purna River Sub-basin

The slope map of the area is given in the Figure 6.3. Hydrogeomorphologically the Karpara sub-basin is divided into areas occupied by valley fill, pediplains, pediments, highly dissected plateau and denudational hills (Table 6.1). By studying the hydrogeomorphological conditions of the basin, it is possible to decipher the groundwater potentiality. It is moderate to good, in most part of the watershed while the fractured zones in the weathered pediplains and valley fills are very good potential zones.

The pediplains and pediments with fractures are having moderate groundwater potentiality (Figure 6.4).

Dug wells are recommended in the weathered pediplains. Most of the dug wells in the pediments and in pediplain dry up during pre monsoon period hence bore wells and dug-cum bore wells are recommended for these areas. The surface water bodies found in the area are Karpara Project, Niwali Project, Jogwada, Pachlegaon percolation tank and various water harvesting structures.

Acknowledgement

The author gratefully acknowledges the encouragement and guidance of Prof. R.D. Kaplay, School of Earth Sciences, S.R.T.M. University, Nanded.

References

Adyalkar, P.G., Ayyangar, R.S., Tikekar, S.S. and Khare, Y.D., 1996. Groundwater potential of Deccan Flood Basalt of Nagpur District in Maharashtra: An imprint derived from satellite imagery in Deccan basalt, Gondwana. *Geol. Soc. Sp.*, 2: 485–492.

Agashe, R.M., 1990. Scope for artificial groundwater recharge in Deccan trap areas of Maharashtra: An over view. In: *Proc. All India Seminar on Modern Techniques of Rainwater Harvesting, Water Conservation and Artificial Recharge for Drinking Water, Afforestation, Horticulture and Agriculture.* G.S.D.A. Publ., Pune, pp. 121–192.

Arya, A.S., Bhandari, R.J., Pathan, S.K., Patel, A.V. and Patel, S.S., 2002. Remote sensing and GIS for micro-watershed development: A grass root level approach. In: *Proceeding Volume of the International Symposium of ISPRS Commission VII on Resource and Environmental Monitoring,* December 3–6, 34(7): 671–674.

Babar, Md., 2005. *Hydrogeomorphology: Fundamentals, Techniques and Applications.* New India Publication Agency, New Delhi, pp. 25–56.

Babar, Md. and Kaplay, R.D., 2003.Groundwater fluctuation in Purna river basin, Parbhani District, Maharashtra. *J. of Applied Hydrology*, 16(1): 56–6l.

Bhan, S.K., 1998. Remote sensing for national development and trends in remote sensing. In: *Proc. of Workshop on Watershed Management: Problems and Prospects (WMPP–98),* Centre for Water Resources, J.N. Technological Univ., Hyderabad, pp. 14–23.

Bhagavan, S.V.B.K., 1998. Remote sensing techniques for watershed programmes in Andhra Pradesh. In: *Proc. of Workshop on Watershed Management: Problems and Prospects (WMPP–98),* Centre for Water Resources, J.N. Technological Univ., Hyderabad, pp. 24–33.

Bhushan, Brij, 2002. Using GIS for crop forecasting and crop estimation in India. In: *Spatial Information Technology: Remote Sensing and GIS–ICORG,* (Ed.) I.V. Murali Krishna, 1: 197–203.

Goswami, D.C., Goswami, I.D., Duarah, B.P. and Deka, P.P., 1996. Geomorphological mapping of Assam using satellite remote sensing technique. *Indian J. Geomorph.*, 1(2): 225–235.

Gupta, R.D., Garg, P.K. and Arora, M.K., 2002. A GIS based spatial modeling for developmental planning. In: *Spatial Information Technology: Remote Sensing and GIS–ICORG,* (Ed.) I.V. Murali Krishna, 1: 265–271.

Jain, K. and Reddy, N., 2002. Map generalization using GIS. In: *Spatial Information Technology: Remote Sensing and GIS–ICORG,* (Ed.) I.V. Murali Krishna, 1: 254–260.

Karanth, K.R. and Seshu Babu, K., 1978. Identification of major lineaments on satellite imagery and on aerial photographs for delineation of possible potential groundwater zones in Penukonda and Dharmavaram talukas of Anantpur district. In: *Proc. of Joint Indo US Workshop on Remote Sensing of Water Resources*, NRSA, Hyderabad, pp. 188–197.

Kulkarni, H., Deolankar, S.B., Lalwani, A. and Lele, V.A., 1994. Integrated remote sensing as an operational aid in hydrogeological studies of Deccan basalt aquifer. Asian-Pacific remote sensing. *J. (ESCAP)*, 6(12): 9–18.

Kumar, Ashok, Prasad, B.B. and Sinha, Ranjan, 1996. Sustainable utilisation and management of groundwater in Churchu watershed, Hazaribagh, Bihar: A remote sensing, geophysical and GIS approach. In: *ISRS Symp.*, 4–6 Dec., Pune.

Muley, R.B., Kulkarni, P.S. and Babar, Md., 2000. Integrated approach of geomorphologic and hydrological studies for watershed development: A case study. *J. Applied Hydrology*, 15(1): 31–36.

Nayak, T.R., 2002. GIS for soil erosion modeling: A case study of Bebas Watershed, Sagar District (M.P.) India. In: *Spatial Information Technology: Remote Sensing and GIS–ICORG*, (Ed.) I.V. Murali Krishna, 1: 191–197.

Patil, B.S., Khadilkar, A.K. and Zambre, M.K., 1999. Shallow groundwater zones mapping by using remote sensing techniques: A case study around Pishore, Aurangabad district, Maharashtra. In: *Seminar on Groundwater and Watershed Development*. Jai Hind College, Dhule, pp. 63–65.

Raju, K.C.B., Rao, G.V.K. and Kumar, B.J., 1985. Analytical Aspects of remote sensing techniques for groundwater prospecting in hard rocks. In: *Proc. Sixth Asian Conference on Remote Sensing*, pp. 127–132.

Satyanarayana, R., 1991. Remote sensing studies on the land and water resources of Hyderabad City and environs. *Ph.D. Thesis*, S.V. Univ., Tirupati, India (Unpublished).

Srinivasa Rao, Y., Krishna Raddy, T.V. and Nayudu, P.T., 1997. Hydrogeomorphological studies by remote sensing application in Niva River basin, Chittor district, Andhra Pradesh, Photonirvachak. *J. Indian Soc. Rem. Sensing*, 25(3): 187–194.

Sara Naaz and Venkateshwara Rao, B., 1998. Application of geographic information system for the identification of soil conservation sites in the parts of Sriramsagar catchment. In: *Indian Proc. of Workshop on Watershed Management: Problems and Prospects (WMPP, 1998)*. Centre for Water Resources, J.N. Tech. Univ., Hyderabad, pp. 43–53.

Vazir Mahamood and Durga Rao, K.H.V., 2002. Groundwater modeling using remote sensing and GIS: A case study of Visakhapatnam (A.P.) India. In: *Spatial Information Technology: Remote Sensing and GIS–ICORG*, (Ed.) I.V. Murali Krishna, 1: 162–168.

Chapter 7

Large Dams: More Harmful than Beneficial

☆ *V.B. Sakhare*

Due to increased demands for reliable supplies of electric power, irrigation and drinking water, the number of new hydropower reservoirs is increasing dramatically, especially in Asia. According to last report of the International Commission on Large Dams, the total number of dams on earth grew about 5000 in 1950 to more than 40,000 in 1986, with China the home to about 50 per cent of these (Mc cully,1996).There are 45,000 large dams in the world. Although the first hydel dam was built in 1890, by 1949, the world had 5000 large dams. By the end of the 20th century, this figure had increased to 45000 in 140 countries. Of these 22000 are in China alone, a country that had only 22 large dams in 1949.The other four in the top five 'large dam' nations are the United States with 6,390,India with 4000 and Spain and Japan with between 1000-1200 each. An Estimated 160 to 320 new large dams are built each year. Between 1989 and 1993, an average of 4 million people was displaced annually by 300 large dams. According to the World Commission on Dams, large dams have displaced an estimated 40 to 80 million people. Large hydropower reservoirs have caused massive social disruption, increased incidences of water–borne diseases, erosion and other social and environmental degradation (Peter, 1978). It has been estimated that the annual inland fish production in Asia's is 5.5 million tons, comrising 57 per cent of the world's inland fish production.However, fish yields from Asian reservoirs comprises just 0.5 million tons of this 5.5 tons.

India has over 4000 large dams as defined by International Commission on Large Dams (ICOLD). Many of these are lower than 15m in height and are classified as large dams based on ICOLD criteria. The majority of Indian dams have been built for irrigation. An examination of the state wise picture of the distribution of large dams shows that nearly half the large dams are in Maharashtra and Gujarat.

Large dams have affected 60 per cent of the world's rivers, often altering ecosystem irreversibly. The world commission on dams has concluded that on the whole large dams have had a negative impact on ecosystems, most of which cannot be mitigated through corrective measures. When river is tilled behind a dam, the sediments it contains sink to the bottom of the reservoir. The proportion of a river's total sediment load captured by a dam-known as its 'trap efficiency'-approaches 100 per cent for many reservoirs, especially in large reservoirs. As the sediments accumulate in the reservoir, so the dam gradually loses its ability to store water for the purposes for which it was built. Every large reservoir loses storage to sedimentation, although the rate at which this happens varies widely. Sedimentation is still the most serious technical problem faced by the large reservoirs. The rate of reservoir sedimentation depends mainly on the size of a reservoir relative to the amount of sediment flowing into it. Large reservoirs in the United Sates lose storage capacity at an average rate of around 0.2 per cent per year,with regional varieties ranging from 0.5 per cent per year in the Pacific states to just 0.1 per cent in reservoirs in the northeast.Major reservoirs in China lose capacity at annual rate of 2.3 per cent (Gleick,1995).

Hirakud dam in Orissa was built to control the floods. Yet extreme floods in the Mahanadi delta between 1960 were 3 times more frequent than before Hirakud was built. In September 1080, hundreds of people were killed after releases from Hirakud breached downstream embankments. Many destructive floods have been recorded on emergency releases from Indian reservoirs. In southwest monsoon period of year 2002, due to sudden releases of water from Yeldari dam in Parbhani district of Maharashtra several peoples were made homeless.

It is a well-known fact that large dams can trigger earthquakes. Reservoirs are believed to have induced five out of nine earthquakes on the Indian peninsula in the 1980s, which were strong enough to cause damage (seeber,1995). The earthquakes caused by large reservoirs are related with the extra water pressure created in the micro cracks and fissures in the ground under and near a reservoir. When the pressure of the water in the rocks increases, it acts to lubricate faults, which are already under tectonic strain, but are prevented from slipping by the friction of the rock surfaces. The most powerful earthquake thought to have been induced by a large reservoir is a magnitude 6.3 tremor which flattened the village of Koyana Nagar in Maharashtra on 11th December 1967,killing around 180 people, injuring 1500 and rendering thousands homeless. The dam was seriously damaged and the power supply to Mumbai and causing panic among its populace, who were able to feel the quake although 230 kilometers from its epicenter. The epicenter of the tremor and numerous foreshocks and aftershocks were all either near the 103 meter high Koyana or under its reservoir

Information on few African and Asian reservoirs revealed that the catch per unit area of water can vary by a factor of almost 200 between different reservoirs (Wellcome,1979).Some reservoirs may increase the total fish catch from a river,many,especially those which wipeout regular seasonal floods downstream and degrade estuarine ecosystems, will reduce the yield. In almost all cases, the fish diversity will drop. Furthermore, local people invariably have less access to fish than before, as reservoir fishing requires different skills and equipment and more capital than river fishing.

Different 459 plant species has been disappeared due to construction of Jaikwadi (Nathsagar) reservoir at village Paithan in Aurangabad district of Maharashtra.Among 1,165 plant species of Marathwada region of Maharashtra 48 species are totally destroyed due to different dams constructed in the region. When vegetation and soils are flooded by a reservoir, they release huge amounts of nutrients, which nourish a fish population, which is suddenly able to expand into a greatly increased habitat. Fishermen can therefore reap a bonanza in new reservoirs. After a number of years, however, when the flush of nutrients from rotting biomass has declined and the specific characteristics of river habitats start to die out, fish catches decline rapidly. In some cases, reservoir waters become depleted of oxygen and clogged with aquatic plants which both decrease fish productivity in the water they smother, and make it almost impossible to catch whatever fish do exist. After creation of a lucrative fishery, there is no guarantee that the local fishermen who have their livelihood affected by the dam will be able to reap the benefits. Often it is only outside entrepreneurs and those with experience of open water fishing that generally requires bigger boats and more expensive gear than river fishing who has the capital and knows how to exploit a newly created reservoir fishery and get the fish to the market. In this case local people lose their previously free access to the fish in the river, and then have to pay to eat fish from a much narrower range of species-from the reservoir. Fishes living many kilometers downstream from a dam who suffer a reduction in catches may be unable to take advantage of a reservoir fishery just because it is so far away. In many Indian states local people are deliberately prevented from taking full advantage of reservoir fisheries because commercial fishing rights in reservoirs are auctioned off to contractors, to who the local people have to sell their catches, usually at very low prices (Vivekanand, 1995).

Groundwater levels and salinity increases in the vicinity of reservoirs. This happened in Nagarjuna sagar reservoir area of Andhra Pradesh. The Bhakra canals in Rajasthan caused water logging and soil salinity (Sreenivsan, 2000). The Bhakra Canals (Rajasthan) caused water logging and soil salinity. In Pakistan, salinity problems cause a loss of 24,000 ha of fertile croplands each year. In Punjab seepage from canals raised the water table by 7 to 9 meters in 10 years. Constitution of a dam of Mae Klong river in Thailand resulted in a loss of thousands of sq km of fertile fields.

It has been stated on the basis of scientific research that maximum quantity of methane is being emitted from large reservoirs of India. As per the new research of Brazilian scientists there are 52,000 reservoirs in the whole world which emit 11.5 crore tones of methane in which maximum contribution is from 450 reservoirs of India, releasing 3.35 crore tones of methane in the atmosphere.Added to this, the reservoirs of Brazil themselves emit 2.18 crore tones of methane which is followed by the reservoirs of china, which contribute 0.44 crores tones of methane in the air. It is also mentioned that India should not be held responsible for this act as a policy outcome, but because of its physical status as a tropical country, where the temperature is normally high. The high range of temperature in the atmosphere is invariably increasing the emission of methane gas from Indian reservoirs. Scientists also believe that the problem of methane may be more in old reservoirs. It is due to high rate of sedimentation and accumulation of silt causing increased organic load through detritus.Inter-govermental Panel of Climatic Change (IPCC) has also stated that the figures of Brazilian research are somewhat threatening.

When an artificial lake is created, the shallow reservoir or shallow portions of deep reservoirs as well as downstream releases could be warmed to undesirable extent, preventing spawning of coldwater fishes prevailing remained by providing outlet works capable of selectively withdrawing coldwater from the lower reservoir depths with desired temperature conditions. In United States, Bureau of Reclamation have provided for selective withdrawal of coldwater at many dams to create favourable temperatures for fish spawning downstream of the dams.At Toa Vaca in Puerto Rica, six gates at different levels have been provided for selective withdrawal of specific temperature water throughout the year.

The dams and barrages form physical obstruction to the migratory fishes causing physiological strain and breeding failure. The classic example is Indian shad (*Hilsa ilisha*) that has migratory range of 1500 km upstream of the estuary. Construction of Farakka barrage, 476 km from the river mouth has nearly eliminated the lucrative *Hilsa* fishery above the barrage. Gone are the days when up to 303.6t of Hilsa were caught at Allahabad (1956-57).The impact of Farakka on Hilsa had been very severe. The average Hilsa landing after commission of Farakka barrage at Allahabad, Buxar and Bhagalpur get reduced by 94.61, 98.12 and 83.05 per cent respectively (Chandra,1989). Another bad effect has been found in the case of *Pangasius pangasius* in Ganga, Brahmaputra, Mahanadi, and Godavari ribers.Dams located in the lower and middle reaches of these rivers obstructed the migration of this fish and adversely affected its population. Torrential fishes like *Glyptothorax, Leptognathus* etc cannot survive in reservoirs and there is a chance of disappearance of their races in nature. The Gandak and Kossi valley projects have also adversely affected the fisheries of North Bihar to a very significant degree. They have affected not only the fisheries of rivers, mauns and chaurs, but indirectly culture fisheries as well as the spawn production, on which culture fisheries lean heavily, has been drastically reduced due to loss of breeding grounds. This has adversely affected the fishery as well as the seed resources of these water bodies, as also those of the main rivers.

In Volta dam the population of bivalve mollusc (*Egeria radiata*) decreased due to increased salinity, affecting the fisheries too. Edible oysters were also affected. Increase of fluoride content of groundwater is a result of impoundments. The belt of land between Mecheri and Morappur in Tamil Nadu is reported to have high fluoride because of Mettur reservoir (Sreenivasan, 2000)

References

Chandra, Ravish, 1989. Riverine fisheries resources of the Ganga and the Brahmputra. In: *Conservation and Management of Inland Capture Fisheries Resources of India,* (Ed.) A.G. Jhingran and V.V. Sugunan. Inland Fisheries Society of India, Barrackpore, pp. 33–39.

Chunhong, H., 1995. Controlling reservoir sedimentation in Chija. *International Journal of Hydropower and Dams.*

Desai, V.R., 2008. Emission of methane from reservoirs. *Fishing Chimes,* 28(4): 10.

Gleick, P.H., 1995. *Water in Crisis: A Guide to the World's Freshwater Resources,* Oxford University Press, Oxford, 1993, p. 367.

Peter, T., 1978. Tropical man-made lakes: Their ecological impact. *Arcives for Hydrobiologie,* 81: 368–385.

Sakhare, V.B. *Reservoir Fisheries and Limnology.* Narendra Publishing House, Delhi.

Sakhare, V.B., 2007. *Applied Fisheries.* Daya Publishing House, New Delhi.

Seeber, L., 1995. Lamont–Doherty Earth Obervatory, Pers.Comm., 18 January.

Sharma, Kalpana, 2001. Large dams: Rights and risks approach. *The Hindu Survey of the Environment, 2001,* pp. 87–87.

Sreenivasan, A., 2000. Large dams be damned: They are not eco-frendly. *Fishing Chimes,* 20 (1): 81–83.

Vivekanandan, K., 1995. A Dam fine effort. *Samudra Madrs,* April.

Wellcome, R.L., 1979. *Fisheries Ecology of Floodplain Rivers.* Longman, London, p. 251.

Chapter 8

Zooplankton Species Richness as Indicator of Environmental Condition in Chilean Lakes

☆ *Patricio De los Ríos, Luciano Parra and Marcela Vega*

Introduction

The zooplankton assemblage in Chilean lakes is characterized by its low species number and high dominance of calanoids copepods that is due mainly to the oligotrophic environment (Soto and Zúniga, 1991; De los Ríos and Soto, 2007). A different situation occurs in northern Chile where the low species number is associated to high salinity (De los Ríos and Crespo, 2004). An opposite situation occurs in northern hemisphere lakes that have high species number (Dodson, 1992; Dodson *et al.*, 2000, 2005, 2008; Dodson and Silva-Briano, 1996; Waide *et al.*, 2003; Willing *et al.*, 2003; Pinto-Coelho *et al.*, 2005).

The northern Chilean lakes (18-27°S) are associated to saline deposit of volcanic origin, and in this scenario, the conductivity is the main regulator factor of species number (De los Ríos and Crespo, 2004), in a second scenario, the trophic status has similar role in central Chile and Patagonian water bodies (33-51° S; Soto and Zúniga, 1991), and finally in southern Patagonian lakes and ponds (51° S), the regulator factors are trophic status and conductivity (Soto and De los Ríos, 2006).

The aim of the present study is analyze the available literature of zooplanktonic crustaceans in Chilean lakes between 23-51° S, for determine the role of conductivity and chlorophyll as regulator factors of species number, considering that the species number would indicate environmental condition.

Material and Methods

It was revised the literature of Chilean lakes and ponds between 23-51° S (De los Ríos, unpublished data; Schmid-Araya and Zúniga, 1992; Campos *et al.*, 1983, 1988, 1990, 1992a, b; Villalobos, 1999; Soto and De los Ríos, 2006), also the conductivity, latitude, surface, maximum depth, chlorophyll concentration and species number. To these data were applied a Principal Component Analysis (PCA) with the aim of determine the regulator factors for discriminate the studied sites. This statistical analysis was applied using the software Xlstat 5.0.

Results and Discussion

The results denoted the existence of a first group that included norhtern Chilean lakes with high conductivity and low species number (Miniques, Miscanti and Toconao), whereas a second group included two central Chilean small lakes (Penuelas and Rungue) with high species number and high chlorophyll concentration (Table 8.3 and Figure 8.1). A third main group joined large Patagonian lakes and ponds that are an heterogeneous group of large and deep oligotrophic and oligo-mesotrophic lakes, with low conductivity and low species number, also are included small shallow mesotrophic ponds with a wide conductivity gradient and high species number (Table 8.3 and Figure 8.1).

The correlation matrix, revealed an inverse relation between chlorophyll concentration and latitude and direct association between surface and maximum depth (Table 8.1). The results of PCA revealed that the variables contributed with 59.81 per cent, for the first axis the most important variables were latitude, maximum depth, and surface that contributed with a 34.82 per cent, whereas in the second axis the most important variables were the conductivity, chlorophyll "a" concentration, and species number contributed with a 24.99 per cent (Table 8.2).

Table 8.1: Correlation Matrix for Variables Considered in the Present Study

	Surface	*Maximum Depth*	*Chlorophyll "a"*	*Conductivity*	*Species number*
Latitude	−0.071	0.098	−0.405	−0.231	0.293
Surface		0.588	−0. 199	−0.219	−0.035
Maximum depth			−0.332	−0.363	−0.159
Chlorophyll "a"				0. 198	−0.146
Conductivity					−0.206

The results obtained about the role of conductivity and chlorophyll concentration agree with descriptions for lakes and ponds of Torres del Paine National Park that are characterized by its environmental heterogeneity (Soto and De los Ríos, 2006; De los Ríos and Soto, 2009). The results of northern Chilean lakes and the role of conductivity as regulator of zooplankton assemblage agree with the first preliminary literature descriptions where the calanoid *Boeckella poopoensis* can be exclusive

component at salinities lower than 90 g/l (Hurlbert *et al.*, 1986; Williams *et al.*, 1995; De los Ríos and Crespo, 2004). A different situation occurs in central Chilean water bodies and northern Patagonian lakes where the trophic status is the main regulator factor, and under low food availability is associated to few species number (Campos, 1984; Soto and Zúniga, 1991; De los Ríos and Soto, 2007, 2009). These results are similar with descriptions for Northern Andean and Patagonian lakes (Quiros and Drago, 1999), mainly in Patagonian water bodies, where

Table 8.2: Percentage of Contribution of Variables for Studied Sites

	F1	*F2*
Latitude	10.378	28.720
Surface	18.816	19.166
Maximum depth	28.153	14.472
Chlorophyll "a"	21.135	4.452
Conductivity	19.559	0.927
Species number	1.960	32.263

Table 8.3: Latitude, Surface, Maximum Depth, Chlorophyll "a" Concentration, Conductivity and Species Number for Studied Sites

	Latitude	Surface (km²)	Maximum Depth (m)	Chlorophyll "a" (µg/l)	Conductivity (mS/cm)	Species Number	Reference
Toconao	23°11' 68°00'	1.00	3.00	173.20	0.40	1	De los Ríos unpublished data
Miscanti	23°44' 67°48'	13.40	9.00	1.00	7.73	3	De los Ríos unpublished data
Miniques	23°45' 67°48'	1.60	5.00	1.00	15.42	1	De los Ríos unpublished data
Rungue	33°00' 70°53'	0.48	15.00	65.90	0.33	4	Schmid-Araya and Zúñiga 1992
Peñuelas	33°10' 71°29'	19.00	15.00	57.40	0.12	11	Schmid-Araya and Zúñiga 1992
Villarrica	39°18' 72° 07'	175.80	185.00	0.40	0.06	7	Campos et al., 1983
Pirihueico	39°50' 71°48'	30.50	145.00	0.60	0.05	5	Wölfl, 1996
Riñihue	39°50' 72°20'	77.50	323.00	1.20	0.05	7	Wölfl, 1996
Ranco	40°13' 70°22'	442.60	199.00	0.80	0.07	9	Campos et al., 1992a
Puyehue	40°40' 72°30'	165.40	123.00	2.10	0.08	7	Campos et al., 1989
Rupanco	40°50' 72°30'	235.00	273.00	1.20	0.06	4	Campos et al., 1992b
Llanquihue	41°08' 72°50'	870.50	317.00	0.50	0.11	3	Campos et al., 1988
Todos los Santos	41°08' 72°50'	178.50	335.00	0.40	0.04	4	Campos et al., 1990
Los Palos	45°19' 72°42'	59.00	5.00	0.80	0.02	4	Villalobos, 1999
Riesco	45°39' 72°20'	14.70	130.00	0.90	0.02	4	Villalobos, 1999
Escondida	45°49' 72°40'	43.00	7.00	0.48	0.01	5	Villalobos, 1999
Larga	51°01' 72°52'	0.10	5.00	0.88	3.45	6	Soto and De los Ríos, 2006
Cisnes	51°01' 72°52'	0.10	1.00	93.70	16.51	6	Soto and De los Ríos, 2006
Redonda	51°01' 72°54'	0.10	3.00	2.25	1.49	8	Soto and De los Ríos, 2006
Juncos	51°01' 72°52'	0.10	3.00	1.82	2.29	6	Soto and De los Ríos, 2006
Nordensjold	51°01' 72°56'	25.00	200.00	0.25	0.13	2	Soto and De los Ríos, 2006
Paso de la Muerte	51°02' 72°54'	0.10	3.00	10.00	0.80	7	Soto and De los Ríos, 2006
Jovito	51°02' 72°54'	0.10	3.00	3.96	1.38	5	Soto and De los Ríos, 2006
Melliza Oeste	51°03' 72°57'	0.13	25.00	3.24	0.66	8	Soto and De los Ríos, 2006
Melliza Este	51°03' 72°57'	0.12	16.00	5.32	0.80	8	Soto and De los Ríos, 2006
Sarmiento	51°03' 72°37'	86.00	312.00	0.34	0.85	5	Campos et al., 1994b
Grey	51°07' 72°53'	15.00	200.00	3.18	0.20	2	Soto and De los Ríos, 2006
Del Toro	51°12' 72°38'	196.00	300.00	0.35	0.07	4	Campos et al., 1994a

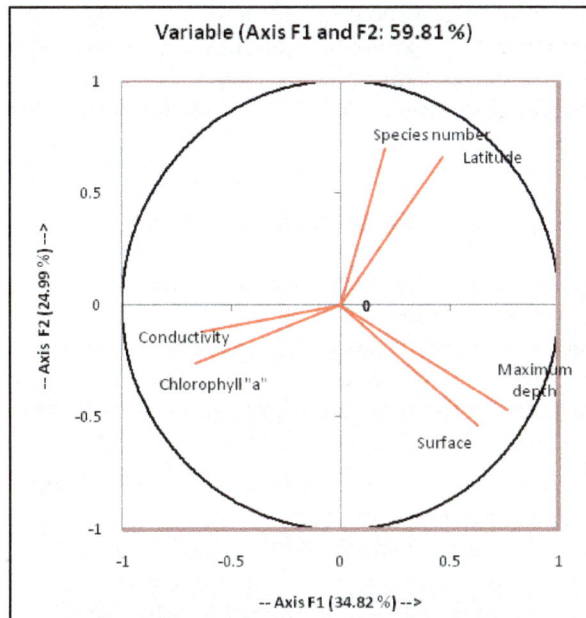

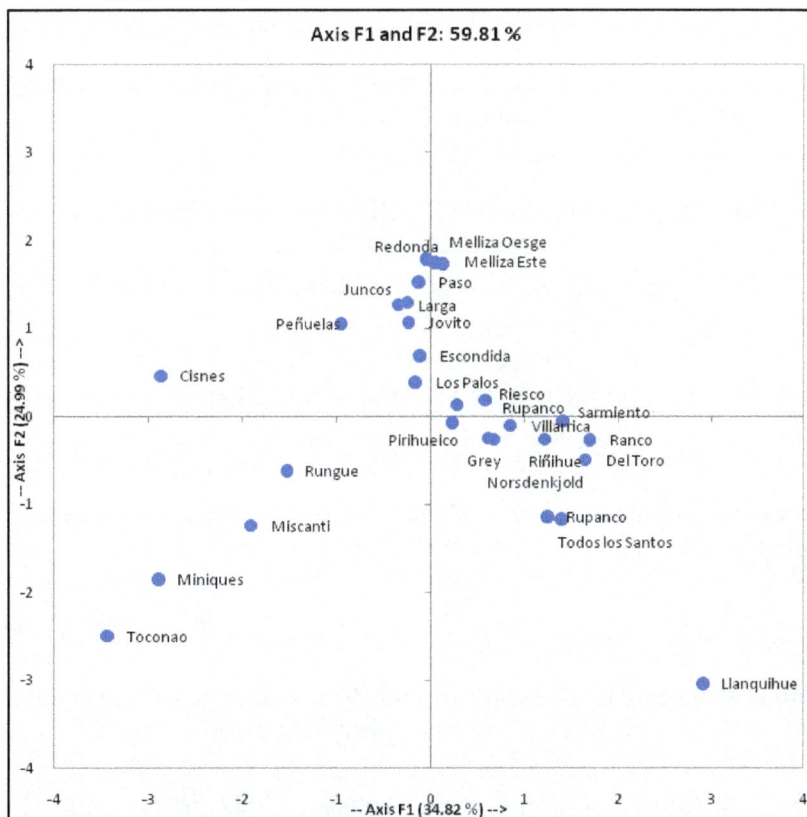

Figure 8.1: Results of PCA for Studied Sites

it is possible found water bodies with marked environmental heterogeneity (Modenutti *et al.*, 1998). These results agree with descriptions for New Zealand lakes and ponds (Jeppensen *et al.*, 1997, 2000). These results agree with integrative descriptions of northern hemisphere lakes, where the chemical characteristics mainly trophic status are the main regulator factors of species number (Dodson, 1992; Dodson *et al.*, 2000, 2005, 2008; Dodson and Silva-Briano, 1996; Waide *et al.*, 2003; Pinto-Coelho *et al.*, 2005; Karatayev *et al.*, 2008).

Other important factor mainly for Patagonian lakes, is the role of natural ultraviolet radiation, that has increase its penetration due ozone depletion, and it can penetrate into water column (Morris *et al.*, 1995). In this scenario, only the tolerant species can occurs (Marinone *et al.*, 2006), and under oligotrophic status, the low species number is enhanced due ultraviolet radiation effects, whereas under mesotrophic status there are high species number due high food availability (De los Ríos and Soto, 2005, 2007; Marinone *et al.*, 2006). Also, in southern Patagonian lakes, the depth provides protection against natural ultraviolet radiation that generates migrations of vulnerable species to zones without natural ultraviolet radiation exposure (Villafane *et al.*, 2001; Alonso *et al.*, 2004).

The exposed results revealed the existence of a notorious geographical gradient that marks environmental heterogeneity with the consequent effects in species diversity that agrees with observations for northern hemisphere ecosystems (Willing *et al.*, 2003). Nevertheless, it is necessary do more integrative and systematic studies for understand ecological process at macro-scales, and compare with other similar descriptions for northern hemisphere or for other kind of environments.

Acknowledgements

This study was funding by the General Research Direction of the Catholic University of Temuco (Project Funding for Development of Limnology, DGI–DCA 2007–01).

References

Alonso C., Rocco, V., Barriga, J.P., Battini M.A. and Zagarese, H., 2004. Surface avoidance by freshwater zooplankton: Field evidence on the role of ultraviolet radiation. *Limnol. Ocean*, 49(1): 225–232.

Campos, H., 1984. Limnological study of Araucanian lakes (Chile). *Verh. Int. Ver. Angew. Limnol.*, 22: 1319–1327.

Campos, H., Soto, D., Steffen, W., Agüero, G., Parra, O. and Zúniga, L., 1994a. Limnological studies in lake del Toro (Chile): Morphometry, physics and plankton. *Arch. Hydrobiol (Suppl.)*, 81: 217–234.

Campos, H., Arenas, J., Steffen, W., Román, C. and Agüero, G., 1982a. Limnological study of lake Ranco (Chile): Morphometry, physics and plankton. *Arch. Hydrobiol.*, 94: 137–171.

Campos, H., Steffen, W., Agüero, G., Parra, O. and Zúniga, L., 1983. Limnological studies in lake Villarrica. Morphometry, physics, chemistry and primary productivity. *Arch. Hydrobiol (Suppl.)*, 81: 37–67.

Campos, H., Steffen, W., Agüero, G., Parra, O. and Zúniga, L., 1987. Limnology of lake Rinihue. *Limnologica*, 18: 339–357.

Campos, H., Steffen, W., Agüero, G., Parra, O. and Zúniga, L., 1988. Limnological study of lake Llanquihue (Chile): morphometry, physics, chemistry and primary productivity. *Arch. Hydrobiol. (Suppl.)*, 81: 37–67.

Campos, H., Steffen, W., Agüero, G., Parra, O. and Zúniga, L., 1989. Estudios limnológicos en el lago Puyehue (Chile): morfometría, factores físicos y químicos, plancton y productividad primaria. *Med. Amb.*, 10: 36–53.

Campos, H., Steffen, W., Agüero, G., Parra, O. and Zúniga, L., 1990. Limnological study of lake Todos los Santos (Chile): morphometry, physics, chemistry and primary productivity. *Arch. Hydrobiol. (Suppl.)*, 117: 453–484.

Campos, H., Steffen, W., Agüero, G., Parra, O. and Zúniga, L., 1992a. Limnological study of lake Ranco (Chile). *Limnologica*, 22: 337–353.

Campos, H., Steffen, W., Agüero, G., Parra, O. and Zúniga, L., 1992b. Limnological studies of lake Rupanco (Chile): Morphometry, physics, chemistry and primary productivity. *Arch. Hydrobiol. (Suppl.)*, 90: 85–113.

De los Ríos, P. and Soto, D., 2007. Crustacean (copepoda and cladocera) zooplankton richness in Chilean Patagonian lakes. *Crustaceana*, 80: 285–296.

De los Ríos, P. and Soto, D., 2005. Survival of two species of crustacean zooplankton under to two chlorophyll concentrations and protection or exposure to natural ultraviolet radiation. *Crustaceana*, 78: 163–169.

Dodson, S.I., 1992. Predicting crustacean zooplankton species richness. *Limnol. Oceanogr.*, 37: 848–856.

Dodson, S.I., Arnott, S.E. and Cotttingham, K.L., 2000. The relationship in lakes communities between primary productivity and species richness. *Ecology*, 81: 2662–2679.

Dodson, S.I. and Silva-Briano, M., 1996. Crustacean zooplankton species richness and associations in reservoirs and ponds of Aguas Calientes, México. *Hydrobiologia*, 325: 163–172.

Dodson, S.I., Lillie, R.A. and Will-Wolf, S., 2005. Land use, water chemistry, aquatic vegetation and zooplankton community structure of shallow lakes. *Ecol. Appl.*, 15: 1191–1198.

Jeppensen, E.T., Lauridsen, L., Mitchell, S.F., Christoffersen, K. and Burns, C.W., 2000. Trophic structure in the pelagial of 25 shallow New Zealand lakes: changes along nutrient and fish gradients. *Journ. Plankt. Res.*, 22: 951–968.

Karatayev, A.Y., Burlakova, L.E. and Dodson, S.I., 2008. Community analysis of Belarusian lakes: correlations of species diversity with hydrochemistry. *Hydrobiologia*, 605: 99–112.

Marinone, M.C., Menu-Marque, S., Anón-Suarez, D., Diéguez, M.C., Pérez, A.P., De los Ríos, P., Soto, D. and Zagarese, H.E., 2006. UV radiation as a potential driving force for zooplankton community structure in Patagonian lakes. *Photochem. Photobiol.*, 82: 962–971.

Pinto-Coelho, R., Pinel-Alloul, B., Méthot, G. and Havens, K.E., 2005. Crustacean species richness in lakes and reservoirs of temperate and tropical regions: variations with trophic status. *Can. J. Fish. Aquat. Sci.*, 62: 348–361.

Schmid-Araya, J.M. and Zuniga, L.R., 1992. Zooplankton community structure in two Chilean reservoirs. *Arch. Hydrobiol.*, 123: 305–335.

Soto, D. and De los Ríos, P., 2006. Trophic status and conductivity as regulators in daphnidd dominance and zooplankton assemblages in lakes and ponds of Torres del Paine National Park. *Biologia Bratislava*, 61: 541–546.

Soto, D. and Zúniga, L.R., 1991. Zooplankton assemblages of Chilean temperate lakes: A comparison with North American counterparts. *Rev. Chil. Hist. Nat.*, 64: 569–581.

Villalobos, L., 1999. Determinación de capacidad de carga y balance de fósforo y nitrógeno de los lagos Riesco, Los Palos, y Laguna Escondida en la XI región. Technical Repor Fisheries Research Foundation-Chile, FIP–IT/97–39.

Villafane, V.E., Helbling, E.W. and Zagarese, H.E., 2001. Solar ultraviolet radiation and its impacts on aquatic ecosystems in Southern Patagonia. *Ambio*, 30: 112–117.

Villalobos, L., Woelfl, S., Parra, O. and Campos, H., 2003. Lake Chapo: A base line study of a deep, oligotrophic North Patagonian lake prior to its use for hydroelectricity generation. II. Biological properties. *Hydrobiologia*, 510: 225–237.

Waide, M.R., Willing, M.R., Steiner, C.F., Mittelbach, G., Gough, L., Dodson, S.I., Juday, G.P. and Parmenter, R., 1999. The relationship between productivity and species richness. *Ann. Rev. Ecol. Evol. Syst.*, 30: 257–300.

Willing, M.R., Kaufman, D.M. and Stevens, R.D., 2003. Latitudinal gradient of biodiversity: Pattern, process, scale and synthesis. *Ann. Rev. Ecol. Evol. Syst.*, 34: 273–309.

Wölfl, S., 1996. Untersuchungen zur Zooplanktonstruktur einschliesslich der mikrobiellen Gruppen unter Berücksichtigung der mixotrophen ciliaten in zweiüdchilenischen Andenfubseen: 1–242 (Universität Konstanz, Konstanz).

Chapter 9

Toxicity Evaluation of Dimethoate to Fish, *Macrones vittatus*

☆ *M.V. Lokhande, D.S. Rathod, V.N. Waghmare and V.S. Shembekar*

Intorduction

The aquatic environment is continuously being contaminated with toxic chemicals generated by human, industrial, agricultural and domestic activities. Pesticides are the xenobiotic substances that have been used in India for a longer period for management of pests in agricultural fields and control of vectors in public health operations.

The most of the insecticides are so hydrophobic that they can be easily absorbed by soil particles and can migrate to natural water systems such as river, lake and pond through the runoff, causing severe aquatic pollution (Ding and Wu 1993, Odonaka *et al.*, 1994). Consequently these xenobiotic molecules have been found in natural water systems and they have a great impact on the environmental quality (Li and Migita 1992). They become accumulated in aquatic organisms and can enter the food chain (Svensaon *et al.*, 1994). Evaluation of the toxicity of pesticides contribute much towards arriving at better means of use and also involving new formulation, which would give results. Chemicals other than required by organisms, especially toxic ones like pesticides pose a health hazards to non-target species and thus gain much importance in screening of pesticide. Hence the present investigation is carried out to toxicity of pesticides on aquatic organism like fish.

Materials and Methods

In present investigation, fishes were collected from Manjara river, Latur and brought to laboratory. These fishes were observed for any pathological symptoms and then placed in a dilute bath of 0.1 per cent potassium permagnate ($KMnO_4$) for 2 minutes so as to avoid any dermal infection. The fish were then washed with water and acclimatized to laboratory conditions for few days in glass aquaria.

During acclimatization, the fish were provided with a diet consisting of live earthworms. Food supply was withdrawn 24 hours prior to the experimentation. Fishes of almost same size measuring 9±2 cm and weighing about 7±2 gms were selected for experimentation.

Different concentrations were made from stock solution as per dilution method suggested by APHA (1998). Fresh stock solutions were used for each exposure.

Static bioassay experiments were conducted as suggested by Doudoroff *et al.* (1951). During exposure period, the animals were starved. Thirteen (13) concentrations were tried. For each concentration, 10 fishes were exposed in 50 liter test solution. This arrangement was made to maintain almost similar ratio of fish weight to water volume in the experiments. Each experiment was repeated six times. The number of fish killed in each concentration was recorded at regular intervals of 24, 48, 72 and 96 hours. The average mortality in each concentration was calculated and LC_{50} values for different intervals of time for dimethoate were calculated by three different methods. (*i*) Statistical (probit analysis Finney, 1971) (*ii*) Dragstedt and Behren's (1975) (*iii*) Graphical method.

Statistical Method

This method makes use of probit analysis (Finney, 1971). The percent mortality was converted into probit mortality and the values were plotted against pesticide concentration in a double logarithmic grid. To fit the straight passing through line which is the pesticide concentration at which there was a probit kill of 50 per cent was noted to represent LC_{50} for that exposure period.

Graphical method

The dose-response curves were fitted by plotting percent mortality Vs pesticide log concentration for fix period in single graph. Lines were fitted and concentration of pesticide at which there was 50 per cent mortality was noted to represent the LC_{50} at 24, 48, 72 and 96 hr of exposure.

Dragstedt–Behren's Method (1975)

In this method cumulative mortality was determined at different concentrations of pesticide and per cent mortalities were calculated from cumulative mortalities values. LC_{50} values were calculated by adopting the formula

$$LC_{50} = A + \frac{50-a}{b-a} \times 2$$

The application of the log to this formula will make the formula as follows:

$$Log\ LC_{50} = LogA + \frac{50-a}{b-a} \times Log2$$

where,

    A: Concentration of the pesticides having the percentage of mortality below 50 per cent

    a: Percentage of mortality below 50 per cent.

    b: Percentage of mortality immediately above 50 per cent.

The values 'A', 'a' and 'b' were obtained after subjecting the recorded observations to cumulative mortality at 24, 48, 72 and 96 hours for two pesticides.

The cumulative percentage mortality is calculated by adding the live and dead number of cumulative mortality and the number of dead animals is divided by live and dead number which is further multiplied by 100.

The LC_{50} values were calculated making use of the formula at 24, 48, 72 and 96 hours for two pesticides.

Determination of safe level or Safe concentration or presumably harmless concentration:

The LC_{50} values are useful in the final evaluation of classifying 'Safe level' or 'Tolerable level' of pollution to the aquatic biota and this will pave the way in establishing 'limits and level of susceptibility' by the biotic components. An estimate of the presumably harmless (safe) concentration of the pesticide can be calculated using the following formula (Hart *et al.*, 1945).

$$C = \frac{LC_{50} / 48hrs \times A}{S^2}$$

where,

C: Presumably harmless concentration.

A: Application factor, (0.3)

$$S = \frac{LC_{50} / 24hrs}{LC_{50} / 48hrs}$$

Results and Discussion

In present investigation static bioassay test was selected to see the toxicity of dimethoate on *Macronus Vittatus*. In this bioassay test mortality was found at 5.0 ppm in dimethoate. The LC_{50} of dimethoate were determined three methods which represented in Table 9.1. In present investigation the average LC_{50} values and safe concentration of dimethoate were 7.73, 7.15, 6.58 and 5.99 ppm for 24, 48, 72 and 96 and safe concentration is 1.657 ppm and 0.615 ppb represent Table 9.2.

Table 9.1: LC_{50} and Average LC_{50} Values Calculated by Three Methods to *Macronus vittatus*

Sl.No.	Name of Toxicant	Time of Exposure	LC_{50} Values			Average LC_{50} Values
			Graphical Method	Probit Analysis	Drasted and Behren's Method	
1.	Dimethoate	24	7.0 ppm	6.998 ppm	2.191 ppm	7.73 ppm
		48	6.5 ppm	6.468 ppm	8.486 ppm	7.15 ppm
		72	6.0 ppm	5.965 ppm	7.778 ppm	6.58 ppm
		96	5.5 ppm	5.392 ppm	7.071 ppm	5.99 ppm

The similar results were reported Pickering *et al.* (1966), Sivaprasad Rao *et al.* (1980), Swarup *et al.* (1981), Arora *et al.* (1971a), Arora *et al.* (1971b), Pankaj *et al.* (2004), Joyti and Narayan (1996), Singh and Narain (1982), Anandswamp *et al.* (1981), Reddy (1977, Vasait *et al.* (2005) studied toxic evaluation of organophosphate insecticide monocrotophos on the edible fish species *Namacheilus botai* for a period of 7 and 14 days and showed that the LC_{50} values were 49.6 and 42.0 ppm respectively. The observed result indicates that the mortality of the test fish to monocrotophos was dose dependent.

Sivaprasad Rao *et al.* (1980) studied toxicity of methyl parathion on freshwater teleost, *Tilapia mossambica* and reported that the LC_{50} value was 0.266 ppm.

Table 9.2: Relative Toxicity of Pesticides on the *Macronus vittatus*
(Calculated by D.J. Finney 1971)

Name of Toxicant	Time of Exposure	Regression Equation $Y= = \bar{y} +b(x- \bar{x})$	LC_{50} Values	Safe Concentration or Presumably Harmless Concentration
Dimethoate	24	Y = 5.9284986X −0.0135217	6.998ppm	1.657ppm
	48	Y = 7.219508X −0.8391218	6.468 ppm	
	72	Y= 7.018011X −0.436984	5.965 ppm	
	96	Y = 7.3625516X −0.3827795	5.392 ppm	

Prashanth (2006) studied impact of cypermethrin on protein metabolism of freshwater fish, *Cirrhinus mrigala* and showed that the LC_{50} value was 5 mg/l for days.Naveed, *et al.* (2006) studied toxicity of lihocin on the activities of glycolytic and gluconeogenic enzyme of fish, *Channa punctata* and showed that the LC_{50} values was 19.19 ppm for 48 hours. Sivakumar *et al.* (2006) studied acute toxicity of chromium on behavioural changes in freshwater fish, *Mystus vittatus*, for period of 24, 48, 72 and 96 hours and reported that the LC_{50} values were 82.79, 72.11, 64.42 and 61.67 mg/l respectively.

Prashanth *et al.* (2003) studied effects of cypermethrin on toxicity and oxygen consumption in the freshwater fish, *Cirrhinus mrigala* and reported that the LC_{50} value was 5.13 mg/l for 96 hours. Prabhakara Rao and Radhakrishnaiah (2006) studied pesticidal impact on protein metabolism of the freshwater fish, *Cyprinus carpio* (Lin) and reported that the 48 hour LC_{50} values of furadan, endosulfan, chloropyrifos and mixture of this three were 20.5, 2.2, 0.12 and 7.5 mg/l respectively. Paraskar *et al.* (2005) studied effects of cypermethrin on three selected freshwater fish, *C. orientalis, C. batrachus* and *H. fossilis* and reported that the LC_{50} values were 1.5, 2.5. and 3.5 ppm respectively. Veena Sakthivel and Gaikwad (2002) studied tissue histopathology of *Gambusia affinis* Baird and Girad under demecron toxicity and reported that the LC_{50} value was 0.34 ppm. The above literature of toxicology also clears that LC_{50} values decreases with increase in exposure period suggesting that with increase in duration of exposure the pesticide becomes toxic even at lower concentration

References

APHA, 1998. *Standard Method for the Examination of Water and Wastewater,* 20th edn. American Public Health Association Washington, D.C.

Anandswarup, P., Mohan Rao, D., Murthy, A.S. 1981.Toxicity of endosulfan to freshwater fish, *Cirrhinus mrigala Bull. Environ. Contam. Toxicol.,* 27(6): 850–855.

Ding, J.Y. and Mu, S.C., 1993. Laboratory studies on the effect of dissolved organic material on the absorption of organochlorine pesticides by sediments and trasnport in rivers. *Wat. Sci. Tech.,* 28(8–9): 199–208.

Doudoroff, P., Anderson, B.O., Burdick, G.E., Galtsoff, P.S., Hart, W.E., Parick, R. Strong E.R., Surber, E.W. and Vanhorn, W.M., 1951. Bioassay methods for the evaluation of acute toxicity of industrial wastes to fish, sewage. *Industr. Wastes,* 23: 1380–1397.

Finney, D.J., 1971. *Probit Analysis*, 3rd edn. Cambride University Press.

Jyoti, B. and Narayan, G., 1996. Effect of organophosphrous insecticide phorate on Gonads of freshwater fish, *Clarias batrachus* (Linn). *Poll. Res.*, 15(3): 293–296.

Kumar, Pankaj, Sharma, B. and Mishra, A.P., 2004. Efficiancy of malathion on mortality of a freshwater air breathing catfish, *Heteropneustes fossilis* (Biotech) during different developmental stages. *Ecol. Env. and Cons.*, 10(1): 47–52.

Li, S. and Migita, J., 1992. Pesticide run-off from paddy field and its impact on receiving water. *Water Sci. Tech.*, 25(11): 67–76.

Macek, K.J., Hutchinson, C. and Cope, C.B., 1969. Effects of temperature on the suceptibility of blue gills and rainbow trout to selected pesticides. *Bull. Environ. Contam. Toxicol.*, 4: 174.

Naveed, Abdul, Venkateshwarlu, P. and Janailah, C., 2006. Toxicity of lihocin on the activities of glycoytic and glycogenic enzymes of fish, *Channa punctatus. Nature Environ. and Poll. Techn.*, 5(1): 79–88.

Odonaka, Y., Taniguchi, T., Shimamaura, Y., Iijima, K., Koma, Y., Techechi, T. and Matano, O., 1994. Runoof and leaching of pesticides in Gulf coast. *J. Pestic. Sci.*, 19: 1–10.

Prashant, M.S., David, M. and Kuri, Riveendra C., 2003. Effects of cypermethrin on toxicity and oxygen consumption in the freshwater fish, *Cirrhinus mrigala. J. Ecotoxicol. Environ. Monit.*, 13(4): 271–277.

Prabhakar Rao, K. and Radhakrishnaiah, K., 2006. Pesticidal impact on protein metabolism of freshwater fish, *Cyperinus carpio. Nature Environment and Pollution Technology*, 5(3): 367–374.

Prashanth, M.S., 2006. Impact of cypermethrin on protein metabolism in freshwater fish, *Cirrinus mrigala. Nature Environ. and Poll. Tech.*, 5(2): 321–325.

Paraskar, P.S., Deshmukh, S.P., Kulkarni, K.M. and Jadhav, R.G., 2005. Effects on three selected freshwater fishes exposed to cypermethrin. *J. Aqua. Biol.*, 20(2): 187–192.

Reddy, T.G.K. and Gomathy, S., 1977. Toxicity and respiratory effects of pesticide thiodan on catfish, *Mystus vittatus. J. Environ. Hlth.*, 19: 361–363.

Svensson, B.G., Hallberg, T., Nilson, A., Schutz, A. and Hagmar, L., 1994. Parameters of immunological competence subjects with high consumption of fish contaminated with persistant organochlorine compounds. *Ind. Arch. Occup. Environ. Health*, 65: 351–358.

Siva Prasad Rao, K., 1980. Studies on Some aspects of metabolic changes with empasis on carbohydrate utility in the cell free systems of the teleost, *T. mossambica* (Perters) under methyl parathion exposure. *Ph.D. Thesis,* S.V.University, Tirupati, India.

Swarup, P.A., Rao, P.M. and Murthy, A.S., 1981.Toxicity of endosulfan to the freshwater fish, *Cirrhinus mrigala. Bull. Environ. Contam. Toxicol.*, 27: 850–854.

Singh, Braj Bhusan and Narain, Arun Shanker, 1982. Acute toxicity of thiodan to catfish *Heteropheustes fossilis. Bull. Environ. Contam. Toxicol.*, 28: 122–127.

Vasait, J.D. and Patil, V.T., 2005. The toxic evaluation of organophosphorous insecticide monocrotophos on the edible fish species *Nemacheilus botia. Eco. Env. and Cons.*, 8(1): 95–98.

Veena, Sakthival and Gaikwad, S.A., 2002. Tissue histopathology of *Gambusia affinis* (Baird and Girard) under Dimecron Toxicity. *Ecol. Env. and Cons.*, 8(1): 27–31.

Chapter 10

Limnological Studies on Kham River at Aurangabad (M.S.) with Special Reference to Larval Chironomids

☆ *P.R. Bhosale, R.J. Chavan and A.M. Gaikwad*

Introduction

The Chironomidae is the most widely distributed group of insect larvae and occurs in high density and diversity in most types of freshwater ecosystem. The range of environmental conditions under which chironomids are found is more extensive than that of any other group of aquatic insects. Chironomids are the only free-living holometabolous insects which have four stages (egg, larva, pupa and adult) in the life cycle. Chironomidae (Diptera), is a family of small flies whose larval and pupal stages are mainly aquatic. For the larvae the common name "Blood worms" was derived due to the presence of red blood pigment, haemoglobin. The chironomids are one of the most useful group in assessing the quality of running water.

In addition to the changes in the composition and community of the species larval deformities such as mentum, mandible, pectin epipharynx, etc. can be used as sensitive indicators of water quality (Bhattacharya, 2006). The current work was undertaken to know the limnology of Kham River through composition of chironomidae.

The historic metropolitan city Aurangabad is situated on latitude 19° 53'59'' and longitude 70° 20' east along the bank of river Kham. Kham river originates near Harsool dam and flows towards Paithan (Nathsagar). While passing through the urban area of Aurangabad it carries domestic sewage of Aurangabad city.

Materials and Methods

Sampling was done at three spots of the river, viz: A; Sant Asaram Bapu Ashram, B; Pan-Chakki and C; Karnpura. The water samples were collected once a month from May 2009 to October 2009. Water samples were analyzed for some physical and chemical characteristics by method given in Trivedi *et al.* (1995), APHA (1997). The larval chironomids were hand sorted and permanent slides were prepared by treating them with 10 per cent KOH and mounted in Canada Balsam. Each specimen was examined for finite taxonomic features under 450 and 1000 magnifications. Species identification was done using identification manual prepared by Epler (2001).

Results and Discussion

Physico-chemical Status of Kham River

The results of physico-chemical characteristics are given in Table 10.1. Temperature of water varied from 15.9°C to 29.4°C. The pH value in all samples were found approximately 8.00, indicates slightly alkaline nature. Minimum dissolved oxygen concentration (*i.e.* 0 mg/l) was recorded at site C in September and it was ranges from 0mg/l to 8.8mg/l throughout the river. Carbon dioxide levels and total alkalinity was higher in October. Phosphate and nitrate generally peaked during September The hardness values were found in range of 360 to 480mg/l and chloride ranges from 84.5 to 133 mg/l. The physico-chemical analysis of the water from the Kham river reveals that it is highly eutrophic. Syed Quadri (2004). He also studied the impact of physico-chemical variables on zooplankton population. The parameters like temperature, dissolved oxygen and water current also affects the diversity of chironomids (Bhattacharya *et al.*, 2006). Kimio Hirabayashi *et al.* (2006), noted that the

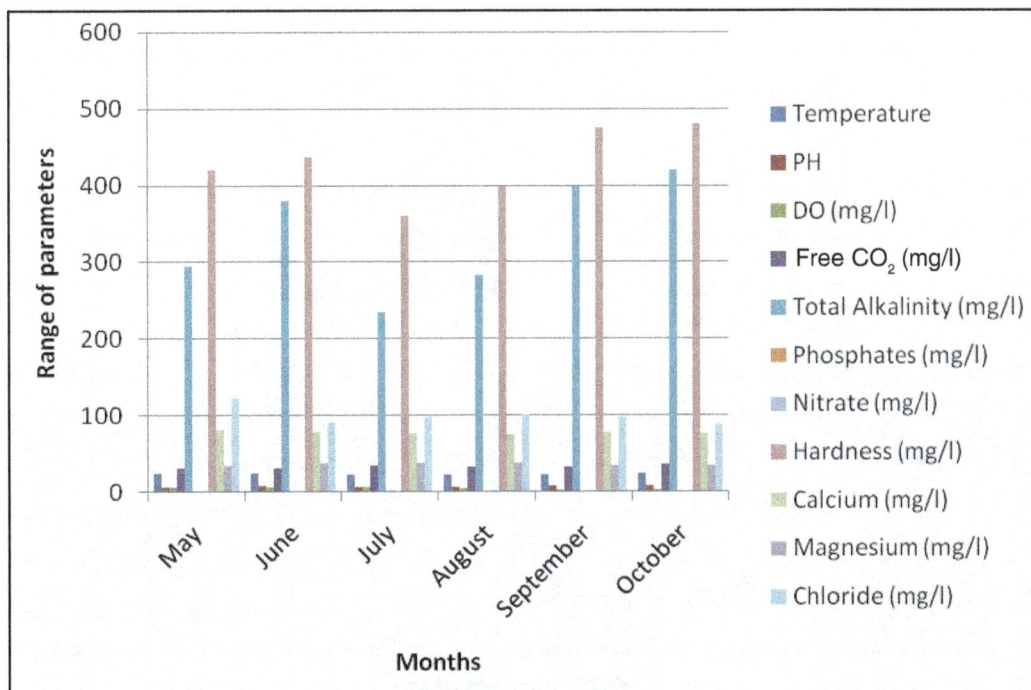

Figure 10.1: Range of Physico-Chemical Parameters of Kham River at Aurangabad

Figure 10.2: Chironomid Larvae Recorded in Kham River at Aurangabad

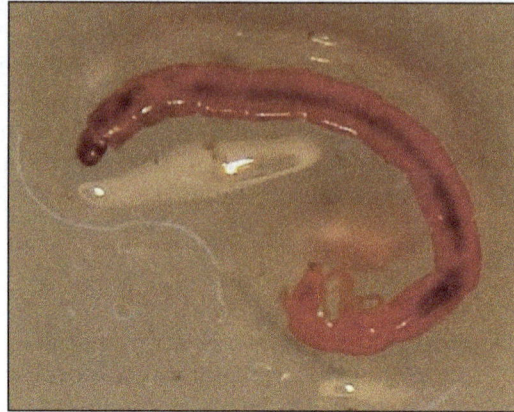

Larva of *Chironomus riparius*

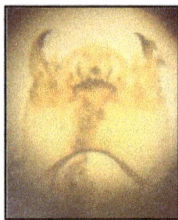

Head capsule

Mentum

Mandible

Antenna

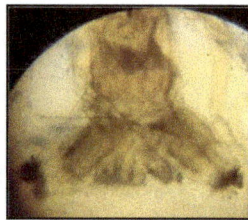

Anal tubule

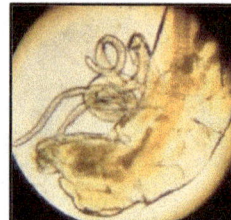

Ventral tuble

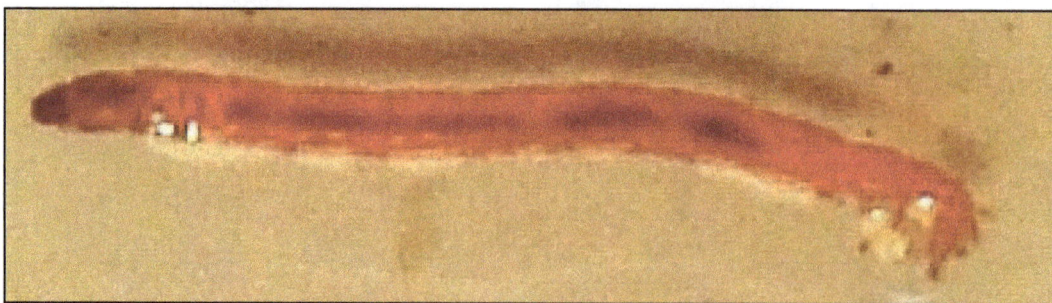

Larva of *Chironomus plumosus*

Contd...

Figure 10.2–Contd...

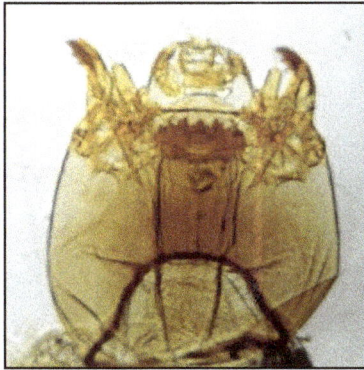  

Head capsule   **Mandible**   **Mentum**

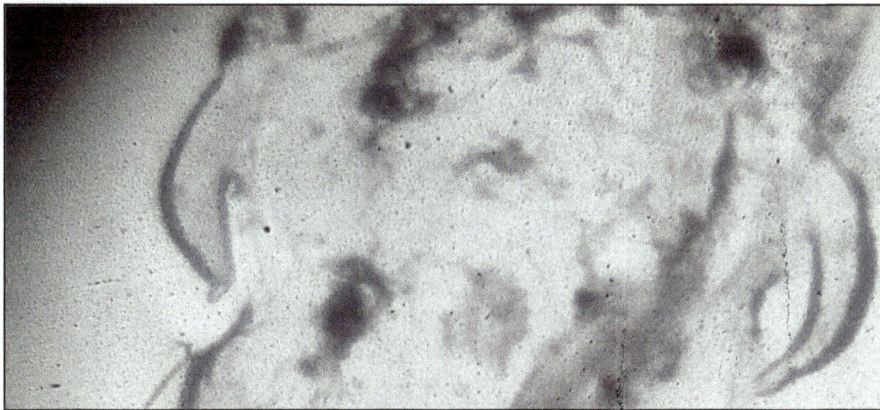

Caudolateral tubule

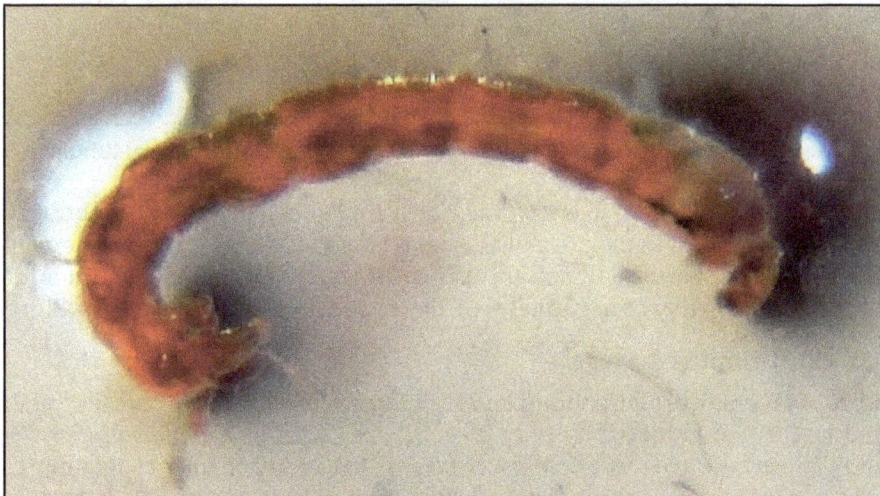

Larva of *Chironomous stigmaterus*

Contd...

Figure 10.2–Contd...

Mandible

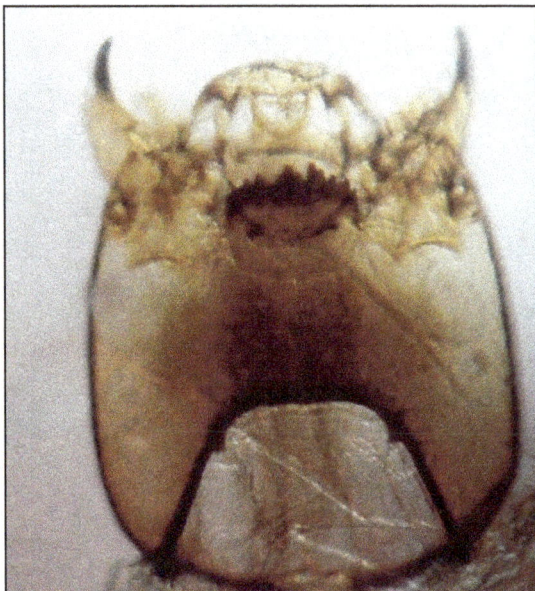

Head capsule

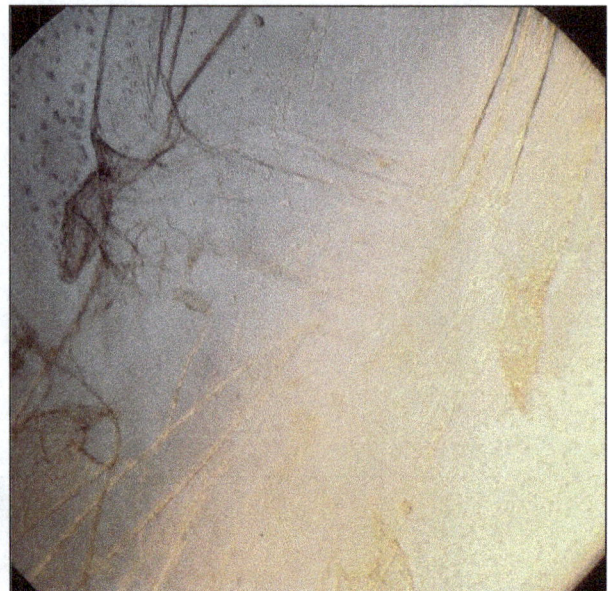

Caudolateral tubule

percentage of *Chironomus plumosus* in eutrophic lake, Kawaguchi, Japan is near about 20.5 per cent of the chironomid communities, In the present studies it was found that it is near about 24.3 per cent at study sites of Kham river. Warusawithana and Yatigammana (2007) noted that different limnological variables were either positively or negatively correlated to species composition of Chironomidae.

Diversity of Chironomus Larvae

The studies of diversity of Chironomid larvae at three sampling spots reveals that the population of chironomid larvae is dominated by only three species such as *Chironomus riparius*, *Chironomus stigmaterus* and *Chironomus plumosus*. These three species were recorded under subfamily Chironominae. *Chironomus riparius* was the dominant species in Kham river

The diversity of identified larvae is shown in Table 10.2.

Table 10.1: Physico-Chemical Parameters of Kham River at Aurangabad

Parameter	May	June	July	August	September	October
Temperature (°C)	25±3.8	25.4±2.7	23.8±2.3	23.1±2.2	24±1.3	25.5±1.2
pH	6.7±0.3	7.7±0.4	7.5±0.6	6.8±0.2	8.5±0.6	8.5±0.6
DO (mg/l)	6.8±3.8	5.9±1.8	6.3±1.6	4.3±3.6	2.6±2.6	3.8±2.0
Free CO_2(mg/l)	32±3.4	31.6±2.8	34.4±4.2	32.8±2.8	33.2±1.6	36.2±2.8
Total Alkalinity (mg/l)	295±7.3	380±9.3	235±6	283±8.3	398±6.2	420±7.5
Phosphates (mg/l)	0.7±0.2	0.7±0.3	0.8±0.4	0.7±0	0.8±0.3	0.8±0.2
Nitrate (mg/l)	0.6±0.1	0.7±0.2	0.7±0.1	0.4±0.1	0.8±0.2	0.7±0.3
Hardness (mg/l)	420±16	437±9	360±15	399±8	475±18	480±20
Calcium (mg/l)	81.8±3.2	78.2±1.3	76.9±6.9	74.5±1.4	77.6±3.1	76.8±1.9
Magnesium (mg/l)	35.2±0.3	38.9±1.1	38.1±2.4	38.2±1.8	34.7±0.6	35.1±0.8
Chloride (mg/l)	123.5±10.2	90.9±6.3	98.6±5.1	101±3.1	98.7±12.6	88.5±4

Table 10.2: Relative Abundance of Identified Chironomus Larvae

Diptera–chironomidae	Relative Contribution (Per cent of total mean density)
Chironomus riparius	53 per cent
Chironomus plumosus	24.3 per cent
Chironomus stigmaterus	22.7 per cent

Key to the Chironomid Larvae

Chironomus riparius
- ☆ Caudolateral tubules not present, ventral tubules present.
- ☆ Length of antennal segment 1/segment 2 > 3.5mm.
- ☆ Mandibal with three dark inner teeth; 2 pairs of ventral tubules present; fourth instar larvae much less than 30mm.

Chironomus plumosus
- ☆ A pair of caudolateral tubules present on the abdominal segment anterior to the segment bearing the ventral tubules.
- ☆ Mentum with trifid median tooth and 12 lateral teeth; premandible apically bifid with at most one additional small tooth near center; widespread.
- ☆ Anteromedial margin of ventromental plate with fine teeth.
- ☆ Mandible with 3 dark inner teeth.

Chironomus stigmaterus
- ☆ A pair of caudolateral tubules present on the abdominal segment anterior to the segment bearing the ventral tubules.

☆ Mentum with trifid median tooth and 12 lateral teeth; premandible apically bifid with at most one additional small tooth near center; widespread.

☆ Anteromedial margin of ventromental plate smooth.

☆ Head usually with central dark dorsal stripe; central tooth of mentum usually basally constricted.

Acknowledgement

The authors are thankful to University Authorities and Professor and Head, Dept. of Zoology, Dr. Babasaheb Ambedkar Marathwada University, Aurangabad for providing laboratory facilities undertaking the present research.

References

Aguiaro, T. and Caramaschi, E.P., 1998. Trophic guids in fish assemblages in three coastal lagoons of Rio de Janeiro State (Brazil). *Verh. Internat. Verein. Limnol.*, 26: 2166–2169.

APHA, 1997. *Standard Methods for the Examination of the Water and Wastewater*, 21st edn, New York.

Bhattacharya, Gautam *et al.*, 2006. Assessment and impact of heavy metals on the communities and morphological deformities of Chironomidae larvae in the river Domoda (India, West Bengal). *Supplimenta ad Acta Hydrobiologica*, 8: 21–32.

Epler, J.H., 2001. Identification manual for the larval Chironomidae of North and South Carolina, *Spixiana Suppl.*, 14: 1–471.

Freimuth, P. and Bass, D., 1994. Physico-chemical conditions and larval Chironomidae (Diptera) of an urban pond. In: *Proc. Okla. Acad. Sci.*, 74: 11–16.

Kimio Hirabayashi *et al.*, 2006. Change of Chironomid fauna (Diptera: Chironomidae) in eutrophic lake Kawaguchi Japan. *Bol. Mus. Mun. Funchal* Sup., 13: 109–117.

Panatta, A., Stenert, C. and Maltchik, L., 2006. Diversity of chironomid larvae in palustrine wetlands of the coastal plain in the south of Brazil. *Limnology*, 7: 23–30. DOI 10.1007/s 10201-005-0160-y.

Quadri, Syed Atharuddin, 2004. Studies of biodiversity of zooplankton from Aurangabad district. *Ph.D. Thesis*, Dr. B.A.M.University, Aurangabad (Dr. M.M. Shaikh).

Trivedy, R.K., Goel, P.K. and Trisal, C.L., 1995. *Practical Methods in Ecology Environmental Science.* Environmental Publications, Karad.

Warusawithana, L. and Yatigammana, S.K., 2007. Species composition of the Family Chironomidae and their relationship to limnological variables in Kandy lake. *Proceedings of the Paradeniya University Research Sessions*, Sri Lanka, 12(1).

Chapter 11

Assessment of Biocontamination of Bennetura Tank in Osmanabad District, Maharashtra

☆ *C.V. Pawar, R.R. Jadhav and D.V. Menkudale*

Introduction

Pure water is a basic need of life. Day by day, pure water is a problem of the word it is polluted by various ways *i.e.* chemical and biochemical ways.

Biocontamination of the water resources were studied throughout the world, some worker like Kulshrestha *et al.* (1992), Kodarkar (1998) and Goel and Trivedy (1984) worked out on this problem. However no such work was recorded on Benetura tank in Osmanabad District of Maharashtra. Therefore the present work was under taken to study the biocontamination of Benetura tank.

Study Area

Location of the Benetura tank is, its longitude 76°-25" and latitude 17°-20". It is earthen dam having maximum height 11.99 meter, catchments area 12.69 meters this water tank were constructed for irrigation, agriculture and drinking purposes.

Material and Methods

Monthly samples were collected from the four sampling stations for a period one-year from January to December 2008. Surface water sample collected directly in two lit. Capacity container. The analysis of water temperature, air temperature and dissolved oxygen was made on site, for other parameter, samples brought to laboratory and analyzed for bacteriological analysis, pH was measured with the help of field pH. Meter hannomodel champ. The methods were used for the analysis of various physico-chemical parameters are as given in methodology for water analysis (Trivedy and Goel 1984, APHA 1980, and Kodarkar *et al.*, 1998).

Results and Discussion

Temperature

Air

Temperature ranged between 21° to 32°C in the month of January and maximum in the month of may.

Water

The water temperature various between 19 to 40°C. The minimum temperature was recorded in the month of January and maximum in the month of May.

Water temperature exhibited positive correlation with MPN, helminthes eggs, protozoa, rotifers and arthropods whereas negative correlation with dissolved oxygen, pH, free CO_2 and alkalinity.

pH

The pH of water varied from 7.2 to 8.2. The pH of water of Benetura tank water was less alkaline. The minimum pH was 7.2 recorded in the month of May and maximum 8.2 recorded in the month of December.

The pH values shows positive correlation with dissolved oxygen, CO_2, alkalinity and negative correlation with MPN, Helminthes eggs, Protozoa, Rotifer, Temperature, Alkalinity and Arthopod.

Free CO_2

The values of free CO_2 ranging between 8 mg/lit and 13 mg/lit. The seasonal variations in the values of free CO_2 were also observed. Free CO_2 were also observed. Free CO_2 exhibited negative correlating with temperature, dissolved oxygen, Alkalinity, Helminthes eggs, Protozoa and MPN where as positive correlation with pH.

Alkalinity

The water of the tank was moderately alkaline throughout the year. Total alkalinity was ranged between 80 to 111 mg/lit.

Total alkalinity showed correlation with temperature, MPN, protozoa, helminthes eggs, rotifer, and arthropod.

Dissolved Oxygen

The values of dissolved oxygen ranging between 6.9 mg/lit and 9.50 mg/lit. The Seasonal variations in the values of dissolved oxygen were also observed. Dissolved oxygen value is higher in rainy season and lower in summer. Dissolved oxygen exhibitive negative correction with water temperature free CO_2, alkalinity, MPN, helminthes eggs, protozoa, rotifer and arthropod whereas positive correlation with pH.

MPN of Coliform

During the course of investigation on M.P.N. of coliform was detected. It was ranged between 50-550/100 ml of sample. It was detected maximum during the month of May and minimum during the month of May and minimum during the month of January.

MPN of coil from exhibited positive correlation with temp, helminthes eggs; protozoa rotifer and arthropod were as negative correlation with free CO_2, D.O. pH and Alkalinity.

Protozoa

The protozoan represented by cysts of *Balantidium coli* and cysts of *Entamoeba histolytica*. The total pollution was highest in 12/lit in the month of May and lowest in the 6/lit in the month of November.

Protozoa of coliform exhibited positive correlation with Temperature, Helminthes eggs, MPN, Rotifer and Arthropods where as negative correlation with free CO_2, D.O, pH and Alkalinity.

Helminthes Eggs

The Helminthes eggs, identified belongs to *Ascaris lumbricoides, Enterobius vermicularis,*

Table 11.1: Physico-chemical and Biological Profile of Benetura Tank (Jan to Dec, 2008)

Sl.No.	Parameter	Range
1.	Temperature	
	a. Air	21° to 32° C
	b. Water	19° to 40° C
2.	pH	7.2 to 8.2
3.	Free CO_2	8 to 13 mg/lit
4.	Alkalinity	80 to 111 mg/lit.
5.	D.O.	6.9 to 9.50 mg/lit.
6.	MPN	50 to 550/100 ml
7.	Protozoa	6 to 12/lit.
8.	Helminthes eggs	11 to 25/lit.
9.	Rotifer	23 to 70/lit.
10.	Arthropods	18 to 55/lit.

Fasciola hepatica and *Trichurus trichore*. The helminthes eggs found maximum 1/lit in the month of Sept. Eggs of Ascaris lumbricoides were most prevalent being found partilly in all months followed by Trichurus trichure observed for 10 month.Fasciola hepatica and Hymenolepis nana observed for 9 months Enteribius Vermicularis observed for 8 months.

Helminthes eggs exhibited negative correlation with pH, D.O, CO_2 where as positive correlation with water temperature, Alkalinity, MPN, Protozoan, Rotifer and Arthropod.

Rotifer

The Rotifer was represented by 5 genus *i.e. Branchionus keratella, Chrochrometogahtre, Epiphanies* and *Filinia* in the summer season rotifer population was maximum where as during the winter season minimum. The highest density of Rotifers is 70/lit in the months of May and lowest density *i.e.* 30/lit. In the months of Dec. Rotifer exhibited negative correlation with pH, Dissolved oxygen, Alkalinity free CO_2 whereas positive correlation with water temp, MPN, protozoan, Helminthes eggs and Arthropod.

Arthropoda

In Arthropods *Cyclops, Daphnia* and *Nauplius* were studied. The arthropod population was higher 55/lit in the month of May and lowest 18/lit.in the month of November. Arthopod population was dominated by *Nauplius*.

Arthropod exhibited positive correlation with temperature free CO_2 Alkalinity, Protozoa, MPN, Helminthes eggs and negative correlation with D.O.., free CO_2 and pH.

Acknowledgement

The authors are thankful to Principal, S.D. Peshwe, A.S.C. College, Naldurg Dist. Osmanabad for providing laboratory facilities.

References

Agarkar, S.V., Bhosale, A.B. and Patil, P.M., 1998. Physico-chemical analysis of drinking water from Buldhana District, M.S. *J. Aqua. Biol.*, pp. 62–63.

APHA, AWWA, WPCF, 1985. *Standard Method for the Examination of Water and Wastewater*, 2nd edn. American Public Health Association, Washington D.C.

Battish, 1992. *Freshwater Zooplankton of India*. Oxford and IBH Publishing Co. Pvt. Ltd., pp. 223.

Bankar, J.T. and Deshmukh, A.M., 2004. Bacteriological characteristics of dringking water from public places in Satara district Maharashtra. *J. Aqua. Biol.*, 19(2): 1–6.

Chavan, R.J., Hiware, C.J. and Tat, M.B., 2004. Studies on water quality of Manjra project reservoirs in district Beed, Maharashtra. *J. Aqua Biol.*, 19(2): 73–76.

Chattrji, K.D., 1980. *Parasitology*, 12th edn. Chatterji Medical Publisher, Kolkata, p. 1–238.

Dhanpathi, M.V.S.S., 2004. On the occurrence of the Trichotria Similes (Stebnroos, 1998). *J. Aqua. Biol.*, 19(2): 11–13.

James, A. and Evison, L., 1979. *Biological Indicators of Water Quality*. John Wiley and Sons, New York, 2: 28–38.

Kodarkar, M.S. *et al.*, 1998. *Methodology for Water Analysis: Physico-chemical, Biological and Microbiological*. IAAB Publication, Hyderabad.

Patil *et al.*, 2002. Limnological investigations of Ujani Wet land. *ZSI Wetland Ecosystem Series*, 3: 27–61.

Salajkar, P.B. and Yeragi, S.G., 2003. Seasonal fluctuations of plankton population correlated with physico-chemical factors in powas lake, Mumbai, Maharashtra. *J. Aqua. Biol.*, 18(12): 19–21.

Thorat, 2000. Pollution status of Salim Ali lack Aurangabad. *Poll Res.*, 19(2): 307–309.

Trivedy, R.K., Goel and Trisal, L.L., 1987. *Practical Methods in Ecology and Environmental Science*. Enviro-media Publications, Karad, Maharashtra.

Vyas, N. and Nama, Pankaj, 1988. Studies on bioecology of water bodies of Jodhpur Rajsthan. *Hydrobiol.*, 4(1): 9–14.

Wellborn *et al.*, 1996. Mechanisms creating community structure across a freshwater habitat gradient. *Annual Review Ecology Systematic*, 27: 337–363.

Chapter 12

Studies on Biological Contamination of Kurnur Tank in Solapur District, Maharashtra

☆ *P.L. Sawant, M.G. Babare and H.K. Jadhav*

Introduction

The biological contamination of water bodies includes disease producing Bacteria, Protozoan, Helminthes eggs, Zooplanktons and some Arthropods fauna. Water borne disease is arising due to bacterial contamination through sewage entering in water body.

Considerable biocontamination investigations are carried out on the reservoirs in India. The workers like Krishnamurthy and Subramaniam (1999, Khan (1978), Sexena and Sharma (1981), Hag (1988), Narsinha rao and Jaya Raju (2001), Batish and Kumar (1986). However no such work was carried on Kurnur water tank in Solapur District of Maharashtra.

Materials and Methods

Monthly samples were collected from the four sampling stations for a period one year from January to December, 2008. The water samples collected with the help of sampler in late morning hours.

Sample were brought to laboratory and analyse the paratmets some parameters were checked on four sampling stations. The methods used for the analysis of physico-chemical and biological parameters are as given in APHA (1980),Trivedi and Goel(1984) and Kodarkar *et al.* (1998).

Results and Discussion

Temperature

Air

The monthly variation in air-temperature ranged between 20 and 39°C. The minimum temperature was recorded in the month of January at all sites and maximum in the month of May at all sites.

Water

The water temperature ranged between 21°C to 41°C. The minimum temperature was recorded in the month of January and maximum in the month of May at all sampling stations. The water temperature exhibited positive correlation with MPN, helminthes eggs, protozoa, rotifers and arthopods whereas negative correlation with dissolved oxygen, pH, free CO_2 and Alkalinity.

pH

The pH of water ranged between 7.2 and 8.6. The pH of water of Kurnur was less Alkaline throughout the year. The minimum was 7.25 in the month of June and highest in the month of December.

The pH was found to be minimum in rainy season and maximum in winter season. pH values shows positive correlation with CO_2 and negative correlation with MPN, helminthes eggs, protozoa, rotifer, temperature, alkalinity and arthopods.

Free CO_2

The values of free CO_2 ranging between 7mg/lit and 26 mg/lit. The free CO_2 is recorded lowest in the summer season, highest in the rainy season and average in the winter season. Highest in the winter season. Free CO_2 exhibited negative correlation with temperature, dissolved oxygen, and protozoa where as positive correlation with alkalinity, helminthes eggs, pH, MPN, and arthropods.

Alkalinity

The water of the tank was moderately alkaline throughout the year. Total alkalinity was ranged between 81 mg. to 130 mg/lit.

Total alkalinity showed positive correlation with temperature, MPN, protozoa, helminthes eggs, rotifer, and arthropod, whereas negative correlation with pH and dissolved oxygen.

Dissolved Oxygen

The dissolved oxygen ranging between 7 mg/lit and 10 mg/lit. Seasonal variations in the values of dissolved oxygen were also observed. In rainy season average D.O. is 8.55 mg/lit., in winter average dissolved oxygen was 9.13 mg/lit and in summer the average D.O. was 8.84 mg/lit

Dissolved oxygen exhibited negative correction with water temperature Free CO_2 alkalinity, MPN, helminthes eggs, protozoa, rotifer and arthropod whereas positive correlation with pH.

MPN of Coliform

In present investigation on MPN, of coliform was detected. It was ranged between 15-125. It was detected maximum during the month of May and minimum during the month of January. MPN of coliform exhibited positive correlation with Temperature, MPN, and helminthes eggs, protozoa, rotifer and arthropod where as negative correlation and D.O.

Protozoa

The protozoan was represented by cysts of *Balantidium coli* and cysts of *Entamoeba histolytica*. The total pollution was highest 16/lit in the month of May and lowest 9/lit in the month of the November.

Protozoa exhibited negative correlation with Free CO_2, pH, dissolved oxygen, where as positive correlation with Water, Temperature, Alkalinity, MPN, helminth eggs, Rotifer and Arthropod.

Helminthes Eggs

The Helminthes eggs, identified belongs to *Ascaris lumbricoides, Enterobius vermicularis, Fasciola hepatica, Hymenolepis nana* and *Trichurus trichore*. The helminthes eggs found maximum 20/lit in the month of April, and minimum 8/lit in October. *Ascaris lumbricoides* found throughout the year but, *Fasciola hepatica* observed 11 months.

Table 12.1: Physico-chemical and Biological Profile of Kurnur Tank (Jan to Dec, 2008)

Sl.No.	Parameter	Range
1.	Temperature	
	a. Air	20° to 38° C
	b. Water	21 to 41° C
2.	pH	7.2 to 8.6
3.	Free CO_2	7 to 26 mg/lit
4.	Alkalinity	81 to 130 mg/lit.
5.	D.O.	7 to 10 mg/lit.
6.	MPN	15 to 125
7.	Protozoa	9 to 16/lit.
8.	Helminthes eggs	8 to 22/lit.
9.	Rotifer	45
10.	Arthropod	15 to 37/lit.

Enterobius Vermicularis and *Trichurus trichure*. observed for 10 month and *Hymenolepis nana* observed for 9 months. Helminthes eggs exhibited negative correlation with pH, dissolved oxygen, where as Positive correlation with water temperature, free CO_2, alkalinity MPN, protozoan, rotifer and arthropods.

Rotifers

The Rotifer was represented by *Branchionus keratella, Chromatogastre, Epiphanies* and *Filinia*. *Branchionus* was dominated in the month tank. The highest density of Rotifer (45/lit) observed in the of May and lowest in the month of January.

Rotifer exhibited positive correlation with Temperature, free CO_2, alkalanity,MPN protozoan, helminthes eggs and arthropods and negative correlation with dissolved oxygen and pH.

Arthropods

In Arthropods *Cyclops, Daphnia* and *Nauplius* were studied. The arthropod population was higher 39/lit in the month of May and lowest 18/lit.in the month of November. The density winter season. A minor peak of the population of arthropod population was dominated by nauplius.

Arthropod exhibited positive correlation with temperature free CO_2 Alkalinity, Protozoa, MPN, Helminthes eggs and negative correlation with dissolved oxygen and pH.

Acknowledgements

The authors are thankful to Principal, S.D. Peshwe, A.S.C. College, Naldurg Dist. Osmanabad for providing necessary laboratory facilities.

References

APHA, 1980. *Standard Method for the Examination of Water and Wastewater*, 15th edn. New York, pp. 1134.

Kodarkar, M.S., Diwan, A.D., Murugan, N., Kulkarni, K.M. and Ramesh, Anuradha, 1998. *Methodology for Water Analysis: Physico-Chemical Biological and Microbiological*. Indian Association of Aquatic Biologist, Hyderabad, pp. 102.

Arora, H.C., 1996. Studies on Indian Rotifer Part–V on species of some genera of the family Brachionidae arch. *Hyrobiol.*, 61: 482–483.

Arjariya, Amita, 2003. Physico-chemical profile and plankton diversity of Ranital lake Chahatarpur M. P. *Nature, Environmental Pollution Tech.*, 2(3): 327–328.

Agarkar, S.V., 2000. Evaluation of physico-chemical and microbiological parameters of Vyzadi reservoir, water. *Indian Hydrobiol*, 3(1): 3–5.

Batish, S.K. and Kumar, 1986. Effect physico-chemical factors on the seasonal abundance of Cladocera in tropical pond at village of Ragba, Ludhiana, India.*J.Ecol.*, 13(1): 146–151.

Bankar and Deshmukh, 2004. Bacteriological characteristics of Drinking water from public places in Satara Dist. Maharashtra. *J. Aqua. Biol*, 19(2): 1–6.

Sakhare, V.B. and Joshi, P.K., 2002. Ecology of palas Nilegaon reservoir in Osmanabad district, Maharashtra. *J. Aqua. Biol.*, 17(2): 17–22.

Shastri, Yogesh and Pendse, D.C., 2001. Hydrobiological study of Dahikhuta reservoir. *Environ. Biol.*, 22(1): 67–70.

Subbamma, D.V. and Rama Sharma, D.V., 1992. Studies on the water quality characteristics of a temple pond near Machillipatanam, Andhra Pradesh.*J. Aqua. Biol.*, 7(1 and 2): 22–27.

Thomas, Sabu and Abdul Azij, P.K., 2002. Physico-chemical limnology of tropical reservoir in Kerla, S. India. *Eco. Env. and Cons.*, 6(2): 160–167.

Trivedy, R.K. and Goel, P.K., 1984. *Chemical and Biological Methods for Water Pollution Studies*, 2nd edn. Environmental Publication, Karad.

Chapter 13

Physico-chemical Parameters of Groundwater in Parli (V.) in Beed District, Maharashtra

☆ *K.S. Raut, S.E. Shinde, T.S. Pathan, and D.L. Sonawane*

Introduction

Water has the ability of dissolve a greater range of substances than any other liquid. The ability of water to dissolve minerals determines the chemical nature of the groundwater. Therefore, it needs a constant monitoring of chemical parameters throughout the year in all seasons. For any regional hydro-chemical studies, a set of observation of tube-well water is to he selected and sampling has to be done periodically. The hydrological cycle is responsible for our weather; it makes our river run and balance the level of groundwater (Yadav *et al.*, 2009).

Groundwater is by far the most abundant and readily available sources of freshwater, followed by lakes, reservoirs, rivers and wetlands, groundwater represents over 90 per cent of the world's readily available freshwater resources (Boswinkel, 2000). About 1.5 billion people depend upon groundwater for their drinking water supply (WRI, UNEP, UNDP, World Bank, 1998).

Groundwater contains salts in solution that are derived from the location and movement. All groundwater contains salts in solution; reported salt contents range from less than 25 mg/l in quartzite to greater than 300,000 mg/l in brines. Soluble salts in groundwater originate from solution of rock materials. Water passing through the root zone of cultivated areas contains significant amount of salt concentration. High salinity found in soils and groundwater in dry or semiarid regions indicates the poor leaching effects of rainwater. Groundwater pollution is defined as the artificially induced degradation of natural groundwater quality. Consumption of the water creates hazards to public health through toxicity or the spread of disease. Sub-surface water pollution is difficult to detect and even more difficult to control and persist for decades. The sources and causes of groundwater pollution are associated with human use of water. Disposal of wastes from municipal, industrial, agriculture and others, in percolation ponds, spreading or irrigation, seepage pits or trenches, dry streambed, landfills, disposal wells and injection wells cause groundwater pollution (Nagarajan, 2006).

Pollution level in the drinking water of the area under investigation has also to be checked up as such water is directly used for human consumption and agricultural pursuits. Some of the pollutants slowly get coagulated and settled with soil sediments mainly by the mechanism of adsorption. In course of time, the pollution effect travels into the soil and hence, is the necessity for the determination of the quantity and monitoring quality variances in groundwater, Parli (V.), India.

Physico-chemical parameters of Groundwater quality in relation to (drinking quality) health has been studied by number of workers throughout India. But less attention has been found to this aspect compared to river, reservoirs, lake, pond's physico-chemical studies and health. Some workers are (Venkata Mohan *et al.,* 1992), (Vijaykumar *et al.,* 1996), (Reddy *et al.,* 2001), and (Nagraja M. *et al.,* 2005) etc.

Parli Vaijnath is situated at 18° 51' 0" North Latitude and 76°27´ 0" East longitudes. The thermal power station and some small scale industries are located in this city.

The present investigation has been undertaken to assess the Groundwater quality of Parli–V. [M.S] India. This study is a consorted effort towards the understanding of various natural and anthropogenic processes influencing the groundwater and to develop effective management strategies for the future.

Materials and Methods

The water samples for physico-chemical analysis were collected from Parli (V.), at 4 stations in those 8 different sites as follow.

Station 1 (South side): Tal and Sneha Nagar,

Station 2 (East side): Madhav Bag and Someshwar Nagar,

Station 3 (North side): Shivaji Nagar and Station Road

Station 4 (West side): Industrial Area and Anand Nagar.

The samples were collected in acid washed five liter plastic container in the early morning between 8 am to 11 am in the first week of every month during January 2007 to December 2007. Separate samples were collected for dissolved oxygen in 250 ml bottles and dissolved oxygen was fixed on the sites by adding alkaline iodide-azide solution immediately after collection. The samples were analyzed immediately returned to the laboratory.

Physico-chemical variations of the bore wells water like temperature, pH, conductivity, dissolved oxygen, total hardness, and chloride were determined seasonally in summer, monsoon and winter according to standard methods (APHA, 1998; Trivedi and Goel, 1987).

Results and Discussion

The Groundwater parameters were studied and recorded in three seasons viz summer, monsoon and winter respectively. The seasonwise physico-chemical parameters data of bore well water, Parli [M.S] India have been presented in Table 13.1 and Figure 13.1. Physico-chemical variations were observed and recorded as follows.

Temperature

Air and water temperature mean ranged between 22.4 to 34.5 0°C and 19 °C to 29.8 0°C. The over all air and water temperature seasonal mean were 29.86±4.87 0°C and 24.63±3.04 0°C. The maximum air and water temperature mean during summer was 33±5.78 0°C and 27.97±3.27 0°C and minimum air and water temperature mean during winter was 24.25±3.20 0°C and 22±3.55 0°C as recorded in Table 13.1 and Figure 13.1. The water temperature was consistently lower than atmospheric temperature.

Electric Conductivity

The electric conductivity mean ranged between 0.8375 to 1.324 micromohs/cm². Electric conductivity mean was maximum during winter 1.267875±0.316 micromohs/cm² and minimum during summer 0.89155±0.035 micromohs/cm². The overall seasonal mean was 1.0460±0.19 micromohs/cm² as show in Table 13.1 and Figure 13.1.

The conductance is beyond the prescribed limits by WHO and ISI. Similar results were recorded by (Satyanarayana *et al.*, 1992).

pH

The pH mean values ranged between 6.1 to 7.9. The maximum pH mean was recorded in winter 7.61±0.79 and minimum was in summer 6.76±0.24. The overall seasonal mean was 7.29±0.46 as show in Table 13.1 and Figure 13.1.

High pH values are recorded in winter season. Similar results were recorded by (Veeramani *et al.*, 2008)

Dissolved Oxygen (DO)

The dissolved oxygen mean values were ranged between 1.6 to 3 mg/l. The DO mean values were maximum in summer 2.83±0.26 mg/l and minimum during monsoon 2.58±1.01 mg/l. The overall seasonal mean was 2.66±0.14 mg/l as recorded in Table 13.1 and Figure 13.1.

The values of Dissolved Oxygen are beyond the permissible limits given by WHO and ISI. Through out year it was less than four. High Dissolved Oxygen was recorded in winter similar results were recorded by (Kumar, 1993).

Chlorides

Chlorides mean were ranged between 103.55 to 634.81 mg/l. The higher values of chlorides mean were recorded in monsoon 536.46±70.42 mg/l and lower in winter 253.51±150.29 mg/l. The overall seasonal mean was 368.49±148.73 mg/l given in Table 13.1 and Figure 13.1.

The high values of chlorides were recorded from April to October. The high values of chlorides are due to pollution of groundwater samples from chloride rich effluent (Karnath, 1989) also recorded same results.

Salinity

Consumption of water with high concentrations of Total Dissolved Salts has been reported to cause disorder of alimentary canal, respiratory system, nervous system, coronary system besides causing miscarriage of cancer (Reddy and Subha Rao, 2001).

Table 13.1: Seasonal Mean Variations in Physico-chemical Parameters of Groundwater, Parli (M.S) India (During January 2007–December 2007)

Parameter	Summer	Monsoon	Winter	Average	WHO	ISI
Atmospheric temperature (0°C)	33±5.78	32.35±3.2	29.86±4.87	29.86±4.87	–	–
Water temperature (0°C)	23.92±4.18	27.97±3.27	22±3.55	24.63±3.04	–	–
Elect. Conductivity (micromohs/cm^2)	0.89±0.03	0.97±0.04	1.26±0.31	1.04±0.19	0–1000	0–1000
pH	7.61±0.79	7.50±1.12	6.76±0.24	7.29±0.46	6.5–9.2	6.5–9.2
Dissolved Oxygen (mg/l)	2.58±1.01	2.588±0.41	2.83±0.26	2.66±0.14	6.2–7	6.2–7
Chloride (mg/l)	315.5±86.99	536.46±70.42	253.51±150.29	368.49±148.73	200–1000	200–600
Salinity (mg/l)	578.94±159.63	988.36±136.61	465.20±275.79	677.5±275.15	–	–
Total hardness (mg/l)	810.24±158.52	690.34±135.22	490.83±83.72	663.79±161.34	200–600	200–500

WHO: World Health Organization; ISI: Indian Standard Institute.

Figure 13.1: Graphs Showing Seasonal Mean Variations in Atmospheric and Water Temperature, Electric conductivity, pH, Dissolved Oxygen, Chloride, Salinity and Total Hardness at Different Seasons of Groundwater in Parli (V)

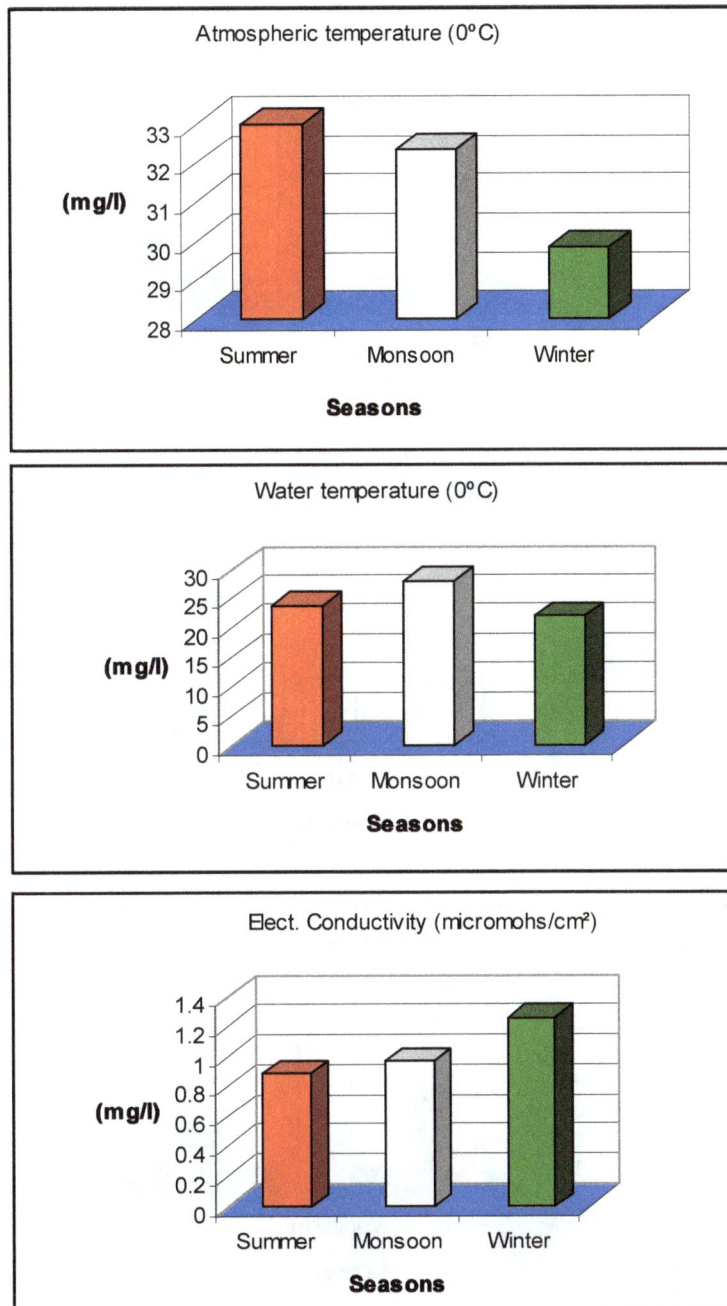

Atmospheric temperature (0°C)

Water temperature (0°C)

Elect. Conductivity (micromohs/cm²)

Contd...

Figure 13.1–Contd...

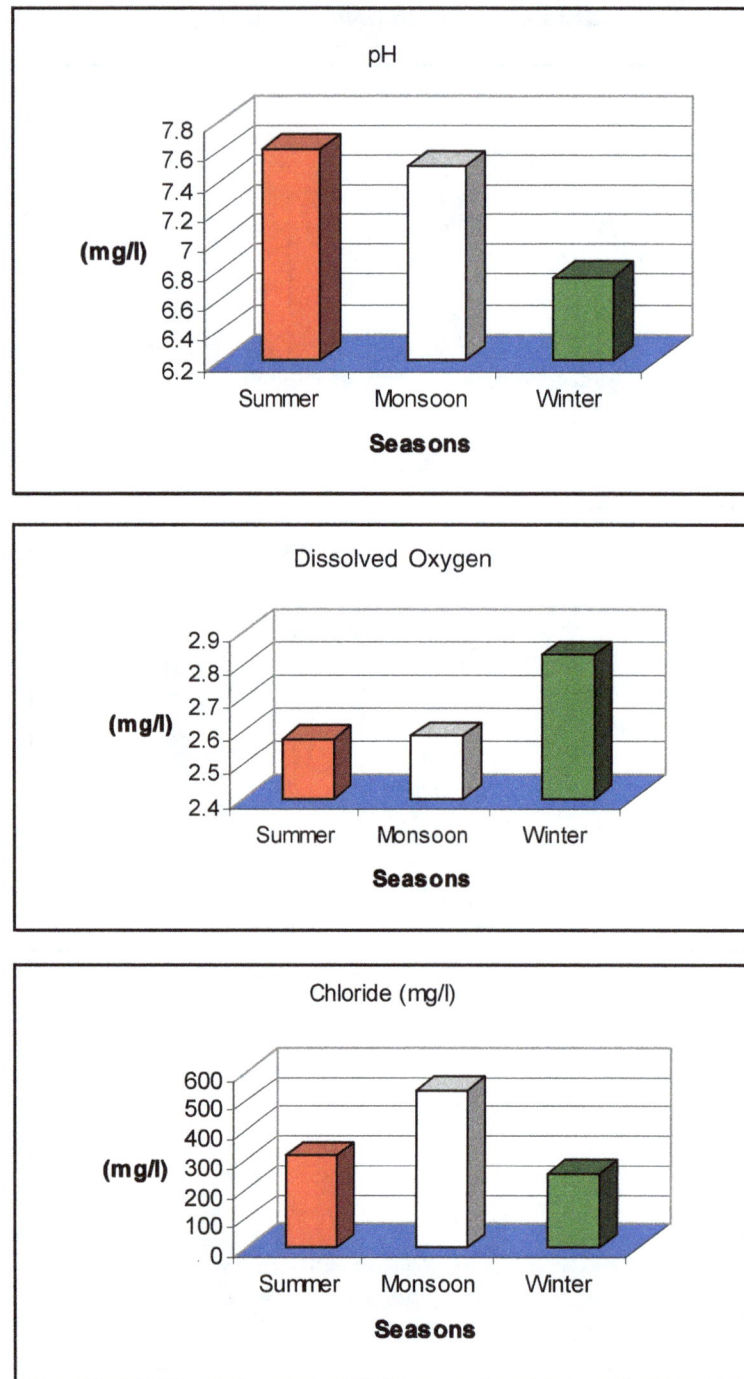

pH

Dissolved Oxygen

Chloride (mg/l)

Contd...

Figure 13.1–Contd...

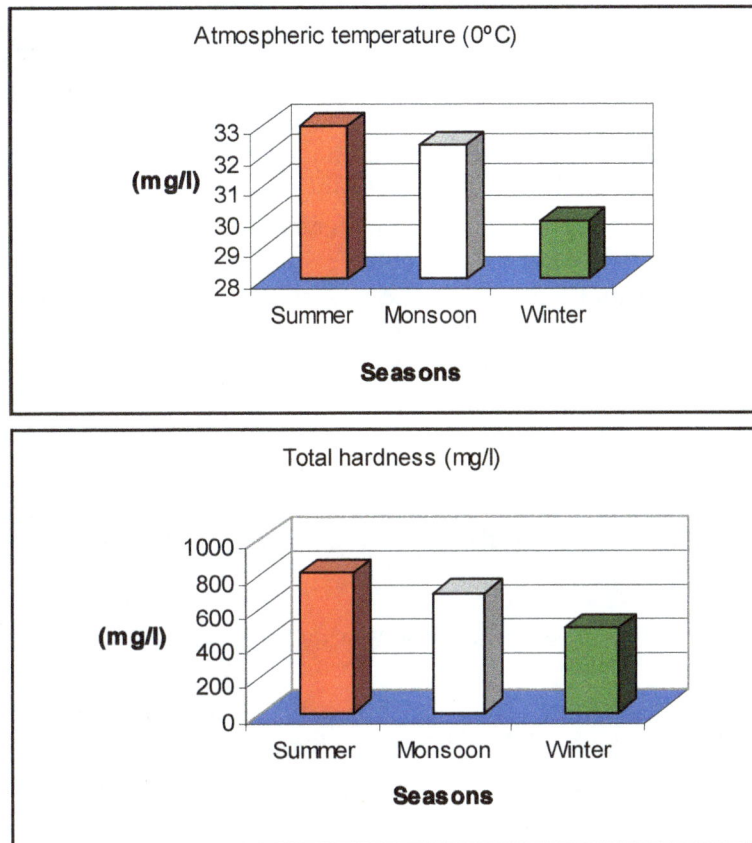

Atmospheric temperature (0°C)

Total hardness (mg/l)

Salinity mean were ranged between 190.01 to 1180.67 mg/l. The higher values of Salinity mean were recorded in monsoon 988.36±136.61 mg/l and lower in winter 465.20±275.79 mg/l. The overall seasonal mean was 677.5±275.15 mg/l given in Table 13.1 and Figure 13.1.

The high values of salinity are recorded from April to October. The high values of salinity levels can make water unfit to use for any purpose and even low levels can create health problems for individuals who may suffer from high blood pressure. High salinity values were also recorded by (Prakash, 1996) and (Vijaykumar *et al.*, 1996).

Total Hardness

The Total Hardness mean ranged between 415.32 to 1027.57 mg/l. Total Hardness mean values were maximum during monsoon 810.24±158.24 mg/l and minimum during winter 490.83±83.72 mg/l. The overall seasonal mean was 663.79±161.34 mg/l as show in Table 13.1 and Figure 13.1.

The high values above permissible limits by WHO and ISI were recorded from summer and monsoon, below limits in winter. Total hardness, calcium and magnesium above limits causes encrustation in water supply structure and adverse effect on domestic use. (Ramteke *et al.*, 1988) (Dhamij *et al.*, 1995) and (Nagraj *et al.*, 2005).

Conclusions

The present study show detailed physico-chemical variations of Groundwater in Parli (V).

1. The summer, monsoon and winter seasons shows different mean seasonal fluctuations in physico-chemical parameters.

2. Dissolved Oxygen, Chloride, Salinity and Total Hardness in this region were beyond the permissible limit according to WHO and ISI standards for drinking purpose in the year. It may be due to addition of industrial effluent and thermal effluent in groundwater.

3. Over exploitation of groundwater and improper management of natural resources led to the unequal distribution of major ions in nature.

4. It can also be seen that the major ion concentration are predominantly influenced by anthropogenic activities than by natural agencies.

5. To improve suitable quality of groundwater for drinking purpose, there should be continuous monitoring of pollution level.

Acknowledgements

The authors are thankful to Head, Dept of Zoology, Dr. Babasaheb Ambedkar Marathwada University, Aurangabad-431004 (M.S) India for providing laboratory Facilities.

References

APHA, 1998. *Standard Methods for the Examination of Water and Wastewater*, 20th edn. APHA, Washington USA.

Boswinkel, J.A., 2000. *Information Note on Freshwater*. International Groundwater Resources Assessment Centre (IGRAC), Institute Applied Geosciences, Netherland.

Dhamija, S.K. and Jain, Yatish, 1995. Studies on water quality index at Jabalpur (M.P.). *Poll. Res.*, 14(3): 341–346.

ISI, 1983. *Indian Standard Specification for Drinking Water*, ISI, p. 105–500.

Karanth, K.R., 1989. *Hydrogeology*. Tata McGraw Hill Publ. Co. Ltd., New Delhi, p. 1–455.

Kumar, S., 1993. Correlation among water quality parameters of groundwater in Balmer district. *Indian J. Environ. Prot.*, 13: 487– 489.

Mohan, Venkata and Reddy, Jayaramo, 1992. Assessment of overall water quality of Tirupati. *Poll. Res.*, 14(3): 275–282.

Nagraj, M., Nagraju, D. and Balasubramaniyam, 2005. Groundwater quality of Mandya Taluk, Karnataka, India. *J. Ecotoxicol. Environ. Monit.*, 15(2): 169–178.

Nagarajan, R., 2006. *Water Conservation, Use and Management for Semi-arid Region*.

Prakash, H.S.M., 1996. Groundwater pollution. *J. Geol. Soc., India*, 48: 595.

Reddy, P.M. and Rao, N. Subba, 2001. Effects of industrial effluent on the groundwater regime in Vishakhapatnam. *Poll. Res.*, 20(3): 385–386.

Ramteke, D.S. and Moghe, C.A., 1988. Manjula on water and wastewater analysis, NEERI Annexure: XV–XVII.

Satyanarayana, Raju, 1992. Groundwater quality of Machaliptanum and total dissolved solids prediction through conductivity measuring. *Poll. Res.*, 11: 65–68.

Trivedi, R.K. and Goel, P.K., 1987. *Chemical and Biological Methods for Water Pollution Studies.* Environmental Publications, Karad, India.

Vijay Kumar, V., Prabhakarachery, M. and Rao, P.L.K., 1996. Studies on quality of groundwater in pedda cherruva sub-basin in P.R. District, A.P. *Poll. Res.*, 15(2): 181–186.

Veeramani, T., Vadivel, S. and Gnanavel, M., 2008. Studies on physico-chemical and benthic characteristic of Vedharanyam Saltpan, Tamil Nadu, India. *J. Aqua. Biol.*, 23(2): 11–16.

WHO, 1999. *Guidelines for Drinking Water Quality, Vol. 9: Surveillance and Control of Community Supplies.* World Health Organisation, Geneva.

WRI, UNEP, UNDP and World Bank, 1998. *World Resources (1998–99): A Guide to Global Environment.* Oxford University Press, New York.

Yadav, M. P., Bijay, K., Rishikesh, K. and Yadav, N.K., 2009. Physico-chemical parameters of tubewell water of Bhagalpur district. *J. Haematol and Ecotoxicol.*, 4(1): 52–56.

Chapter 14

Physico-chemical Properties of Groundwater of Jintur Taluka in Parbhani District of Maharashtra

☆ *V.B.Pawar and Kshama Khobragade*

Introduction

Unplanned urbanization, rapid industrialization and indiscriminate use of artificial chemical in agriculture, causing heavy and varied pollution in aquatic environment leading to deterioration of water quality depletion of aquatic biota The knowing to understand chemical content of water it is difficult to biological phenomenon fully, because chemical nature of water, reveals much about the metabolism of the ecosystem and explain the general hydro biological interrelationship.

The water which moves down word percolates in soil becomes groundwater further reacts with the soil and rock materials, These reactions primarily consist of solution of solid phase in accordance with the solution of chemistry of particular minerals. These minerals range from almost insoluble to very soluble. The solubility also affected by temperature and pressure.

Various workers in our country have carried out extensive studies on fluoride Das *et al.* (2000) reported fluoride hazardous of groundwater in Orissa, Khedkar and Dixit (2003), evaluated suitability of Ambanala water for Irrigation. Recently Kumar and Gopal (2000) published a review article on fluorosis and its preventive strategies. Chand (1999) reported fluoride study on human health and Patra. Diwedi and *et al.*, studied industrial fluorosis on cattle in Udaipur, Rajasthan. Jagdap *et al.* (2002); Hussain *et al.* (2003) and Abbasi *et al.* (2002), have studied water quality in different rivers. Sriniwas *et al.* (2000), Jha *et al.* (2000) and Warma (2003) studied water quality in Hyderabad and Bihar respectively. A survey of literature reveals that there is no systematic study on evaluation of physico-chemical properties in groundwater of Jintur taluka of Parbhani district. Hence present work has been undertaken for the study.

Material and Methods

The study area lies in the Parbhani district of Maharashtra State. Jintur is located at 19.62° N 76.7° E. It has an average elevation of 455 meters (1492 feet). Water samples are collected from 15 numbers of selected Bore wells as well as of open well in sterilized bottles (Kudesia, 1985) of one liter capacity during October 2007 to February 2008. Water samples were collected in morning at 8.00 A.M. to 11 A.M. Sampling has been carried out without adding any preservative in rinsed bottles directly for avoiding any contamination and brought to the laboratory. A.R. grade chemicals and double distilled water used for preparing solutions for analysis. Bacteriological examination was done using standard procedure suggested by Trivedy and Goel (1984) and APHA (1989)

Various physical parameters like pH, EC, DO and TDS which are important to evaluate the suitability of groundwater for portability were determined on the site with the help of digital portable water analyzer kit (CENTURY-CK-710). The chemical analysis was carried out for Calcium (Ca^{2+}), Magnesium (Mg^{2+}), Chloride (Cl^-), Carbonate (CO_3^{2-}) and Bicarbonate (HCO_3^-) by volumetric litigation methods, while fluoride (F-) by spectrophotometer method. Sodium (Na^+) and Potassium (K^+) are by flamephotometery (Kudesia, 1985).

Result and Discussion

The characteristics of groundwater collected from different sites are presented in Table 14.1. the pH values of water samples are varying from 6.2 to 8.28 and these values within the limits except samples S8 (8.28), prescribed by ISI and ICMR but all of these samples within the limits as per WHO (1993) Table 14.3 gives the physico-chemical properties of groundwater of Jintur taluka of Parbhani district.

Table 14.1: Sampling Stations of Groundwater of Jintur Taluka of Parbhani District

Sl.No.	Station Code	Name of Stations
1	S1	Adarsh Colony
2	S2	Anand Nagar
3	S3	Nivriti Mohalla
4	S4	Lecture colony
5	S5	Bori village
6	S6	Bamani
7	S7	Bhogaon
8	S8	Itoli
9	S9	Jamb
10	S10	Yeldari
11	S11	Limbala
12	S12	Pungala
13	S13	Bori Tanda
14	S14	Wazar
15	S15	Warud

Conductivity values varied from 300 to 2986µ siemens/cm for groundwater and these values are very much higher than the prescribed standards limits (1400µ siemens/cm) recommended by WHO except S5, S6 and S10 (Figure 14.1) higher EC and TDS values reflect greater salinity of water and it can not be suitable for drinking and irrigation under ordinary conditions but may be used occasionally under special circumstances. By using the only EC values (Table 14.2) Wilcox (1995) has classified the groundwater for irrigational waters. By considering this classification more than 50 per cent of samples of study area were found to be doubtful to unsuitable and the classification is as follows.

Total Dissolved Solids (TDS) values varied from 120 to 1525 mg/liter Figure 14.2 except S2, S9, S13, S14 and S15 remaining all sampling station TDS values are higher than the standard values recommended by ISI, ICMR and WHO. The samples which have high value of TDS are unsuitable for Drinking and Irrigation purpose.

Table 14.2: Classification of Samples According to their Class of Samples

EC (μ siemens/cm)	Class	No. of Samples
<250	Excellent	–
250–750	Good	5
750–2000	Permissible	7
2000–3000	Doubtful	1
>3000	Unsuitable	2

Table 14.3: Physico-chemical properties of groundwater of Jintur area of Parbhani District

Sampling Station No.	Location	pH	Turbidity	EC	TDS	DO	TA	Ca^{2+}	Mg_2^+	Cl_2^-	F^-	E. coli
1	Adarsh Colony	7.86	5.0	1887*	960	5.1	226	195	65	225	0.20	–
2	Anand Nagar	7.91	4.2	535	370	4.8	186	210*	70	220	0.29	–
3	Nivratti Mohallah	6.64	3.1	969	610	4.6	140	280*	95	124	0.16	1
4	Lecture colony	7.62	3.6	1578*	946	5.8	205	485*	164*	226	0.05	–
5	Bori	7.06	3.2	2748*	1525	4.9	210	385*	185*	180	0.12	2*
6	Bamni	7.5	2.8	2986	874	5.9	246	615*	134	139	0.16	–
7	Bhogaon	6.2	5.3	1395	530	4.6	198	198	65	136	0.28	–
8	Itoli	8.28	6.2	934	510	5.1	230	172	69	121	0.53	–
9	Jamb	6.92	5.6	958	120	4.8	138	126	69	85	0.43	2*
10	Yeldari	7.04	5.3	2542*	1210	4.5	102	340*	185*	160	0.16	1
11	Limbala	7.58	5.0	1162	520	5.1	120	176	93	114	0.29	14*
12	Pungla	7.19	8.0	680	210	4.8	44	56	35	24	0.16	120*
13	Bori Tanda	7.3	6.2	480	160	4.6	78	52	26	28	2.43	81*
14	Wazar	6.4	6.1	300	120	5.1	90	68	42	15	0.15	45*
15	Warud	6.87	6.2	690	150	4.9	172	143	113	40	0.13	19*

*: Exceeding the permissible limit.

Note: All Parameters are Expressed in Mg/lit. Except p^H, Turbidity (NTU), Electrical Conductivity (μ mhos/cm) and *E. coli* (Number/100 ml.)

Hardness of water relates to its reactions with soap and to scale and incrustations accumulating in conduct where the water is heated or transported. Since soap if precipitated by Ca and Mg ions. Water classified according to its hardness as

Soft	0-60
Moderately Hard	61-120
Hard	121-180
Very Hard	>181

ELECTRIACL CONDUCTIVITY

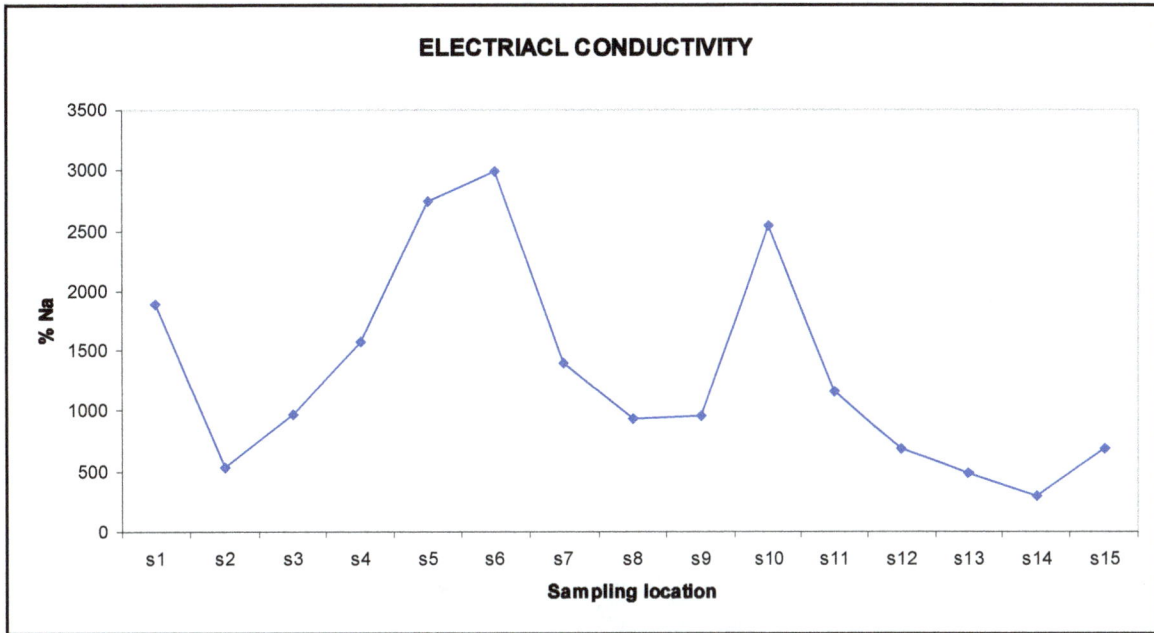

Figure 14.1: Variation of EC in Groundwater of Jintur Taluka of Parbhani District

TOTAL DISSOLVED SOLIDS

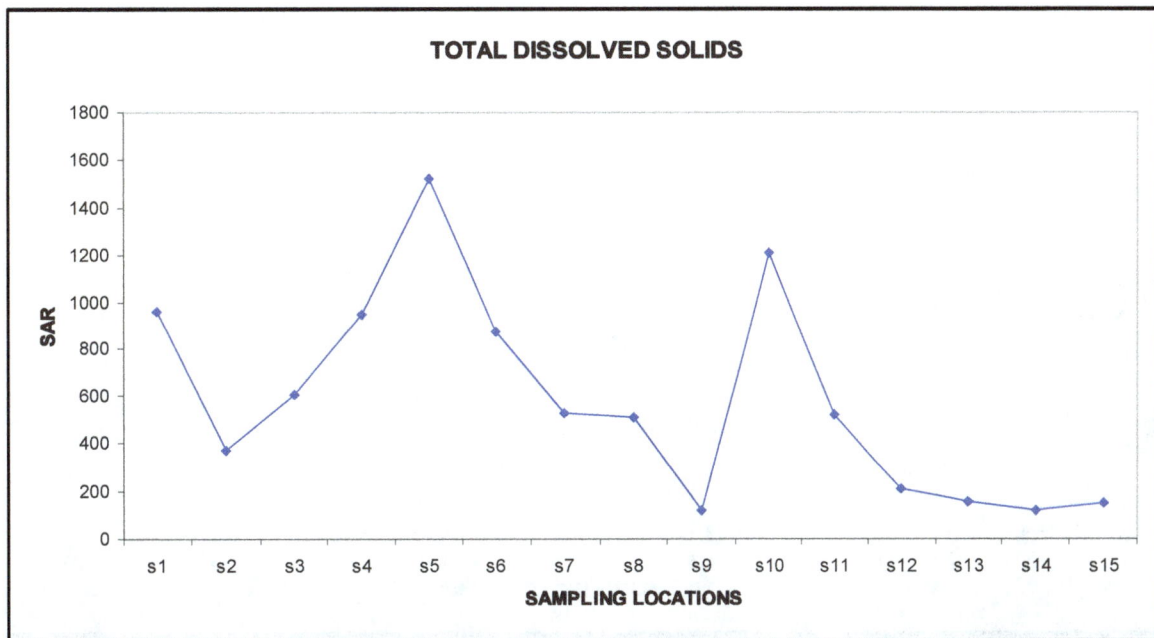

Figure 14.2: Variation of TDS in Groundwater of Jintur Taluka of Parbhani District

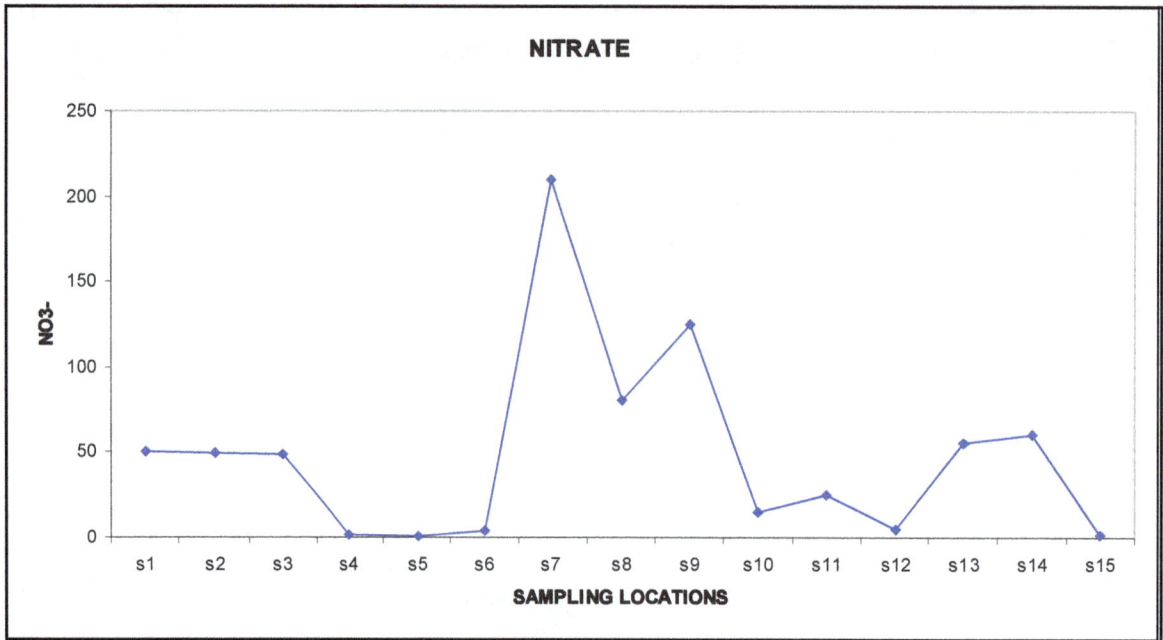

Figure 14.3: Variation of NO$_3^-$ in Groundwater of Jintur Taluka of Parbhani District

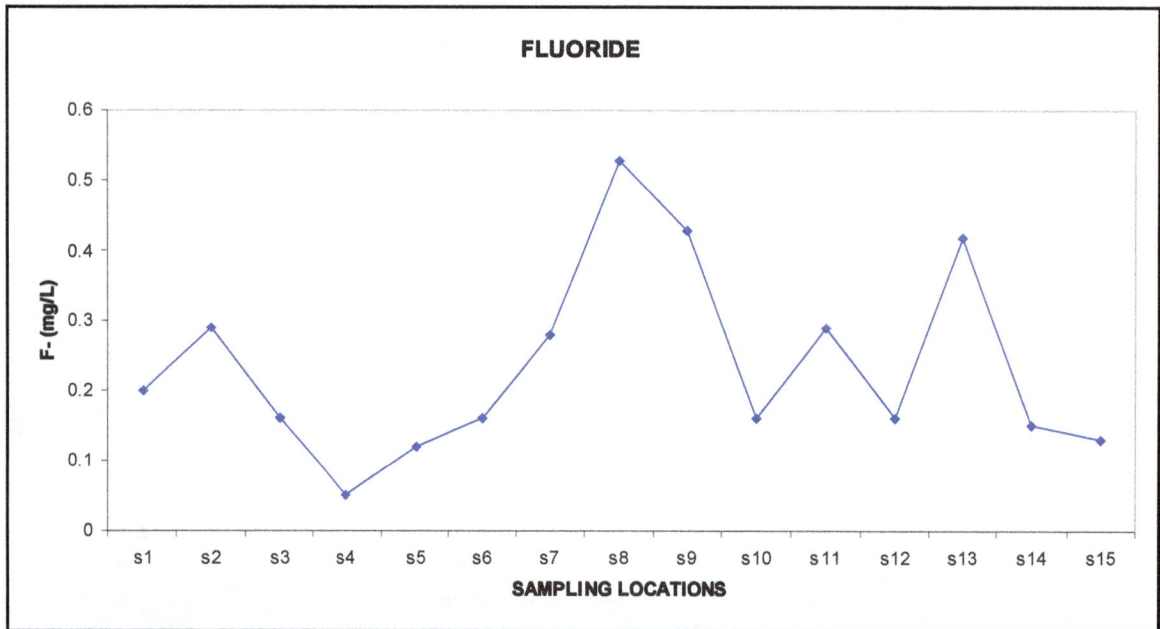

Figure 14.4: Variation of F$^-$ in Groundwater of Jintur Taluka of Parbhani District

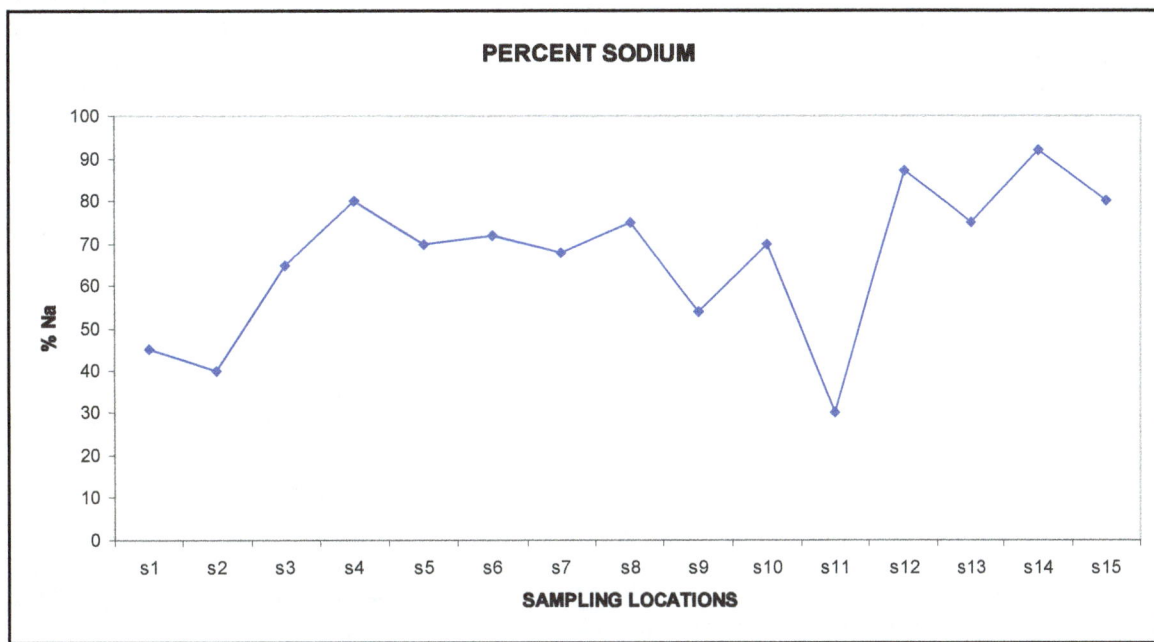

Figure 14.5: Variation of Per cent Na in Groundwater of Jintur Taluka of Parbhani District

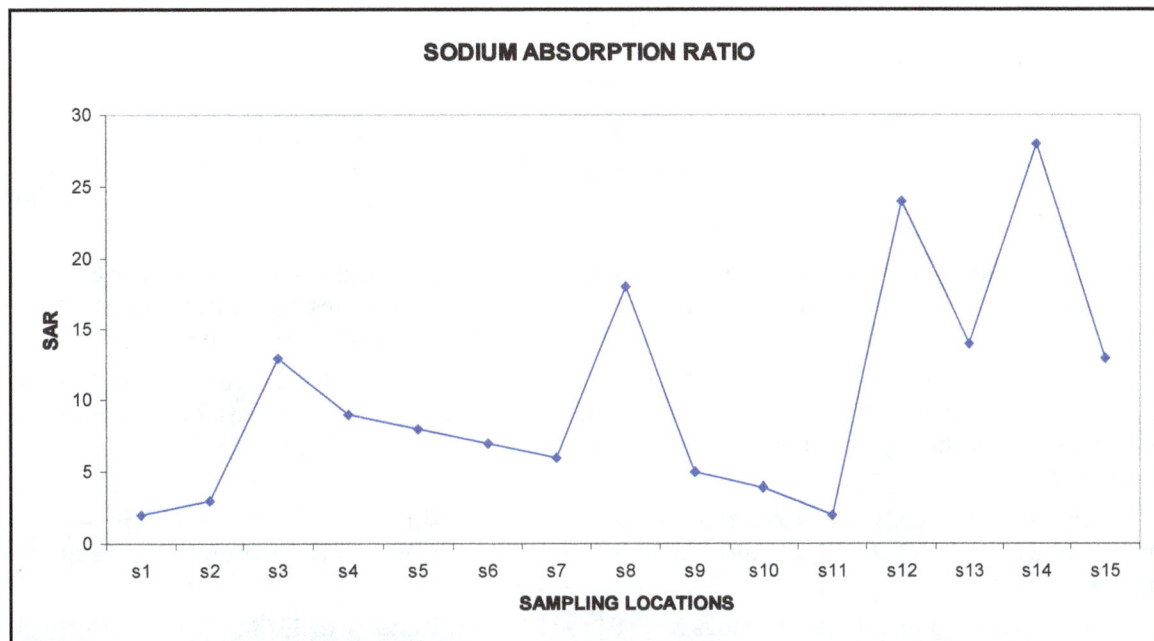

Figure 14.6: Variation of SAR in Groundwater of Jintur Taluka of Parbhani District

Water for domestic use should not contain more than 80 mg/liter total hardness, Hardness of groundwater can be increased if contaminated by acid leachate from garbage waste disposal areas. The disadvantage of hard water is that it precipitate soap thus increasing soap requirements low incidence of heart diseases apparently also occur in areas with very soft water which led Neri et. al (1975).

Potassium values ranging from 0.89 to 24 mg/liter potassium is common in igneous rock. These minerals however are very insoluble so that potassium levels in groundwater normally are much lower than sodium concentration.

Sodium Absorption Ratio is defined as following formula

$$SAR = Na^+/\sqrt{[(Ca^{2+}+Mg^{2+})/2]}$$

Where the individual ions have been expressed as meq/L. there is a significant correlation between SAR values of irrigation water and the extent to which sodium is absorbed by the soil. The SAR values from 1.35 to 28.42 (Figure 14.6) in which samples S3, S8, S12, S13, S14 and S15 have high value of SAR recommended by USSL, so these samples are not suitable for irrigation purpose.

Bicarbonate concentration of water has been suggested as an additional tool for classification of irrigation of water by Richard (APHA,1989). Bicarbonate hazard is evaluated by calculating residual sodium carbonate (RSC) and based on these values groundwater of the investigated area is suffer from bicarbonate hazard and the classification is as follows.

RSC Values	Class	No. of Samples
<1.25	Good	11
1.25-2.5	Medium	01
>2.5	Bad	03

Calcium hardness values are varied from 52 to 615 mg/L and these values are within permissible limit as prescribed by ICMR and WHO. Sources of Calcium are igneous rock minerals like silicate pyroxenes, amphiboles feldspar and silicate minerals produced in metamorphism. Since the solubility of these minerals is low in calcium as well as in TDS.

Magnesium hardness values are varied from 26.0 to 185 mg/L and are within permissible limit as prescribed by ICMR and WHO standards except S4, S5 and S10 magnesium in around water from igneous rock primarily derives from ferromagnesian minerals like Olivine, pyroxenes, amphiboles.

Chlorides values varied from 15 to 226 mg/L and these values are within permissible limits as per prescribed by ICMR, WHO standards except S1 and S4. In reasonable concentration these are not harmful to human beings impart salty taste beyond concentration 250 mg/L which is objectionable to many peoples.

Sulphate values are varied from zero to 210 mg/L and are within permissible limit as prescribed by ISI, ICMR and WHO Drinking water should be not exceed 250 mg/L of Sulphate because water will have a better taste.

Carbonates and Bicarbonates values from 04 to 113.0 mg/L and 77 to 397 mg/L respectively all of the studied samples Bicarbonate values are lesser than the prescribed by US standards and carbonates has no standard values for drinking purpose suggested ISI (1991), ICMR(1975) and WHO. Sources of Carbonate and Bicarbonate include CO_2 from atmosphere CO_2 Sulphate produced by the biota of the soil or by the activity of Sulphate producers and the various carbonate rocks and minerals.

Nitrate values varied from 05 to 215 mg/L all the studies samples nitrate value are lesser than the prescribed by ICMR, WHO standards except S7, S9 and S14 (Figure 14.3). Nitrate occurs naturally in certain water supplies and may also find access through directly or indirectly discharges of wastes and sewage.

Fluoride values of fifteen samples varies from 0.05 to 2.43 mg/l Except samples S13.which are varied from 2.43 mg/L, so the samples are higher than the prescribed limit recommended by ICMR, ISI and WHO, These groundwater very hazardous for Human consumption.

The dissolved oxygen was recorded in the range of 4.5 to 5.9 to mg/L the highest value of dissolved oxygen was encountered in the samples S6. There were not any variations of dissolved oxygen values found and they all are in Permissible limit. All of the samples have lower BOD values except S10.The higher BOD indicates pollution from domestic sources.

Conclusion

The general conclusion can be drawn as the quality of almost all groundwater samples collected from the study area indicate that the concentration of pH, Ca^{2+}, Mg^{2+}, SO_4^{2-}, DO, BOD and F values are within permissible of ISI, ICMR and WHO but Cl^-, TDS, NO_3^- and EC values are that poor water quality in most of the studied groundwater samples. A study area classifies water under moderate category and is not best for house hold and irrigation.

References

Abbasi, S.A., Khan, F.I., Senthilveal, K. and Shobudeen, A., 2002. Modelling of Buckinggham canal water quality. *Indian J. Environ. Health*, 44(4): 290–297.

APHA, 1989. *Standard Methods for Examination of Water and Wastewater*, 17th Edn., (Ed.) Lenore S. Clescrei. APHA, AWWA, WPCE, Washington DC.

Chand, D., 1999. Fluoride and human health: Causes for concern. *Indian Journal of Environmental Protection*, 19(2): 81–89.

Das, S., Mehta, B.C., Samanta, S.K., Das, P.K. and Srivastava, S.K., 2000. *Indian Journal of Environmental Health*, 1(1): 40–46

Hussain, M.F. and Ahmad, I., 2003. Variability in physico-chemical parameters of Pachin River (Hanagar). *Indian J. Env. Health*, 44(4): 329–336.

Jayshri, Jagdap, Bhushan, Kachawe, Leena, Deshpande and Prakash, Kelkar, 2002. Water quality assessment of the purna river for irrigation purpose in Buldhana district, Maharashtra. *Indian J. Environ. Health*, 44(3): 247–257.

Jha, A.N. and Verma, P.K., 2000. Physico-chemical property of drinking water in town area of Godda District under Santal Pargana, Bihar (India). *Pollution Research*, 19(2): 245–247.

Kumar, S. and Gopal, K., 2000. *Indian Journal of Environment Protection*, 20(6): 430.

Patnaik, K.N., Satyanarayan, S.V. and Swoyam, Poor Rout, 2002. Water pollution from major industries in Pradeep area: A case study. *Indian Journal Environment Health*, 44(3): 203–211.

Sriniwas, C.H., Tisko, R.S., Venkateshwar, C., Satnarayan Rao, M.S. and Reddy Ravindra, R., 2000. Studies on groundwater quality of Hyderabad. *Poll. Res.*, 19(2): 285–289.

Sharma, Surendera Kumar and Chandel, C. P. Singh, 2006. Physico-chemical properties of groundwater of Dudo block of Jaipur District. *Ecology, Environment and Conservation*, 12(1): 141–147.

Chapter 15

Aquariculture of Living Jewels: An Opportunity for Aquabusiness

☆ *Indranil Ghosh*

"Aquariculture" is the term, which exclusively means the cultivation of ornamental fishes within the aquarium system. The aquarium is a miniature aquatic ecosystem where all the necessary components are maintained artificially adopting proper biological and engineering technology. The fishes breed there to produce fry and are reared up to the marketable size with the aid of proper feeding and micro-ecosystem management. They are also called "The Living Jewels" for their glory, gorgeousness and gravity. Besides the aesthetic beauty, the 'World of Silence' of Aquarium revealed the hypnotic power of healing, for several psycho-somatic disorders in human being, and hence, the status of aquarium had been raised up to a standard consumer product today. Involvement of a large number of fish and other animal species beautified with species-specific ornamental peculiarities and their ever-growing demand in both domestic and foreign market has created a superb scope of a potential aqua-trade worth millions of dollars. The recognition of the second largest hobby in the world next only to photography has thus been justified. The real opportunity lies in its value addition through selective breeding and genetic manipulations, use of pigmented feed, application of hormones etc. resulting into the development of new attractive colour strains and ornamental characteristics.

International Trade of Ornamental Fish

The trade with a turnover of US $ 7 Billion and an annual growth rate of 6 per cent offers a lot of scope for development. According to FAO (2006), exports of ornamental fish were $250 billion compared to that of food fish of over $55 billion, in 2004, which support the backbone of multibillion dollar ornamental aqua-business incorporating the ancillary units like aquaria, water purification system, air diffuser, medicines etc necessary for proper maintenance of aquarium environment. The top exporting country is Singapore followed by Czech Republic, Japan, Malaysia, Indonesia, and Israel. India ranks 26th amongst the exporter countries. Since the break-up of former USSR, the Czech Republic has successfully and rapidly penetrated European and American Market and has captured the second largest exporter in the new millennium. The largest importer of Ornamental fish is the USA followed by UK, Germany, Japan, France, Singapore, Netherlands, Belgium, Italy, Hong Kong (FAO, 2006).

A couple of the new countries are becoming more interested in the "Ugly–Darlings", *i.e.*, the dull but strange, hideous and odd-looking Indian fishes; and our native fishes constantly making their strong, permanent footings in the foreigners' aquarium which confirms the common proverb "Familiarity Breeds Contempt". Though our contribution to the Global trade is still insignificant, the Ornamental fish export from India is growing continuously as evidenced from export figure (Table 15.1). The present turnover is a meagre Rs. 577.08 lakh. The major contribution of this trade (90 per cent) is that of West Bengal followed by Mumbai and Chennai. About 95 per cent of our export of ornamental fishes is based on wild collection of fish from the eastern and North Eastern states of the country. Since the nearest international Airport to the collection sites is located at Calcutta, the trade of these species have thrived in West Bengal. Few traders/exporters located at Calcutta regulate the market. The earning potential of this sector has hardly been understood and the same is not being exploited in a technology driven manner in harmony with nature. It is shocking to know that our rich faunistic resources in the North Eastern States are being squandered by local inhabitants by exploiting these as food resource instead of harnessing the same in terms of their ornamental values of recreative nature. It is a fact that 20 per cent of those fishes being caught are the potential ornamental fishes

Table 15.1: Export of Ornamental fish From India (Value in lakhs of INR)

Year	All India	West Bengal
1992-93	35.42	35.27
1993-94	55.32	54.86
1994-95	65.03	53.45
1995-96	81.67	67.36
1996-97	108.34	102.7
1997-98	79.42	73.32
1998-99	158.23	142.56
1999–'00	174.27	168.2

It is encouraging to note that in West Bengal where the business is mostly developed, a large number of entrepreneurs have taken to breeding and rearing these fishes for their ornamental value. There are three categories of people practicing aquaculture, of which the amateurs, or mere hobbyists in true sense who don't depend upon this trade for income, are the first; professionals who have this trade as either primary or secondary source of income, are the second; and thirdly, the farmers and entrepreneurs who have their own culture/production unit. There are innumerable units which have come up in almost all the districts of West Bengal, of which Howrah 24 Parganas (south), 24 parganas (north), Hooghly and Maldah are the major contributor to the market. In certain areas these units have come up in clusters providing opportunity for area Development scheme as being promoted by The National Bank for Agriculture and Rural Development. These units would comprise of nucleus seed farm, satellite rearing units and other ancillary units.

Such approach to development of the sector not only promoted export (Table 2) but also contributed substantially towards regional economy by way of providing employment to educated unemployed youth and women folks engaged in the trade. In addition to that, it has also provided the opportunity to the entrepreneurs to diversify their business. The Marine Products Export Development Authority has already postulated their schemes for the interested candidates. DRDA (District Rural Development Agency), NCDC (National Cooperative Development Council), FFDA (Fish Farmers Development Agency, under Fisheries department of the state government), and CADC (Comprehensive Area Development Corporation, Under Panchayat and Rural Development Department) have also come in front with their economic aids.

Table 15.2: Export of Ornamental Fish in Current Millennium (Value in lakhs of INR)

Year	All India Figure
2001-01	229.31
2001-02	314.08
2002-03	254.95
2003-04	307.83
2004-05	443.84
2005-06	517.31

Technology

There are more than 500 varieties of fish which are at present grown for their ornamental importance in the country along with approximately 450 species are of indigenous origin which include loaches, Gouramis, Barbs, Danios, eels, catfishes, air-breathing fishes, glass fishes, shrimps/prawn and many other vertebrates and invertebrates. The commonly grown groups include Molly, Guppy, Platy, Swordtail, Gouramis, Barbs, Danios, Chiclids, Angels, Siamese fighter, Tetras, Gold fish, Manila crap, sharks etc.

The ornamental fish varieties can be broadly grouped into 4 groups as under:

1. Tropical freshwater types
2. Tropical coldwater types
3. Marine tropical types

A fourth category is the "Brackishwater Types' is nothing but a recent introduction since past few years to the farmers as its technology was not at their hands till then, although it needs more care and attention in future. Most of the people keep the first two categories, while very few-advanced aquarists keep marine.

Management of these 4 categories are different in nature. However the species under reference belong to the first category barring Gold fish and Koi carps. These fish species/varieties differ widely in terms of their tolerance to water hardness, breeding characteristics and temperature tolerance etc. (Table 15.3).

As may be evident from the table, the fishes differ considerably in their characteristics, which would limit their rearing to certain water quality regime and their association in community rearing. At the same time they may be complementary as regard succession of breeding breeding is concerned thus allowing combination of species ensuring better utilisation of installed capacity.

However such grouping and associations can be done in relation to a project location considering the water quality, climatic conditions, ease of rearing and other associated factors. Hence a generalisation of project cost has been avoided. Instead, a species wise comparison of economics of rearing has been attempted to give the intended end user a range of species to select from.

Captive Breeding of Tropical Ornamental Fishes

Classification based on reproductive stategies:

The Bearers

External Bearers

1. *Transfer brooders*–eggs carried for some time before deposition
2. *Auxiliary brooder*–adhesive eggs carried in clusters or balls on the spongy skin of ventral side
3. *Mouth brooders*–oral incubation of eggs and or hatchlings
   (*a*) Ovophile mouth brooder
   (*b*) Larvophile mouth brooder
4. *Pouch brooder*–eggs incubated in external marsupium

Internal Bearers

1. Lecithotrophic
2. Matrotrophic
3. Facultative internal bearer–eggs sometimes fertilized internally by accident via close apposition of gonophores in normally oviparous fish.

The Guarders (Mostly egg-depositors)

Substrate Chooser

1. *Pelagophiles*–Pelagic spawner
2. *Aerophiles*–Emergence spawner
3. *Lithophile*–Rock-spawner
4. *Phytophile*–plant spawner

Nest/Cavity Spawner

1. Foam or froth nester–nest formed by mucous bubbles
2. Polyphiles–miscellaneous substrate and material spawner
3. Lithophiles–rock and gravel nesters
4. Phytophiles–plant material nesters
5. Psammophiles–sand nesters, mud and shell nesters
6. Speleophiles–hole nester

The Non Guarders

Open Substrate Spawner

1. *Pelagophile*
2. *Lithopelagophile*–rock and gravel spawner with pelagic larvae
3. *Lithophiles*–rock and gravel spawners with pelagic larvae
4. *Phytolithophiles*–non obligatory plant spawners
5. *Phytophile*–obligatory plant spawners

6. *Psammophiles*–sand spawners

7. *Aerophiles*–terrestrial spawners

Brood-Hiders

1. *Xerophiles/Egg Burriers*–annual or seasonal

2. *Speleophiles*–cave spawners

3. Rock and gravel spawners

4. *Ostracophiles*–spawners in live invertebrates.

Modes of Reproduction in Ornamental Fish

1. *Pond spawners*: –normally spawns in ponds or large tanks(*e.g.*, barbs, gold fish, koi-carps)

2. *Dedicated spawners*: –remain as dedicated mates for life (*e.g.*, angel, diascus)

3. *Spawners oriented to hormone-inducement*: –spawning hormone required (sharks, loaches, some cat fishes)

4. Anytime or short period spawners: come together only for a short period to spawn

Table 15.3: Species Tolerance and Breeding Characteristics

Species	Water Quality	Breeding Type	Egg Type/Care
Molly	Hard water Sp.	Live Bearer	Young Ones
Guppy	Hard water Sp.	Live Bearer	Young Ones
Platy	Hard water Sp.	Live Bearer	Young Ones
Swordtail	Hard water Sp.	Live Bearer	Young Ones
Blue Gourami	Wide Tolerance	Nest Builder	Male Guard eggs
Pearl Gourami	Wide Tolerance	Nest Builder	Male Guard eggs
Rosy Barb	Wide Tolerance	Egg Scatterer	Adhesive
Gold Fish	Wide Tolerance	Egg Scatterer	Adhesive
Z/P/Vl Danio	Wide Tolerance	Egg Scatterer	Non Adhesive/no care
Siamese Fighting Fish	Wide Tolerance	Nest Builder	Male Guard eggs
Catfish	Wide Tolerance	Egg Scatterer	Adhesive/female care
Angel*	Soft Water	Nest Builder	Adhesive/Parents Fan Eggs
FM Cichlid	Hard Water	Egg Depositors	Enclosures Reqd./bi-parental care
R D Cichlid	Hard Water	Egg Depositors	Enclosures Reqd./bi-parental care
Bl W Tetra	Soft Water	Egg Scatterer	Adhesive/no care
B A Tetra	Soft Water	Egg Scatterer	Adhesive/no care
Serpae Tetra	Soft Water	Egg Scatterer	Adhesive/no care
Manila Carp	Soft Water	Egg Scatterer	Adhesive/no care
Shark	Soft Water	Egg scatterer	Adhesive/no care
Oscar	Soft Water	Egg depositor	Adhesive/bi-parental care
T Loach	Soft Water	Rearing Only	Rearing Only
R loach	Soft Water	Rearing Only	Rearing Only

Rearing Economics

For the purpose of working out economics a unit size of 150 sq.m. Rearing unit has been considered with four cemented tanks for stocking fry. Units of this size have been designed considering the small entrepreneurs in view. However the same could be increased on modular basis and the economics can be worked out in project situation keeping combination of species in mind. For the sake of model economics one species has been taken as an example (Table 15.4).

Table 15.4: Model Economics

Project Cost for Rearing Unit		
Unit size: 150 square meters		
A	**CAPITAL COST**	**(Amount in Rs)**
I	Building and Civil work	
1.	Shed with polythene roofing	8000
2.	Cement tanks with Epoxy coating	30000
3.	Valves and pipe lines	2500
4.	Drainage pit in RCC	1000
5.	Flooring in granite stone chips	1000
6.	Borewell and Water supply	8000
II	Machinery and equipments	
1.	Electric pump	5000
2.	Air blower (0.5 HP) and Supply network	5000
3.	Heater (Aqua. Impression healer)–20 Nos.	1000
4.	Electrical fittings and connection	4000
III	Misc. Fixed Assets	
1.	Small glass aquarium 10 Nos.	2000
2.	Laboratory instruments	1000
3.	Netting for tanks	1000
4.	Hand nets	500
	Sub total	**70000**
B.	**RECURRING COST Fish species (Molly)**	
	Fish seed	1400.00
	Feed	3969.00
	Medicines	268.45
	Electricity	268.45
	Miscellaneous	268.45
	Sub total	**6174.35**
	UNIT COST	**76174.35**
	NPW at 15 per cent DF (Rs)	**536.13**
	BCR at 15 per cent DF	**1.01**
	IRR	**15.48**

Chapter 16

Aquatic Pollution

☆ *Meenakshi Jindal and Kavita Sharma*

"A lake is a body of water or other liquid of considerable size surrounded entirely by land. A vast majority of lakes on Earth are freshwater, and most lie in the Northern Hemisphere at higher latitudes. In ecology the environment of a lake is referred to as lacustrine. Large lakes are occasionally referred to as "inland seas." The studies of these inland water bodies and related ecosystems is referred as Limnology. Limnology divides lakes into three zones: the littoral zone, a sloped area close to land; the photic or open-water zone, where sunlight is abundant; and the deep-water, benthic zone, where little sunlight can reach. The depth to which light can reach in lakes depends on turbidity, determined by the density and size of suspended particles. A natural lake is a fairly large body of water occupying an inland basin (low-lying geographic area). Lakes cover only about 1 per cent of the continents, and contain less than 0.02 per cent of the world's water, but they are important ecosystems and may be sources of water supply in certain regions.

Lakes are extremely varied in terms of origin, occurrence, size, shape, depth, water chemistry, and other features. Lakes can be only a few hectares in surface area (*i.e.*, less than a square kilometer), or they can be thousands of square kilometers. Their average depth can range from a few meters to more than a thousand meters. Lakes can be nearly uniformly round, or they can be irregularly shaped. Their water can be highly acidic (as in some caldera lakes), nearly neutral, or highly alkaline (as in soda lakes). Lakes can be low in nutrients (oligotrophic), moderately enriched (mesotrophic), or highly enriched (eutrophic).

Lake size and depth can change over time, owing to various reasons. Through natural processes, lakes will ultimately fill with sediment, thereby "evolving" into a terrestrial ecosystem. But human influences can accelerate the process. For example, Lake Chad, once one of Africa's largest bodies of freshwater, has decreased in size due to an increasingly dry climate and human demands for irrigation water. The Aral Sea in Uzbekistan has decreased from an area of 68,000 square kilometers (26,500 square miles) in 1960, to 29,000 square kilometers (11,200 square miles) in 1998

Water is the essential substance of life; the dominant component of all living organisms. About 75-95 percent of the weight of all living cells is water and there is hardly a physiological process in which water is not of fundamental importance. Covering some 75 per cent of the planet's surface water is also the dominant environment on earth. Aquatic ecosystems are divided into two major categories: –Marine water and Freshwater. These two major categories are further divided into a variety of aquatic ecosystems based on the depth of flow, substrate and the type of organisms that dominate. Aquatic ecosystems, including rivers, lakes and inland seas, flood plains, coastal lagoons and estuaries, coastal shelves and open oceans cover a very large part of the earth's surface and, among other amenities, goods and services, sustain the production of fisheries and aquaculture.

Fisheries and aquaculture exploit a large diversity or organisms ranging from algae, ascidians and sea-cucumbers to molluscs, crustaceans, fish and marine mammals. All marine and freshwater aquatic environments are linked either directly or indirectly as components of the water cycle. The largest reservoir in the global water cycle is the oceans which contain more than 97 per cent of the total volume of water on earth. Although we as humans recognize this fact, we disregard it by polluting our rivers, lakes and oceans. Subsequently we are slowly but surely harming our planet to the point where organisms are dying at a very alarming rate.

Water Pollution

Pollution may be defined as a change in the physical, chemical or biological aspects of environment which makes it harmful for humans, other living organisms and cultural assets. The basic cause of pollution is man itself. When toxic substances enter lakes, streams, rivers, oceans and other water bodies, they got dissolved or lie suspended in water or get deposited on the bed. This results in the pollution of water whereby the quality of the water deteriorates affecting aquatic ecosystems.

Types of Water Pollution

Point Source

It represents those activities where wastewater is routed directly into receiving water bodies for *e.g.* discharge from factories, sewage systems, power plant, underground coal mines and oil wells. These sources are relatively easy to identify and easier to monitor and regulate.

Non-point Source

It delivers pollutants indirectly through environmental changes for *e.g.* fertilizer from a field is carried into a stream by rain in the form of run off which in turns affects aquatic life. These are more difficult to regulate.

Land-based Activities Causing Pollution

Disposal of Domestic Sewage

Demographic pressure in the urban cities and towns has resulted in the production of enormous amounts of domestic waste materials. These materials reach the marine environment either directly or indirectly through rivers, creeks, bays, etc. The domestic sewage contributes to the largest amount of waste and it has been estimated that approximately 18,240 million liters per day (MLD) (as of 1994) reach the coastal environment of the country. These wastes predominantly contain degradable organic matter which utilises enormous amount of oxygen from seawater for its oxidation. Domestic wastes are discharged mostly in untreated condition due to the lack of treatment facilities in most of the cities and towns. It has been reported that only primary treatment facilities are available in cities and towns where the population is more than 100,000 and the capacity of the plants is not adequate for the treatment of the total waste generated in the city. For example, in Mumbai, the treatment facilities are available only for 390 MLD as against 1,200 MLD of domestic sewage which is generated. Due to such partial treatment, the chemical characteristics of the wastewater retain almost their original features and cause damage to the environmental water quality.

Discharge of Industrial Waste

India is one of largest industrialised nations in the world. Major industrial cities and towns of the country such as Surat, Mumbai, Kochi, Chennai, Vishakhapatnam and Kolkata are situated on or near the coastline. While the major industries discharge treated effluents into the sea, numerous small and medium scale industries discharge the untreated effluents into the adjoining wastewater canals, municipal drains, creeks, etc.

Radioactive and Thermal Wastes

Although power generation is mostly thermal in India, nuclear power is also being generated. So far no serious harm has been reported from these sources, but fly ash from thermal power plants invariably creates environmental problems. Radioactive wastes from nuclear power plants are normally disposed of according to strict international conventions. However, their heat generation poses several problems. Nuclear power plants normally release 50 per cent of their generated heat to the coastal marine environment. Localised damage to ambient flora and fauna appears to be unavoidable.

Agricultural Wastes

India, being an agricultural country, uses large quantities of fertilizers and pesticides to sustain its agricultural production. This is essential in order to meet the food requirements of its population. The run-off from the agricultural fields reaches the rivers. Since the riverine flow is very much less during the dry period, the chemical elements present in the run-off undergo biogeochemical changes in the riverine environment itself with minimum input into the sea. Agriculture is both cause and victim of water pollution.

Eutrophication of Lakes and Rivers

"Stimulation of algal growth and a shift in the algal flora to the blue green algae leading to production of obnoxious blooms floating blankets of algae results into phenomenon known as eutrophication". Eutrophic means well nourished, this sometimes results into very undesirable consequences. Process of eutrophication occurs naturally but very slowly often over a period of time. This leads to the ever increasing of the organic matter which gets deposited at the bottom. The thickness of the bottom sediments increases slowly resulting into the formation of swamps, marshes and finally to extinction of water body

Freshwater Pollution

Non point source pollutants irrespective of source are transported over land and through soil by rainwater and melting snow. These pollutants ultimately find their way into groundwater, wetlands, rivers and lakes and finally to the oceans.

All of the world's fresh-water fisheries face some type of threat from pollution. Lake Victoria is threatened by agricultural runoff and industrial waste. The Chang Jiang is the cleanest of China's rivers, but still threatened by pollution. And completion of the Three Gorges Dam (scheduled for 2009) will reduce flow in the lower river, which in turn may allow contaminated sediments to build up. In recent years, fishing on the Chang Jiang has been severely affected by environmental pollution, construction of irrigation projects, and the reclamation of land from lakes.

In North America, chemical pollution is a great concern in the Great Lakes region because of its heavy industrial and urban centres within larger agricultural areas. The Lake Erie catch of whitefish (a desirable species) collapsed in the 1960s, partly due to the effects of agricultural chemical runoff. Whitefish populations in Lake Superior and Lake Michigan already were reduced as a result of parasitism by sea lampreys. Since the 1960s, the white-fish catch has been steadily increasing as efforts to reduce water pollution and control the sea lamprey had been successful. At the same time as the whitefish take was declining, the tons of carp harvested per year steadily increased. This is one of the ironic aspects of this type of pollution. The lake remained very productive, but carp and the other kinds of fish taken were less economically desirable, so the income from commercial fishing dropped.

Marine Pollution

The oceans are the ultimate sink in which all wastes in river and much of that in air are likely to find their way. Some major activities responsible for the pollution of oceans are:

Developmental Activities like Construction of Ports and Breakwaters

The coastal engineering studies have revealed that the construction of breakwaters alter the sediment transport mechanism in the coastal areas, thereby causing erosion and accretion, depending on the direction of the littoral drift. Such impacts have already been noticed due to construction of port of Madras (Chennai), Paradip, etc. It has been found that due to the establishment of Madras port, an accretion of 75 acres occurred in the last 60 years south of the port as well as severe erosion in the north resulting in the loss of 83 acres of land. Serious coastal erosion problems are also experienced in the western part of India due to intense monsoonal activity. The cyclonic weather which is common along the East coast also causes severe erosion problems reported in north of between Point Calimere and Vishakhapatnam.

Accretion

The littoral drift along the coastal area, caused by reversing monsoonal surface circulation, leads to erosion and accretional problems. The beach sand erosion eroded in the area is deposited either offshore or north or south of that area depending on the direction of the drift. Deposition of sand and sediment by perennial rivers also cause sandpits and small islets. For example, deposition of silt load of $9.4\,m^3$ by the Mahanadi river over the years has led to the formation of Hukitola island. Similarly the Chilka lake, the largest brackish water lake in Asia, is facing a heavy siltation problem due to sand deposition at its mouth. Formation of sandbars due to nourishment also leads to poor exchange of tidal water with the adjoining brackish water ecosystem which is being slowly converted to freshwater ecosystem of adverse dimension. This has been affecting the marine life and also the nesting of migratory birds which rely on the marine ecosystem for its food.

Ship-breaking Industries

During the ship-breaking activities, the major components like engine, etc., are removed and offloaded to the shore. The hull and other steel parts are cut into different sizes and transported as scrap. During the operation, the iron particles and the paint containing lead, get into the marine environment while the soluble matter contaminates the surrounding environment. The iron particles settle in the sediment. Increased concentrations of lead in sea water is lethal to organisms and the iron debris settling in the sediment causes damage to the benthic organisms particularly the filter feeders like clams, mussels, etc.

Sea-based Activities Causing Environmental Disturbances and Pollution

Oil pollution is one of the major sources of pollution from the sea-based activities. India imports nearly 30 metric tons of petroleum products every year through its major ports located at Kandla, Mumbai, Kochi, Chennai and Kolkata. India also produces oil through its inshore and offshore oil fields.

Oil Production in Offshore Platforms and Handling in Ports

The total quantity of oil produced from the offshore wells of ONGC along the Western coast of India is approximately 30 million tons per day. This oil is transported mainly through the pipelines and the oil tankers. Additionally, approximately 30 million tons of crude oil imported from foreign countries is being handled at major ports and the total quantity of petroleum products handled in major ports is about 50 million tons per year at present and is likely to increase in future. This crude oil is carried by tankers and ships which number more than 1,600 per year. The Shipping Corporation of India (SCI) also operates more than 24 crude carriers which carry imported oil to the major ports of the country. Considering the large volume of oil transported and high rate of tanker movement the probability of tanker accident is high–once every few years.

Chapter 17

Fisheries and Aquaculture

☆ *Kavita Sharma and Meenakshi Jindal*

Fishery is the occupation or industry of catching, processing or selling fish or shell fish. However, Aquaculture is the cultivation of aquatic animals and plants, especially, fish shell fish and seaweed, in natural or controlled marine or freshwater environments. It is commonly known as 'Fish Farming' and since marine or freshwater animals and plants are grown on a commercial scale, it can be termed as 'Commercial Underwater Agriculture'.

Brief History

Aquaculture is an ancient culture technique. It was first practiced in Asia. Chinese, in 3500 BC raised carps in ponds. They also developed breeding techniques for increased production. Today China dominates in Aquaculture industry. Currently aquaculture practices all over the world are growing as fast as any other type of agriculture. Aquaculture in India is one hundred years old. Till the end of 19th century the method of pond management was mainly confined to West Bengal, Bihar and Orissa. Later, it gradually spread to whole country.

Major Kinds of Aquaculture

According to the variations in salinity of the respective culturing system, the Aquaculture can be broadly classified into

1. Freshwater culture
2. Mariculture
3. Brackish water culture.

Freshwater Aquaculture

Water is considered to be 'freshwater' when the salinity is <0.5ppt (parts per thousand or ppt). Culturing of organisms in this water is known as freshwater aquaculture. Freshwater aquaculture includes earthen pond culture, temple tanks culture, irrigation tanks culture, race way culture; sewage–fed fish culture, re-circulatory system culture, ornamental fish culture, larvivorous fish culture, food organism's culture sport-fish culture and integrated fish culture.

Within the culture fisheries, the major contributors have been the freshwater aquaculture. India is being basically a carp country. Carp culture in India is believed to be as old as carp culture in china. Carp contributes more than 80 per cent of the total aquaculture production of the country. Carps mainly divided into two groups.

Indian Major Carps

Catla (*Catla catla*), Rohu (*Labeo rohita*) and Mrigal (*Cirrhinus mrigala*)

Exotic Carps

Silver carp (*Hypophthalmichthys molitrix*), Grass carp (*Ctenopharyngodon idella*) and Common carp (*Cyprinus carpio*)

The indigenous and the exotic carps account for the bulk of the production. The three Indian major carps (catla, rohu, mrigal), together, contribute a lion's share of two million tones with exotic carp (silver carp, grass carp, common carp), forming the next important group.

Composite Fish Culture

Culture of different species of fishes together is called composite fish culture. It is also known as polyculture. The objective of this type of fish culture is to select and grow fishes of different feeding habits in order to exploit all types of food available in different regions of pond. In this type of fish culture, fast growing fishes namely Indian major carp, rohu, mrigal, silver carp, common carp and grass carp are usually cultured.

Integrated Fish Culture

Culture of fish along with agricultural crops such as paddy, banana and coconut and livestock such as poultry, duck, cattle and pigs is known as integrated fish farming. The integration of aquaculture with livestock and crop farming offers great efficiency in resource utilization, reduces risk by diversifying crop, and provides additional food and income. This system involves recycling of waste or by-products of one farming system as input for another system and efficient utilization of farming space for maximum production.

Culture of Catfishes

In recent years, culture of catfishes, has also received increased interest, due to their high market price and hardy nature. In all tropical regions of the world, oxygen depletion in shallow water is of common occurrence. The atmospheric temperature has a direct effect on the temperature of the water. There occur diurnal pulses of dissolved oxygen values, influenced by seasonal changes. Such water areas become unutilisable for the conventional human activities, pertaining to their commercial prospecting and are regarded as, derelict and swampy. Survival of carps in such inherently catastrophic environment is meagre but there are number of teleosts in tropical region which are able to thrive in the adverse ecological conditions. Air breathing carnivorous fishes such as murrels, koi, singhi and magur are suitable economically important catfishes for culture.

Brackish Water Aquaculture

Brackish water is water that is saltier than freshwater, but not as salty as sea water. The culture of aquatic organisms in saline water having salinity range >0.5–30‰ is known as brackish water culture. In brackish water areas, milk fish, sea bass, mullet, mugil species and prawns are cultured by installing bamboo or nylon enclosures. These enclosures are called pens. In brackish water, mussels and oysters are also cultured by pole, rack, raft and rope culture methods.

Brackish water aquaculture in India is an important activity for increasing the fish production and to provide rural employment. Though brackish water aquaculture is traditionally practiced in the coastal areas technology improvements in this sector in recent years have motivated farming in new areas.

Along the Indian coast line, the potential area of 1,909 million ha is amenable for brackish water. It has been estimated that we have 9 million ha salt affected land which are unfit or marginally fit for agriculture exist in the hot semi-arid and arid eco-region of northern plains in Haryana, Rajasthan, Uttar Pradesh and Gujarat. The salinization is threatening the productivity of agricultural enterprises affecting the fragile ecosystem and damaging the infrastructure facilities in these areas. The saline water resources however can be used as a potential opportunity for the development of inland brackish water aquaculture adopting the technologies followed in the aquaculture practices of the coastal areas with site specific modifications (Aggarwal, 2006).

Marine Fisheries

The mariculture potential of India is vast as there is great scope for developing farming of shrimps, pearl oysters, mussels, crabs, lobsters, sea bass, groupers, mullets, milkfish, rabbit fish, sea cucumber, ornamental fishes, sea weeds etc. Although about 1.2 million ha is suitable for land based saline aquaculture in India, currently only 13 per cent is utilized. In India till date mariculture activities are confined only to coastal brackish water aquaculture, chiefly shrimp farming.

Mariculture is the rearing of the aquatic organisms under control or semi controlled conditions in the coastal and offshore waters where the salinity is maximal and not subject to significant daily or seasonal variations. Mariculture contributes to the production of protein rich food and has been the source of livelihood of millions of coastal villages.

The other activities which can be categorized as artisanal mariculture include green mussel farming, lobster fattening, crab farming, edible oyster culture, clam farming and seaweed culture. A flourishing international trade of marine ornamental fishes is also in vogue which offers scope for the culture of marine ornamental fishes.

Farming of Molluscs	Marine Crustaceans Farming
Oyster farming	Shrimps
Clams farming	Crabs
Mussels farming	Lobsters
Pearl farming	

Aquaculture has already assumed the status of fast expanding industry, not only in India, but also, in several Asiatic countries over the recent past. Now a day, increasing urbanization have reduced the available land for agriculture and agriculture alone, is not sufficient to meet the requirement of protein rich, balanced diet for the population, which is growing at an alarming rate. The demographic flame is likely to consume this planet, unless, the problem of food production is efficiently tackled in time. So, aqua farming is rightly identified as one of the priority areas in food production systems. Aquaculture provides a great opportunity for production of valuable health food, employment, best land/water use upliftment of rural poor, foreign exchange earnings and overall improvement in the economy of the country.

The role of fisheries sector in the national economy is, in general, relatively limited. Fisheries sector as compared to other sectors of the national economy probably comes under the most complex category. The complexity of fisheries sector stems from the interaction between nature, men and technology.

The fisheries sector has been recognized as a powerful income and employment generator as it stimulates growth of a number of subsidiary industries and is a source of low cost animal protein to the people particularly to the economically weaker sections of the society and thereby it is an advantageous position to ensure national food security

Currently, one-third of the world's 6 billion people rely on fish and other aquatic products for at least one-fifth of their annual protein intake, and catches by subsistence and artisanal fisheries make up more than half of the essential protein and mineral intake for over 400 million people in the poorest countries in Africa and south Asia[1]. Fisheries and aquaculture directly employ over 36 million people worldwide, 98 per cent of whom are in developing countries. Taking into account ancillary occupations and their dependents, there are approximately 520 million fisheries-dependent people. Fisheries and aquaculture also support global trade worth over 78 billion dollars in 2008 (FAO, 2008 and World Bank, 2009).

Aquatic Resources of India

Indian fisheries has diverse resources both in terms of aquatic habitats ranging from glaciers lakes to deep seas and a rich biodiversity of fish and shellfish of more than 10 per cent of global biodiversity. The country is endowed with 8,129 km of coastline and 2.02 million Sq.km of perennially, clear and mercury free water, within the exclusive economic zone. In the inland sector, the country has 4, 5000 km of rivers, 1, 26,334 km network of canals, reservoirs and lakes, covering 3.3 million hectares and freshwater ponds and tanks, which are nearly 2.4 million hectares. In India aquaculture has been in progress for the past 5 decades. India's aquaculture industry is highly dependent on exports. World demand exceeds production and per capita consumption is also steadily increasing. Therefore India has great scope in expanding its aquaculture industry. There has been a remarkable improvement in the productivity level of aqua culture ponds.

Production

The fish production of the country has increased from 0.75 million tonnes in 1950 to over 68.69 lakh tonnes at present. Therefore, an overall eight fold increase with production of over 30.24 lakh tonnes from the inland fisheries and 38.45 lakh tonnes from the marine sector. With this production, the country has occupied the 2nd position in the world in terms of inland fish production and third position in overall fish production (Annual report, 2007-08).

Food Security

From the nutritional point of view, aquaculture has a special significance as fish is a rich source of protein for human being and also for other livestock populations. The diet of an average Indian is very low in calories, that's why, fish is considered as rich food for poor people, containing proteins and micronutrients such as iron, iodine, zinc, calcium, vitamins A and B. (Bene, 2005).The abundance of fish is the solution to fight against malnutrition, in developing societies and countries, on account of the high content of essential amino acids and high digestibility. Nutritionally, fish is often presented as an important source of protein, especially where other sources of animal protein are scarce or expensive.

Research and Extension

Research and extension in the field of fish culture are now given high priority in India and the success achieved is highly encouraging in the recent years. The production of fish has increased at a higher rate compared to food grains, milk, eggs and many other food items. Aquaculture occupies an important position in terms of livelihood of millions of people all around the world. From the nutritional point of view as well, aquaculture has a special significance as fish is a rich source of protein for human being and also for other livestock populations. Vast potential of natural resources of water in India offers great scope and possibilities of commercial freshwater ornamental fish production and export. Many fishery by-products such as pearls, isinglass, liver oil, frozen items also play a important role in Indian economy. India recognizes and shares the objective of the importance of management of fishery resources towards sustainable fisheries. Both the central and state governments have undertaken resource management measures, including closed season fishing, regulation of mesh size, earmarking areas for fishing vessels. India is signatory to UNCLOS. India has also adopted various provisions of the FAO Code of Conduct for Responsible Fisheries–though the Code is voluntary, and Government policies and programmes are broadly in tune with the provision of the Code. India is a member of the Indian Ocean Tuna Commission and is actively participating in all the major programmes aimed at conservation and management of Tunas in the Indian Ocean Region.

Major Threat to Fisheries and Aquaculture

Fisheries and aquaculture are threatened by climate change like, higher water temperatures, rising sea levels, melting glaciers, changes in ocean salinity and acidity, more cyclones in some areas, less rain in others, shifting patterns and abundance of fish stocks. Climate change compromises the sustainability and productivity of a key economic and environmental resource, but it also presents opportunities, especially in aquaculture. Developing countries that depend on fish for food and exports will have a real challenge adjusting to the changes.

Climate change stresses will compound existing pressure on fisheries and aquaculture and threaten their capacity to provide food and livelihoods. The Intergovernmental Panel on Climate Change projects that atmospheric temperatures will rise by 1.8–4.0°C globally by 2100. This warming will be accompanied by rising sea temperatures, changing sea levels, increasing ocean acidification, altered rainfall patterns and river flows, and higher incidence of extreme weather events (IPCC, 2007).

The physical, biological and ecological impacts of climate change in aquatic ecosystems are becoming increasingly apparent. Coral reefs are bleaching and their associated fisheries collapsing rapidly. Commercially exploited fishes are moving northward and into deeper waters at rapid rates, invading polar seas, and withdrawing from sub-polar seas, semi-enclosed seas and the tropics. Climate change may affect fisheries, and their contribution to local livelihoods, national economies and global trade-flows, through both direct and indirect pathways.

Recent analysis demonstrated that African and southeast Asian countries are the most economically vulnerable to climate change impacts on fisheries and aquaculture sectors. This vulnerability arises from a relatively high reliance on fisheries combined with low levels of societal capacity to adapt to anticipated temperature increases. Of the 33 nations identified as being most vulnerable to climate impacts on their fisheries sectors, 19 are among the world's least developed countries, whose inhabitants are twice as reliant on fish and fisheries for food as those of more developed nations. Not only are the most vulnerable countries highly dependent on fish for protein, they also rely on fish and fisheries products as a source of income, producing around 20 per cent of the total tonnage of global fish exports, a fraction worth about US$6.2 billion.

The productivity, distribution and seasonality of fisheries, and the quality and availability of the habitats that support them, are sensitive to these climate change effects. In addition, many fishery-dependent communities and aquaculture operations are in regions highly exposed to climate change. Researchers and policymakers now recognize that the climate change impacts on coastal and riparian environments, and on the fisheries they support, will bring new challenges to these systems and to the people who depend on them. Coping with these challenges will require adaptation measures planned at multiple scales.

Mitigation strategies for fisheries include promoting the use of fuel-efficient fishing vessels and methods, removing such disincentives to energy efficiency as fuel subsidies, and reducing overcapacity in global fishing fleets, as there are too many boats burning too much fuel to chase too few fish. Aquaculture technologies that reduce energy consumption and optimize the potential for carbon sequestration provide opportunities for mitigation. Similarly, conserving and restoring mangroves sequesters carbon, protects coastlines, and enhances fisheries and livelihoods. Opportunities for funding adaptation through novel schemes that also contribute to mitigation, such as the Reduced Emissions from deforestation and degradation scheme for mangroves, should be promoted.

High-quality research that involves resource users, builds strong partnerships and harnesses political will is crucial for making fisheries and aquaculture systems more resilient to the challenge of global climate change and securing a bright future for the people that depend upon them.

References

Aggarwal, V.P., 2006. *Aquaculture and Fisheries Science.* S.R. Scientific Publications, Agra, Delhi, p. 1–19.

Annual Report, 2007–2008. Department of Animal Husbandry, Dairying and Fisheries, Ministry of Agriculture, Government of India, New Delhi, p. 60–81

FAO, 2009. *The State of World Fisheries and Aquaculture, 2008.* FAO Fisheries and Aquaculture Department, Rome, 176 p.

ICAR, 2006. *Handbook of Fisheries and Aquaculture.* Directorate of Information and Publications of Agriculture, ICAR, New Delhi.

IPCC, 2007. *Inter-governmental Panel on Climate Change.* http: //www.ipcc. ch/ipccreports/ assessments–reports.htm.

World Bank and Food and Agriculture Organization, 2008. *The Sunken Billions: The Economic Justification for Fisheries Reform.* Agriculture and Rural Development Department, The World Bank, Washington DC.

Chapter 18

Macrophytic Community of Masoli Reservoir in Parbhani District of Maharashtra

☆ *S.U. Kadam and Md. Babar*

Introduction

Macrophytes (aquatic weeds) are those unabated plants, which grow and complete their life cycle in water and cause harm to aquatic environment directly and to related eco-environment relatively. Water is one of most important natural resource and in fact basis of all life forms on this planet. Therefore, appropriate management of water from source to its utilization is necessary to sustain the normal function of life. It is one important part of natural resource management. The area under small tanks and ponds is equally important due to the establishment of many small irrigation schemes and watershed management projects all over the world. For example, India has 1.9 m ha under water in reservoirs and 1.2 m ha under irrigation canals. The area under village ponds and tanks is nearly 2.2 m ha.

Aquatic weeds often reduce the effectiveness of water bodies for fish production. Aquatic weeds can assimilate large quantities of nutrients from the water reducing their availability for planktonic algae. They may also cause reduction in oxygen levels and present gaseous exchange with water resulting in adverse fish production. Although excessive weed growth may provide protective cover in water for small fish growth it may also interfere with fish harvesting. Aquatic weeds in India are one of the crucial problems of pisciculturists. They widely grow in unmanaged reservoirs.

Materials and Methods

Seasonal sampling were conducted around two years of study from three different stations during Jan. 2006 to Dec. 2007. From each sampling station different sites were located to collect Macrophytes.

The collection of samples was done from sub surface. For collection of phytoplanktons large water (100 lit) was sieved through plankton net. A simple type of planktonic net was used, the water sample was filtered through the sieves and the planktons concentrated in a bottle connected at the lower end of the net.

Macrophytes collected from three different sites of the reservoir included aquatic ferns and true seed producing angiosperms. A systematically classified list of aquatic macrophytes is prepared and grouped into four categories; they are submerged weeds, emerged weeds floating weeds and marginal weeds.

Results and Discussion

Aquatic weeds in India are one of the crucial problems of pisciculturists. They widely grow in unmanaged reservoirs. In this field much work has been done by Unni (1976), Misra (1991), Mishra *et al.* (1992), Gaur and Khan (1995) and Shrivastava and Desai (1997). A systematically classified list of aquatic macrophytes is given in Table 18.1.

Table 18.1: Macrophytic Community of Masoli Reservoir

Life form	Name of species	Class	Family
Floating	*Salvinia molesta*	Fern	Salviniaceae
	Eichhornia crassipes	Monocot	Pontederiaceae
	Lemna sp	Monocat	Lemnaceae
	Pistia sp.	Monocot	Araceae
	Potamegaton sp.	Monocot	Potamegatonaceae
	Nelumbo mucifera	Dicot	Nymphaceae
Submerged	*Marsilea quadrifolia*	Fern	Narasiliaceae
	Naja sp.	Monocot	Najadaceae
	Ceratophylum sp.	Dicot	Ceratophyllaceae
	Utricularia exoleta	Dicot	Lentibulariaceae
Emergent	*Trapa sp.*	Dicot	Hydrocaryaceae
	Cyperus sp.	Monocot	Cyperuaceae
	Marsilea quadrifolia	Fern	Narasiliaceae
Marginal	*Ipomea cornea*	Dicot	Convolvulaceae
	Cyperus sp.	Monocot	Cyperaceae
	Marsilea quadrifolia	Fern	Narasiliaceae

The submerged weeds under collection include *Marsilea sp. Hydrilla verticellata, Vallisneria spiralis, Najas sp., Cerotyphyllum sp.* and *Utricularia exoleta.* Submerged aquatic weeds show vegetative growth during monsoon and flourished during post monsoon months. *Hydrilla sp.* showed profuse growth during summer. Sharma and Singhal (1969) reported impact of floating and emergent vegetation on the trophic status of tropical lake. The growth of submerged plants during post monsoon months might have been due to high mineral concentration and better light condition Philipose *et al.* (1970).

Most of the weeds are of emergent nature, which can grow under saturated and emerged soil–water condition. They grow in soil from saturated moisture on the banks of irrigation canals and drainage ditches. They have been observed in seepage areas of canals, depressions containing water along canals, earthen embankments of water reservoirs, tanks and shallow depths of water reservoirs. The emergent weeds includes *Potamegaton sp., Nymphea stellata Nelumbo nucifera., Ipomoea cornea* and *Trapa sp.* Among these *Trapa sp.* and *Potamegaton sp.* were dominant in Masooli reservoir. *Nymphea* and *Nelumbo sp.* Show considerable growth during monsoon and post monsoon periods whereas *Ipomoea sp., Nymphaea sp., Nelumbo sp.* and *Trapa sp.*

The floating aquatic plants have leaves floating on water surface either singly or in group. They have true roots, leaves and flowering parts above the water surface. Some of them are free floating while the roots of few are anchored in mud in the bottom of water body. These plants rise and fall with the level of water in the water body. The floating weeds were represented by *Salvinia molesta, Eichhornia crassipes, Lemna sp., Pistia sp., Potamegaton sp.* and *Nelumbo nucifera.*

The group of marginal weeds was represented mainly by *Ipomea sp., cyperus sp.* and *Marsilea sp.* The marginal weeds were observed to flourish along the marginal areas of the reservoir from post monsoon. Philopse *et al.* (1970), Jhingran (1982), Joshi *et al.* (1987), have also reported that these plants are common inhabitants of freshwater bodies of India.

During the present investigation maximum population of *Hydrilla, Najas, Potamegaton* and *Trapa* were observed. This is in accordance with the statement of Jhingran (1982) that either *Hydrilla sp.* alone or *Hydrilla sp.,* and *Naja sp.* together dominates some water and do not normally permit the establishment of other plant except *Trapa sp.* and *water lilies.*

Aquatic weeds create situations, which are ideal for mosquito growth. The mosquitoes are sheltered and protected from their predators by aquatic weed roots and leafy growth and are responsible for the spread of Malaria, Yellow fever, river blindness and encephalitis. Snails are able to multiply, playing a crucial role in the life-cycle of blood and liver flukes (parasitic worms) as they shelter, and find sustenance among the root zones. Schistosomiasis and fuscioliasis diseases spread as the floating weed carry the snails to new locations. People living close to these areas complain of mosquito problems (Kumar and Singh, 1987).

Fish production is greatly affected by the presence of floating and submerged aquatic weeds. Isolated weed beds may be tolerated, providing shelter and shade for fish, but when the growth becomes thick and covers entire water body, it can be lethal for fish growth. Fish may suffocate from a lack of oxygen and may cause death. When floating and submerged aquatic weeds become extremely dense, many fish species are unable to exist in such environments and vanish. For example, fish production in Harike lake in Punjab is decreasing and is a matter of concern to all (Chandi and Nagpal, 1987).

The decomposition of huge amounts of biological mass creates condition where CO_2 and carbon monoxide are produced and released to the atmosphere. The decomposition period is much less than decomposition of other vegetation on land. The decomposition creates emissions of foul smells, which are unpleasant to public convenience. Mosquitoes and other parasites grow in these situations and affect the life of those living in close proximity.

Aquatic weeds also affect quality of water. These weeds cause taste and odour problems and also increases biological oxygen demand because of organic loading. They increase the organic matter content of water that may affect the strength of the concrete structures when used as curing and mixing water. It is due to the organic matter that combines with cement to reduce bond strength and may cause large amount of air entrained in concrete.

Aquatic weeds impede the free flow of water, which may contribute to increased seepage and may cause rises in water tables in the adjoining areas. It may lead to water logging. This may also create saline or alkaline conditions in the soil and also give rise to many other land weeds.

References

Chandi, J.S. and Nagpal, B.R., 1987. Problems of *Typha* sp. and its control. In: *Proceedings of the Workshop on Management of Aquatic Weeds*, Amritsar, Punjab, India, 21st November.

Gaur, R.K. and Khan, Asif A., 1995. Physico-chemical characteristics on an eutrophic lentic environment with a permanent bloom of a cyanobacterium *Microcystis aeruginosa. J. Ecobiol.*, 7(4) 263–267.

Jhingran, V.G., 1982. *Fish and Fisheries of India*. Hindustan Pub. Corp. India, Delhi.

Joshi, A., Sharma, A.P. and Pant, M.C., 1981. Limnological studies in a subtropical system (Lake Sattal, India). *Acta Hydrochim et. Hydrobiol.*, 9(4): 407–425.

Kumar, M. and Singh, J., 1987. Environmental impacts of aquatic weeds and their classification. In: *Proceedings of the Workshop on Management of Aquatic Weeds*, Amritsar, Punjab, India. 21st November.

Mishra, P.C., 1991. Environmental status of Hirakud reservoir system. *J. Ecotoxicol. Environ. Monit.*, 1(1): 23–30.

Mishra, S.R., Sharma, Sanjay and Yadav, R.K., 1992. Phytoplanktonic communities in relation to environmental conditions of Lentic waters. *J. Environ. Biol.*, 13(4): 291–296.

Philipose, M.T., 1970. Freshwater plankton of Inland fisheries. In: *Proc. Symp. Algal.*, ICAR, New Delhi, p. 272–291.

Sharma, Alka and Singhal, P.K., 1988. Impact of floating and emergent vegetation on the trophic status of tropical lake. I. The macrophytes and physico-chemical status. *J. Environ. Biol.*, 9(3 suppl.): 303–311.

Shrivastava, N. P. and Desai, V.R., 1997. Studies on the bottom macro-fauna of Rihand reservoir, Uttar Pradesh. *J. Environ. Biol.*, 18(4): 325–331.

Unni, K. Sankaran, 1972. An ecological study of the macrophytic vegetation of Doodhadari lake, Raipur, M.P.S. chemical factors. *Hydrobiologia*, 40: 25–36.

Chapter 19

Lipid Changes in Oyster *Crassostrea cattuckensis* during Accumulation and Depuration of Cadmium

☆ *G.D. Suryawanshi, Y.A. Shaikh and U.H. Mane*

Introduction

A small amount of many of relative toxic materials into aquatic environment causes multiple changes in the internal dynamics of aquatic organism at sublethal levels. In aquatic toxicology extensive literature is available on effect of various pollutants on the biochemical composition of tissues of different types of marine bivalve mollusks Regoli, *et al.* (1992). Kumarswamy, and Karthikeyan, (1999), Kumbhar, (2001), The study on biochemical processes is very important to understand the mechanism of metal toxicity to commercially important invertebrates, Rao *et al.* (1987) studied the effect of fluoride and mercuric chloride on the freshwater bivalve *Indonaia caeruleus*. Lipid composition of egg and adductor muscle in giant scallops *Piacopecten magellanicus* was studied Napolitund *et al.* (1992). Cadmium chloride and mercuric chloride induced changes in the biochemical composition of the freshwater bivalve *Lamellidens marginalis* respectively was noticed Kulkarni (1993). Changes in lipid peroxidation in the gills and muscle of *Perna viridis* during exposure of cadmium and copper was reported Arasu and Shreenivasula Reddy (1995). Further, bioaccumulation and metabolic effects of zinc and mercury on marine dreissinid bivalve, *Mytilopsis sallei* was studied Devi (1995). Munsi *et al.* (1997) mixture of heavy metals on the biochemical composition of two penaeid shrimp post larvae was studied. While workers Patil and Mane (1997) studied seasonal changes of biochemical in different body parts of *L. marginalis* during exposure of mercury.

Materials and Methods

The oysters, *Crassostrea cattuckensis* were collected from Bhatye estuary is one of them [situated 73⁰ 15 East and 16"51 North] in south of Ratnagiri coast where Kajvi River meets the sea in post winter

(February). Soon after the fishing they were brought to the laboratory and the shells were brushed to clean the fouling biomass and mud. They were then stocked in continuous aerated filtered seawater pumped in the laboratory from the estuary for 24 to 48 h for depuration. Based on the acute toxicity tests the chronic test were performed on the oysters of (80-90 mm shell length) using sub lethal concentrations, *i.e.* $1/10^{th}$ of LC_0 and LC_{50} values of cadmium metal salts. The experiments were carried for 15 days and 30 days to metal exposure, and then transfer of oysters of 15 days metal exposed to normal seawater to next 15 days for detoxification. The control group in normal seawater was run simultaneously during each experiment. The oysters belonging to control and experimental were sacrificed separately to obtain soft body parts like mantle, gills, gonad, hepatopancreas, adductor muscles and siphons. These tissues were weighed and they were then kept in hot air oven at 92°C till constant weights were obtained. The dried product was ground to obtain fine powder. From the replicates of the three samples lipid were estimated Barnes, and Blackstock (1973). The data obtained were statistically analyzed for confirmation of the results. The results were calculated using regression equations and expressed in mg/100mg dry weight.

Results and Discussion

Lipids play a nutritionally and physiologically important role in marine bivalves by providing an efficient source of high-energy content and essential fatty acids Waldock and Holland (1984). The metabolism and transport of lipids in many bivalve molluscs is of particular important in reproduction and larval development Robinson, *et al.* (1981). On the other hand, Nagabhushanam and Mane (1978) investigated the female mussels contained lipids twice as much as males during gametogenesis and in adult bivalve lipid are stored mainly in gonad. Further, Voogt (1983) stated that lipids in bivalves are multifunctional and in different species one or same of the functions during the maturation of gametes, drastic environmental conditions, starvation, population stress etc. can be more noticeable. In bivalve mollusks the conversion of glycogen into fatty via trios–phosphate entry in glycogen sequence and to the production of pentose sugar for nuclic acid synthesis is well documented by Gabbott (1976). In the view of such a role of lipid in body maintenance metabolism in the bivalves, analysis of lipid, in the present study from *C. cattuckensis* was exposed to cadmium metal for 15 days the lipid content in low concentration when compared with the oysters from experimental control the content increased from siphon (39.08%; P < 0.001), gonad (33.90%; P < 0.001), mantle (29.63%; P < 0.001) but decreased from gill (30.77%; P < 0.01), hepatopancreas (7.15%; N.S.), and adductor muscle (1.59%; N.S.). In high concentration it was decreased from adductor muscle (40.90%; P < 0.001), hepatopancreas (23.29%; P < 0.01) and gill (18.06%; P < 0.01) but increased in gonad (57.54%; P < 0.001), mantle (10.41%; P < 0.01) and siphon (3.13%; N.S.). In high concentration compared to those from low concentration the content was more decreased from adductor muscle (39.95%; P < 0.001) than from siphon (25.85%; P < 0.01), hepatopancreas (17.38%; P < 0.001) and mantle (14.83%; P < 0.01), but it increased from gill (18.36%; N.S.) and gonad (17.66%, P < 0.01).

Further, the content from the oysters when compared from the oysters belonging to respective low and high concentrations on 15 and 30 days exposure showed decreased level from all the tissues from both the concentrations, except gill from high concentration on 30 days exposure. In low concentration more decreased from siphon (66.83%; P < 0.001) followed by gonad (62.48%; P < 0.001), adductor muscle (48.78%; P < 0.01), gill (46.84%; P < 0.01), hepatopancreas (44.40%; P < 0.01) and mantle (9.74%; N.S.). Further in high concentration the content from gill (2.82%; N.S.) almost remained same, but decreased in the content from siphon (46.76%; P < 0.05), hepatopancreas (37.78%; P < 0.05), adductor muscle (37.28%; P < 0.01), mantle (11.46%; N.S.) and gonad (11.35%; N.S.). On the other hand, during detoxification process in low concentration when compared with the 15 days metal

Table 19.1: Changes in the Lipid Content from Different Body Parts of C. cattuckensis during Chronic Test of Cadmium Chloride in Winter

Body Parts	15 Days Exposure			30 Days Exposure			Detoxification	
	Control	LC_0	LC_{50}	Control	LC_0	LC_{50}	LC_0	LC_{50}
Mantle	3.075±0.061	3.986±0.147 (29.63%)***	3.395±0.085 (10.41%)** (14.83%)	2.009±0.061 (49.82%)*** (34.65%)ᴬᴬᴬ	3.598±0.338 (79.09%)* (9.74%)	3.006±0.274 (49.63%)** (16.45%) (11.46%)	5.095±0.484 (153.61%)*** (27.83%)?	4.726±0.169 (135.25%)*** (7.26%) (39.21%)???
Gills	3.195±0.061	2.212±0.294 (30.77%)**	2.618±0.339 (18.06%)** (18.36%)	2.149±0.061 (6.70%)* (32.74%)ᴬᴬᴬ	1.176±0.169 (45.28%)*** (46.84%)ᴬᴬ	2.692±0.129 (25.27%)** (128.93%)ᴼᴼᴼ (2.82%)	3.265±0.167 (51.94%)*** (47.61%)??	3.099±0.509 (44.20%)* (5.0%) (18.38%)
Gonad	2.894±0.112	3.875±0.056 (33.90%)***	4.559±0.179 (57.54%)*** (17.66%)ᴼᴼ	2.129±0.061 (45.70%)*** (26.44%)ᴬᴬᴬ	1.454±0.251 (31.71%)** (62.48%)ᴬᴬᴬ	4.042±0.673 (89.86%)** (78.11%)ᴼᴼ (11.35%)	4.096±0.111 (92.40%)*** (5.71%)?	2.286±0.279 (7.38%) (44.23%)ᴼᴼᴼ (49.88%)???
Hepatopancreas	3.316±0.126	3.079±0.251 (7.15%)	2.544±0.199 (23.29%)** (17.38%)ᴼᴼᴼ	2.211±0.061 (33.06%)*** (33.33%)ᴬᴬᴬ	1.712±0.242 (22.57%)* (44.40%)ᴬᴬᴬ	1.583±0.485 (28.41%)* (7.56%) (37.78%)ᴬ	3.635±0.209 (64.41%)*** (18.06%)	5.447±0.139 (146.36%)*** (49.85%)ᴼᴼᴼ (144.12%)???
Adductor muscle	2.773±0.61	2.729±0.169 (1.59%)	1.639±0.139 (40.90%)*** (29.95%)ᴼᴼᴼ	2.049±0.061 (12.34%) (26.11%)ᴬᴬᴬ	1.398±0.335 (31.78%)** (48.78%)ᴬᴬ	1.028±6.13 (49.83%)** (26.45%) (37.28%)ᴬᴬ	3.062±0.169 (49.44%)*** (12.21%)	3.413±0.452 (66.57%)** (11.46%) (108.24%)???
Siphon	2.109±0.61	2.933±0.111 (39.08%)***	2.175±0.169 (3.13%) (28.58%)ᴼᴼ	1.347±0.061 (54.41%)*** (36.14%)ᴬᴬᴬ	0.973±0.251 (99.28%) (66.83%)ᴬᴬᴬ	1.158±0.547 (49.83%) (19.01%) (46.76%)ᴬ	4.264±0.529 (216.56%)*** (45.39%)?	1.528±0.538 (13.44%) (64.18%) (29.74%)

Bracket values represent percentage differences, (*,O,Δ,?–P<0.05, **,OO,ΔΔ,??–P<0.01, ***,OOO,Δ ΔΔ,???–P<0.001. *–compared to control oysters, O–compared to LC_0 group, Δ,?–compared to 15 days exposure group).

exposed oysters, the content increased from gill (47.61%; P<0.01), siphon (45.39%; P < 0.05), mantle (27.83%; P < 0.05), hepatopancreas (18.06%; N.S.), adductor muscle (12.21%; N.S.) and gonad (5.71%; P < 0.05). In high concentration showed that the content increased from adductor muscle (108.24%; P < 0.001) hepatopancreas (144.12%; P < 0.001), mantle (39.21%; P < 0.001), and gill (18.38%; N.S.) but decreased from gonad (49.86%; P < 0.001) and siphon (29.74%; N.S.).

The decrease in the lipid content of tissues indicates the possibility of pronounced lipolysis and its likely utilization to meet high-energy demand during the metal stress. Sivaprasad Rao and Ramanrao (1979) found considerable decrease in the total lipid in muscle might be due to drastic decrease in glycogen content in the tissue of freshwater fish, *Tilapia mossambica*. Similar decrease in the lipid level was also observed by Muley and Mane (1987) in the freshwater mussel *L. marginalis* when exposed to mercuric chloride. Jalaluddin and Shariff (1987) observed a decline in lipid content in the muscle of fish and this decrease might be due to compensate the energy deficiency in animal. A significant decrease observed of content in some tissues of the oysters. However, increase in the lipid content in the body parts of the oysters was also seen which suggest the inhibition of lipase activity and lipid synthesis likely due to impairment in carbohydrate metabolism and to the inhibition of enzyme activity in lipid metabolism (as suggested by Coley and Jonson). Jaiswal (1986) observed a rise in lipid content during exposure to sublethal concentration of naphthalene on freshwater prawn *Macrobranchium kistnensis*. Where as, Arasu and Shreenivasula, Reddy (1995) also observed a lipid content to increase in the gill and muscle of marine bivalve *P. viridis* during exposure to cadmium and copper. On the other hand, lipid content decreased in both the groups from gills, gonad, hepatopancreas and adductor muscle. In addition, heavy demand of energy exerted upon the oyster body parts was noticed especially by utilization of the lipid from gill, hepatopancreas and adductor muscle where in the content decreased.

On the other hand, upon 30 days exposure in both the groups of cadmium, lipid increased from adductor muscle, gonad, hepatopancreas and siphon. While the content decreased from gill, this showed greater demand of energy over the utilization of body reserves on gill. This further, showed synthesis of lipid in most of the tissues due to metabolic shifts in different body parts due to longer exposure to metal, however, it appears that mantle and adductor muscle showed protein synthesis due to cadmium irrespect of time of exposure period and similarly it appears that gonad and siphon showed lipid synthesis due to cadmium irrespect of time of exposure period, 15 or 30 days. Further, the depletion of lipid from the tissues suggest the possibility of its much utilization to provide excess energy for cellular biochemical processes probably through lipolysis in the fact of increased anaerobic due to metal toxicity. In present study corroborates with Mohite (2002) who mentioned the glycogen content decreased in some body parts with an increased exposed time during 15 days, 30 days exposure period and also next 15 days detoxification process when *P. viridis* were exposed to cadmium.

References

Arasu, S.M. and Shreenivasula, Reddy P., 1995. Changes in lipid peroxidation in the gills and muscle of marine bivalve *P. viridis* during exposure to cadmium and copper. *Chem. Ecol.*, 11(2): 105–112.

Barnes, H. and Blackstock, J., 1973. Estimation of lipids of marine animals in tissue. Detailed investigation of the sulphophosphovanillin method for total lipids. *J. Exp. Mar. Biol. Ecol.*, 12(1): 103–111.

Coley and Jonson, 1973. *Lipids*, 8: 43.

Devi, V.U., 1995. Bioaccumulation and metabolic effects of zinc on marine fouling dreissinid bivalve, *Mytilopsis sallei. Water, Air, Soil, Pollut.* 81(3): 295–304.

Gabbott, P.A., 1976. *Energy Metabolism in Marine Mussels,* (Ed.) B.L. Bayne. Cambridge University Press, London, New York, Melborn, p. 293–355.

Jaiswal, K.B., 1986. Physiology responses of the freshwater prawn *Macrobrachium kistnensis* to the hydrocarbon Naphthalene. *Ph.D. Thesis,* Marathwada Univ., Aurangabad (M.S.), India.

Jalaluddin and Shariff, D., 1987. Modulation in the biochemical nature of the body muscle of a freshwater teleost, *Sarotherodon mossambicus* (Peters) due to 1 DETOL (A synthetic detergent). In: *Proc. Nat. Cont. Env. Impact Biosystem,* p. 279–282.

Kulkarni, S.D., 1993. Cadmium toxicity to freshwater bivalve molluscs *Lamellidens marginalis* from Godavari river near Aurangabad. *Ph.D. Thesis,* Marathwada University, Aurangabad, p. 1–338.

Kumarswamy, S. and Karthikeyan, A., 1999. Effect of cadmium on oxygen consumption and filtration rate at different salinities in an estuarine clam *Meretrix casta* (Chemnitz). *J. Envi. Biol.,* 20(2): 99–102.

Kumbhar, S.N., 2001. Cadmium induced toxicity to esturine clams Katelysia opima, *Meretrix meretrix. Ph.D. Thesis,* Shivaji University, Kohlapur, p. 1–325.

Mohite, V.T., 2002. Base levels of heavy metal and their detoxification mechanisms in green mussel *Perna viridis* from coast Maharashtra. *Ph.D. Thesis,* Institute of Science, Mumbai, p. 1–259.

Muley, D.V. and Mane, U.H., 1987. Sublethal effects of $HgCl_2$ on the tissue compositions of a bivalve molluscs *Lamellidens marginalis. Biol. Bull., India,* 9(1): 31–40.

Munsi, A.B., Su-Young-quan, Li, Shao-Jing, Hong-Li-Yu, 1997. Effect of Cu, Cd and Cu: Cd mixture an the biochemical composition of two penaeid shrimp post larvae. *Chin. J. Ocean. Limnol.,* 15(1): 46–51.

Nagabhushanam, R. and Mane, U.H., 1978. Seasonal variaton the biochemical composition of *Mytilus edulis* at Ratnagiri on the west coast of India. *Hydrobiologia,* 57: 69–72.

Napolitund, G.E., MacDonald, B.A., Thomson, R.J. and Ackman, R.G., 1992. Lipid composition of egg and adductor muscle in giant scallops *Piacopecten magellanicus* from different habitats. *Mar. Biol. (BERL),* 113(1): 71–76.

Patil, S.S. and Mane, U.H., 1997. Tissue biochemical levels in different body parts of the bivalve molluscs, *Lamellidens marginalis* (L.) exposed to mercury in winter season. *J. Aqua. Biol.,* 12(2): 47–52

Rao, K.R., Kulkarni, D.A., Pillai, K.S. and Mane, U.H., 1987. Effects of fluoride on the freshwater bivalve molluscs, *Indonaia caeruleus* in relation to the effect of pH: Biochemical appraoch. *Proc. Nat. Symp. Ecotoxic.,* p. 13–20.

Regoli, F., Nigro, M. and Orlando, E., 1992. Effects of copper and cadmium on the presence of renal concentrations in the bivalve, *Donacilla cornea. Comp. Biochem. Physiol.,* 102(1): 192–198.

Robinson, W.E., Wehling, W.E., Morse, P.M. and McLeod, G.C., 1981. Seasonal changes in soft body component indices and energy reserves in the Atlantic deep sea scallop *Placopecten magellanicus. Fish. Bull. U.S.,* 79: 449–458.

Sivaprasad Rao, K. and Ramanrao, K.V., 1979. Effect of sublethal concentration of methyl parathion on selected oxidative enzymes and organic constituents in the tissue of freshwater fish, *Tilapia mossambica* (Peters). *Curr. Sci.*, 48: 526–528.

Voogt, P.A., 1983. *Lipids: Their Distribution and Metabolism in the Mollusca.* (Ed.) P.W. Hochachka, Academic Press, New York and London, Volk, 1: 329–370.

Waldock, M.J. and Holland, D.L., 1984. Fatty acid metabolism. *Lipids,* 19: 332–336.

Chapter 20

The Mudskipper,
Boleophthalmus boddarti
(Pallas, 1770) (Pisces: Gobiidae)
Distribution along Tamil Nadu,
Southeast Coast of India

☆ *V. Ravi*

Introduction

Mudskippers (Gobiidae: Oxudercinae) are the only fishes known to burrow and reside in the intertidal mudflats or mangrove swamps of the Indo-West Pacific region (Murdy, 1989) and these fishes are uniquely adapted to a completely amphibious lifestyle (Graham, 1997). They are quite active when out of water, feeding and interacting with one another, for example to defend their territories. Compared to the fully aquatic gobies, these fishes have a range of peculiar behavioral and physiological adaptations to an amphibious lifestyle. These include: anatomical and behavioral adaptations that allow them to move effectively on land as well as in the water (Harris,1961), the ability to breathe through their skin and the lining of their mouth (the mucosa) and throat (the pharynx) by means of cutaneous air breathing (Graham, 1997), digging of deep burrows in soft sediments that allow the fish to thermo regulate (Tytler and Vaughan, 1983), avoid marine predators during the high tide when the fish and burrow are submerged (Sasekumar *et al.*, 1984). The mudskipper, *Boleophthalmus boddarti* (Pallas, 1770) (Figure 20.1) is a residential fish inhabiting the mudflats of the estuary and the waterways of mangrove forests, Tamil Nadu Southeast coast of India. It is edible and fishermen consume it during the lean season. It is also used in traditional medicine and fishermen use it as a cure for frequent urination by children.

Figure 20.1: The Mudskipper, *B. boddarti* (Pallas), Tamil Nadu, Southeast Coast of India

World Distribution of Mudskippers and Systematics

Mudskippers are found distributed in temperate to tropical mudflats and mangrove forests along the east and west coasts of India besides Andaman and Nicobar islands, African coasts (Nigeria, Tanzania, Madagascar), Arabian Gulf, Pakistan, Sri Lanka, Bangladesh, Malaysia, Singapore, Indonesia, the Philippines, Thailand, China, Vietnam, Korea and Australia.

About 34 species belonging to genera as *Periophthalmus* (Ps), *Boleophthalmus*, *Periopthalmodon* (Pn.) and *Scartelaos*, which are called as 'mudskippers'. In India, the occurrence of mudskipper species as *B.boddarti* (Pallas), *B. dussumieri* Valenciennes, *Ps. koelreuteri* (Pallas), *Pn. schlosseri* (Pallas) and *Scartelaos viridis* (Hamilton) have been reported (Subramanyam, 1974; Rema Devi, 1992; Mukherjee, 1995; Barman *et al.,* 2000; Rao *et al.,* 2000; Ravi, 2000; Rathod *et al.,* 2002). Among them, the giant mudskipper, *Pn. schlosseri* is one of the world's largest oxudercine gobiid fish (TL 24cm).

Etymology

Boleophthalmus is from the Greek bole, dart or arrow, and ophthalm, eye; *Periophthalmus* is from the Greek, peri, around, and ophthalm, eye, in reference to the prominent, dorasally-placed eyes that appear to provide a broad range of vision; and Periophthalmodon is from the Latin suffix odon, meaning toothed.

Abundance

The population density of the mudskipper (*B. boddarti*) was 4 to 10 and 4 to 7 animals/sq.m in Mudasalodai and Vellar estuary, Southeast coast of India respectively in 1999 (Ravi, 2000) and suddenly got reduced 0 to 1 animal and 2 to 3 animals/sq.m in the above places respectively during Jan 2005 *i.e.* after tsunami (Ravi, 2005). In the west coast of India, Rathod *et al.* (2002) found that the mudskipper population was affected due to severe pollution and increased siltation which made the substratum not conducive for these species in Thane creek, Mumbai.

In 2003, the average population density in 95 Japanese habitats was 24.3 individuals per 100 m² of mudflat (Takegaki, 2008). Although, extremely high densities were observed in some habitats (>50 individuals/100 m2), the densities in many other habitats were found to be much lower.

Habitat and Biology

B. boddarti is an amphibious and gobiid fish which mainly inhabits in the mudflats around the estuaries and mangroves of Tamil Nadu. They are active on the mudflats during the daytime low tides and remain stayed in the burrows during the high tide and at night. This species is herbivorous feeding mainly on diatoms on the mudflat surface (Ravi, 2000). The life span for *B. boddarti* appears to be 2.5 years or above (Ravi and Rajagopal, 2007).

Burrow Dynamics

Mudflat is a large area of mud in the intertidal region and a characteristic feature of an estuary and mangroves. Generally the mudflats are predominantly constituted by muddy substratum and are exposed twice a day during low tides. Habitat selection could play a major role in the maintenance of genetic polymorphism in natural populations (Powell and Taylor, 1979). Soil texture is an important factor determining the distribution of benthic organisms in general and mudskippers in particular.

The mudskipper, *B. boddarti* is restricted to the mudflat zone of the study areas. The distribution of mudskippers is commonly assumed to be based on the presence of suitable muddy substratum (Macnae, 1968). Burrow is tunnel composed of a single or many aperturers extending/branching with chambers in the muddy substratum made by the organisms. Mudskippers build burrows (Figures 20.3a and b) the depth of which range from 50 to 150 cm with one or many entrances on the surface of the mudflats (Clayton and Vaughan, 1986; Ravi *et al.*, 2004; Hong and Zhang, 2004). Similarly, Ishimatsu *et al.* (1998) observed that these fishes construct burrows in the substrata of the high intertidal zone and transport air for storage in its burrow, which is always filled with water. In general, burrowing animals aerate the soil (Lavelle *et al.*, 1995) and form a labyrinth of interconnected tubes through which water can flow possibly providing an extremely efficient pathway for the transfer of nutrients and oxygen besides many others across the swamp-bed interface (Ridd, 1996). Burrow also serves as an important refuge from piscivorous predators (Milward, 1974). Mudskippers build inside the burrow an egg chamber whose depth range from 20 to 30 cm beneath the surface of mudflats (Hong and Zhang, 2004). After spawning, the females leave the chamber and the male protects the eggs.

Burrow width varied from 3-21cm (Table 20.1). Burrow width and number of burrows recorded in different seasons are given in Table 20.2. More number of burrows (14÷0.82) measured in 6-9cm width during premonsoon and summer. During post monsoon, burrows with 12-21cm width were not found which would be due to construction of new burrows (Ravi *et al.*, 2004).

Table 20.1: Burrow Type, Width and Nature of the Mudskipper, *B. boddarti*

Burrow Type	Burrow Width (cm)	Burrow Nature
Single aperture burrow	3-9	Newly constructed and existing burrow
Double aperture burrow	9-15	Existing burrow
Multi-aperture burrow	15-21	Existing/Collapsed burrow

Classification of burrows was confirmed only after carrying out molding analysis in the field. The burrow nature and its construction pattern (Ravi *et al.*, 2004) are shown schematically in Figure 4. Based on the number of aperture found, burrows were categorized as single aperture burrow (having only one aperture in the burrow), double aperture burrow (having two apertures) and multi aperture burrow (having more than two apertures). Furthermore, based on the construction and survival status of the burrows, they were categorized as newly constructed burrow(*i.e*, freshly constructed burrow), existing burrow(living burrow *i.e*, burrow in use) and collapsed burrow (burrow unfit to accommodate the mudskippers, *i.e* physically damaged) (Ravi *et al.*, 2004).

Table 20.2: Changes in Burrow Width in Different Seasons (No. of burrow/m²)

Burrow Width (cm)	Pre-monsoon	Monsoon	Post-monsoon	Summer
3–6	4±1.25	5±3.09	7±1.69	4±1.63
6–9	12±2.94	6.3±2.2	10±1.41	14±0.82
9–12	1.3±0.89	1.7±0.9	2±1.4	1±0.9
12–15	1±0.3	1±0.9	–	1.5±0.5
15–18	1±0.4	1±0.8	–	2±0.5
18–21	1±0.5	1±0.7	–	1±0.4

IUCN Status

Anandha Rao *et al.* (1998) listed few species of mudskippers in India as Endangered (EN)(*B. dussumieri* and *S. viridis*) and Vulnerable (VU) (*B. boddarti* and *Ps. koelreuteri*) which is mainly due to loss of habitat and pollution that lead to reduction in population and restricted distribution. Kruitwagen *et al.* (2006) also observed that the early stages of embryonic development of mudskippers (*Ps. argentilineatus*) would be mostly affected by pollutants along the Tanzanian coasts. Similarly, mudskipper *B. pectinirostris* (Linnaeus) occurring along the coasts of Japan has been included as Vulnerable (VU) in the Red Data Book of Japan (Ministry of the Environment of Japan, 2003), and the decimation is attributed to overexploitation of stock, coastal development and wastewater discharges from coastal areas (Takegaki, 2008).

Threats

Threats Observed in the Mudskipper's Habitat
1. Rail route development near the mudflat (Ennore)
2. Industrial effluent release near the mudflats (Ennore estuary)
3. Mudflat erosion along the coast (Muthupettai mangroves)

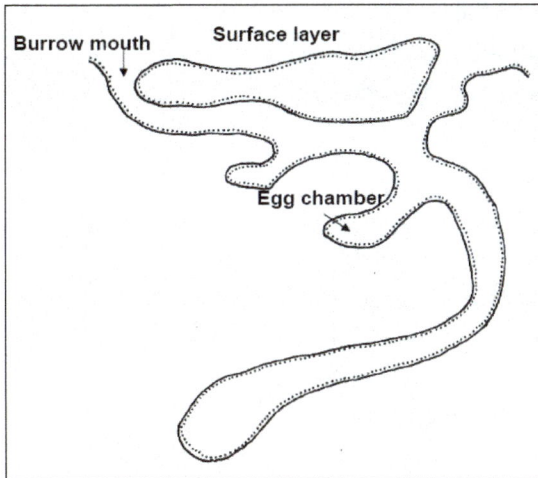

(a) *B. boddarti* (Ravi, 2000)

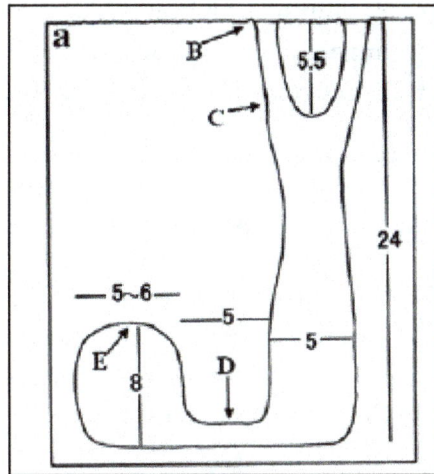

(b) *Ps. magnuspinnatus*
(Courtesy: Baeck *et al.*, 2008)

Figure 20.2: Burrow Structure of the Mudskippers

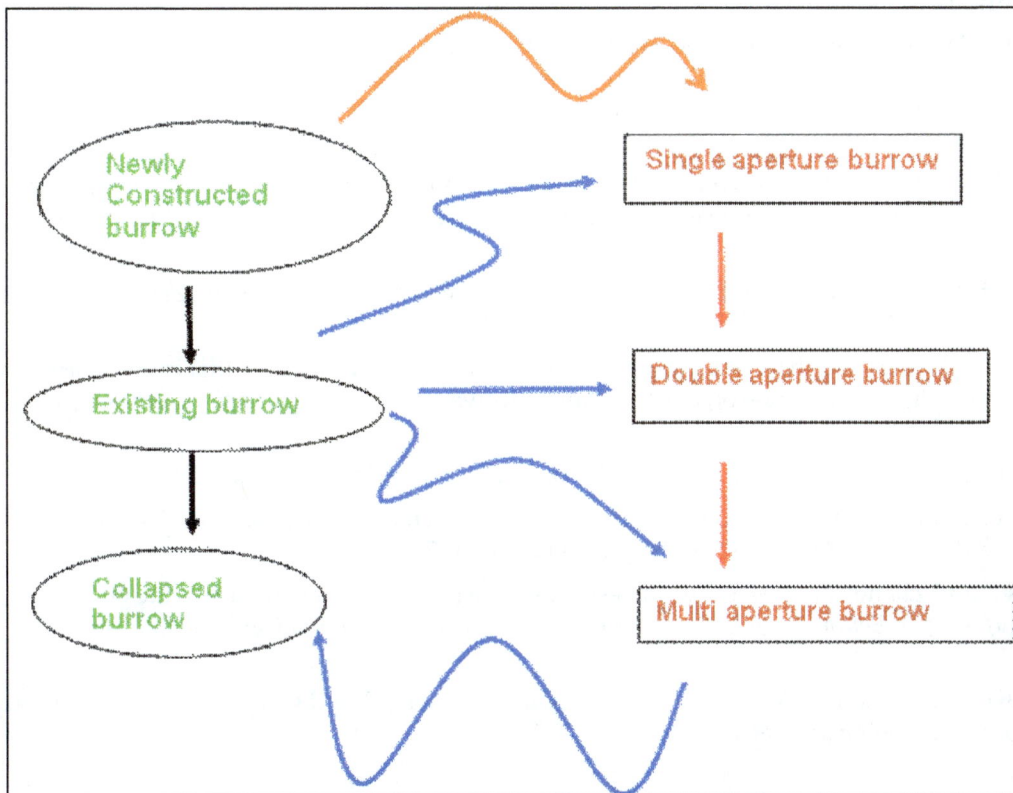

Figure 20.3: Schematic Representation of Burrow Construction and its Condition

4. Ecotourism activity near the mudskipper's habitat (Vellar estuary)

5. Impact of macro algae on mudskipper's habitat (Kollidam estuary)

6. Lack of estuarine water to reach mudflats (Uppanar estuary)

Conservation and Management Issues

☆ Mudflats are important in preventing coastal erosion.

☆ Mudflats worldwide are under threat from predicted sea level rises, land claims for development, dredging due to transport purposes, and pollution.

☆ Ecosystem management towards the improvement of its biodiversity

☆ Enacting strict Regulatory Acts to dumping of waste into the ecosystem

☆ Setting up of waste-water treatment and purification of the ecosystem

☆ Creation of awareness to the users about its importance

Acknowledgements

The author is thankful to the Prof. (Dr.) T. Balasubramanian, Director, CAS in Marine Biology and to the authorities of Annamalai University for the facilities and to the University Grants Commission, New Delhi for the financial assistance.

References

Anandha Rao, T., Molur, Sanjay and Walker, Sally (Eds.), 1998. *Report of the Workshop 'Conservation Assessment and Management Plan for Mangroves of India (BCCP–Endangered Speceis Project)'*, Zoo Outreach Organization, Conservation Breeding Specialist Group, India, Coimbatore, India, 106 p.

Baeck, G.W., Takita, T. and Yoon, Y.H., 2008. Lifestyle of Korean mudskipper *Periophthalmus magnuspinnatus* with reference to a congeneric species *Periophthalmus modestus. Ichthyol. Res.,* 55: 43–52.

Barman, R.P., Mukherjee, P. and Kar, S., 2000. Marine and estuarine fishes. In: *Zool. Surv. India. State Fauna* Series 8: Fauna of Gujarat, Part–I: 311–411.

Clayton, D.A. and Vaughan, T.C., 1986. Territorial acquisition in the mudskipper *Boleophthalmus boddarti* (Pisces: Gobiidae) on the mudflats of Kuwait. *Journal of University of Kuwait (Science),* 15: 115–138.

Graham, J.B., 1997. *Air-Breathing Fishes: Evolution, Diversity and Adaptation.* Academic Press, San Diego.

Harris, V.A., 1961. On the locomotion of the mudskipper *Periophthalmus koelreuteri* (Pallas) (Gobiidae). *Proceedings of the Zoological Society of London,* 134: 107–135.

Hong, W. and Zhang, Q., 2004. Induced nest spawning and artificial hatching of the fertilized eggs of mudskipper, *Boleophthalmus pectinirostris. Chinese Journal of Oceanology and Limnology,* 22(4): 408–413.

Ishimatsu, A., Hishida, T., Kanda, Takita, Oikawa, S., Takeda, T. and Khoo, K.H., 1998. Mudskippers store air in their burrows. *Nature,* 391: 237–238.

Kruitwagen, G., Hecht, T. and Pratap, H.B., 2006. Changes in morphology and growth of the mudskipper (*Periophthalmus argentilineatus*) associated with coastal pollution. *Marine Biology*, 149: 201–211.

Lavelle, P., Lattaud, C., Trigo, D. and Barois, I., 1995. Mutualism and biodiversity in soils. *Plant and Soil*, 170: 23–33.

Macnae, W., 1968. A general account of the fauna and flora of mangrove swamps and forests in the Indo-West Pacific region. *Advances in Marine Biology*, 6: 73–270.

Milward, N.E., 1974. Studies on the taxonomy, ecology and physiology of Queensland mudskippers. *Ph.D. Thesis*, University of Queensland, 276 pp.

Ministry of the Environment of Japan, 2003. Threatened wildlife of Japan. In: *Red Data Book, 2nd edn., Vol 4: Pisces, Brackish and Freshwater Fishes.* Japan Wildlife Research Center, Tokyo.

Mukherjee, P., 1995. Intertidal fishes. In: *Zool. Surv. India.* Ecosystem Series, Part 2, Hughli–Matla Estuary, p. 345–388.

Murdy, E.O., 1989. A taxonomic revision and cladistic analysis of the Oxudercine gobies (Gobiidae: Oxudercinae). *Records of the Australian Museum, Supplement*, 11: 1–93.

Powell, J.R. and Taylor, C.E., 1979. Genetic variation in ecologically diverse environments. *Am. Scient.*, 67: 590–596.

Rathod, D., Sudesh, N., Patil, N., Quadros, Goldin and Athalye, R.P., 2002. Qualitative study of finfish and shellfish fauna of Thane creek and Ulhas river estuary. *Proceedings of the National Seminar on Creeks, Estuaries and Mangroves: Pollution and Conservation*, November, pp. 135–141.

Ravi, V., 2000. Studies on eco-biology the mudskipper *Boleophthalmus boddarti* (Pallas, 1770) (Pisces: Gobiidae). *Ph.D. Thesis*, Annamalai University, India, 193 pp.

Ravi, V., Rajagopal, S., Khan, S. Ajmal and Balasubramanian, T., 2004. On the classification of burrows of the mudskipper, *Boleophthalmus boddarti* (Pallas)(Gobiidae: Oxudercinae) in the Pichavaram mangroves, southeast coast of India. *Journal of the Annamalai University (Science)*, Platinum Jubilee Special Issue, p. 235–239.

Ravi, V., 2005. Post tsunami studies on the mudskipper *Boleopthalmus boddarti* (Pallas, 1770) from Mudasalodai, Tamil Nadu, Southeast coast of India. *J. Int. Goby Soc.*, 4(1): 9–17.

Ravi, V. and Rajagopal, S., 2007. Age and growth of the mudskipper, *Boleopthalmus boddarti* (Pallas, 1770). *J. Aqua. Biol.*, 22(1): 123–128.

Rao, D.V., Devi, Kamala and Rajan, P.T., 2000. An account of Ichthyofauna of Andaman and Nicobar Islands, Bay of Bengal. *Rec. Zool. Surv. India*, Occ. Paper No. 178, 434 pp.

Rema Devi, K., 1992. Gobioids of Ennore estuary and its vicinity. *Rec. Zool. Surv.*, India, 90(1–4): 161–189.

Ridd, P.V., 1996. Flow through animal burrows in mangrove creeks. *Estuarine Coastal and Shelf Science*, 43(5): 617–625.

Sasekumar, A., Ong, T.L. and Thong, K.L., 1984. Predation of mangrove fauna by marine fishes. In: *Proceedings of the Asian Symposium on the Mangrove Environment? Research and Management.* University of Malaya and UNESCO, Kuala Lumpur, p. 378–384.

Subramanyam, K., 1974. Biosystematic studies in Gobioidea (Pisces: Teleostei) of Porto Novo waters (S. India). _Ph.D. Thesis_, Annamalai University, India, 361 pp.

Takegaki, T., 2008. Threatened fishes of the world: _Boleophthalmus pectinirostris_ (Linnaeus 1758) (Gobiidae). _Environ. Biol. Fish._, 81: 373–374.

Tytler, P. and Vaughan, T., 1983. Thermal ecology of the mudskipper, _Periophthalmus koelreuteri_ (Pallas) and _Boleophthalmus boddarti_ (Pallas) of Kuwait Bay. _J. Fish Biol._, 23: 327–337.

Chapter 21

Statisticial Relationship between Body Measurement of *Gobius biocellatus*

☆ *N.R. Jaiswal, M.S. Kadam and N.V. Sunnap*

Introduction

The study on the statistical relationship helps in the determination of the degree of association between body measurements and to establish the equations, enabling conversion of one measurement into another.Some work on statistical relationship between body measurements have been carried out by Sugan and Sankaran (1972), Madalapure (1973), Jadhav (1974) and Somvanshi (1976). Studies on the morphology of *Gobius giuris* have been carried out by Patil (1982). However no account have been given on the statistical relationship between body measurements of *G. biocellatus*.

Materials and Methods

The present investigation was carried out on *G.biocellatus* from Kayadhu river near Hingoli, with a view to establish the relationship between various body measurements of the fish and its total length. A series of measurements of 615 individuals ranging between 80 to 210 mm in total length were taken. The statistical analysis of the relationship between 1) Total length and Standard length, 2) Total length and Head length, 3) Total length and Orbit diameter, 4) Total length and Inter-orbital space, 5) Total length and length of the snout and 6) Total length and depth at the origin of the dorsal fin have been carried out. All the measurements were taken with the help of engineering dividers with a screw.

The various terms regarding the body measurements are explained as follows:

☆ *Total length*: From the tip of the snout to the tip of caudal lobe.

☆ *Standard length*: From the tip of the snout to the origin of the caudal fin.

☆ *Head length*: From the tip of the snout to the pointed edge of the operculum.

☆ *Orbit length*: Diameter of right eye was measured.

Table 21.1: Relation Between Total Length and Standard Length in *G. biocellatus*

Sl.No.	Length Group in mm	Total Length X in mm	Standard Y Length in mm	X2	XY	Calculated Values of Y
1.	80-89	86.66	77.33	7509.95	6701.41	75.18
2.	90-99	94.66	85.33	8964.3000	8077.33	80.76
3.	100-109	105.75	88.75	11183.06	9385.31	88.50
4.	110-119	116.33	93.33	13532.66	10857.07	95.88
5.	120-129	126.33	99.00	15959.26	12506.67	102.86
6.	130-139	135.00	107.5	18225.00	14512.50	108.90
7.	140-149	145.66	110.33	21216.83	16070.66	116.34
8.	150-159	156.00	125.00	24336.00	19500.00	123.56
9.	160-169	165.66	131.66	27443.23	21810.79	130.30
10.	170-179	177.50	141.00	31506.25	25027.50	138.56
11.	180-189	185.00	144.33	34225.00	26701.05	143.79
12.	190-199	195.00	150.33	38025.00	29314.35	150.77
13.	200-209	200.66	156.33	40264.43	31369.17	154.72
		$\Sigma X=1890.21$	$\Sigma Y=1510.22$	$X^2=292390.97$	$\Sigma XY=231833.81$	

$X=145.40$, $Y=116.17$, $Y=14.72+0.6977 X$

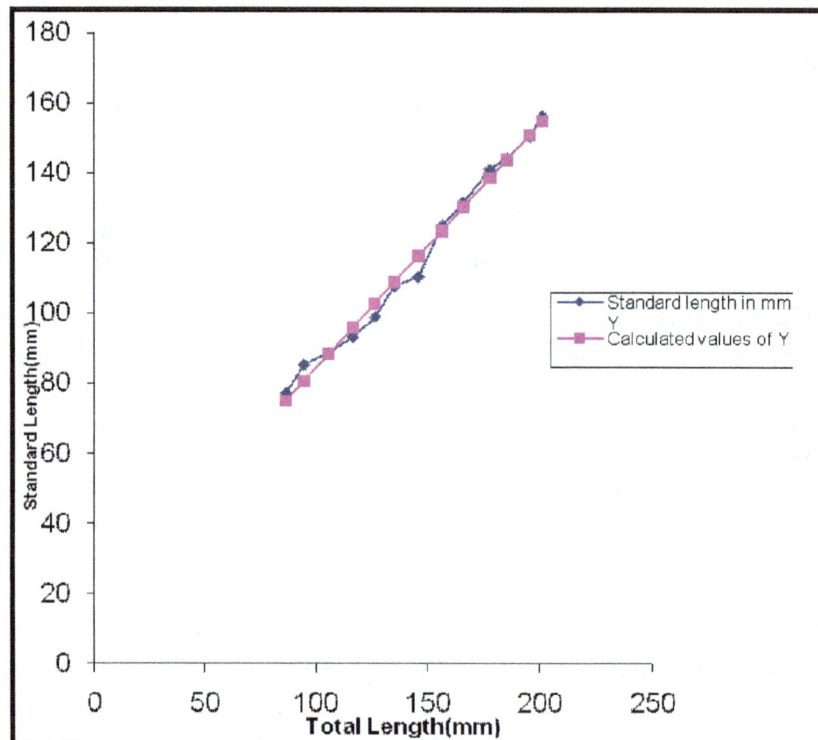

Figure 21.1

Table 21.2: Relationship Between Total Length and Head Length in *G. biocellatus*

Sl.No.	Length Group in mm	Total Length X in mm	Head Length in mm Y	X2	XY	Calculated Values of Y
1.	80-89	86.66	22.3	7509.95	1935.11	20.4
2.	90-99	94.66	24.00	8964.3000	2271.84	22.1
3.	100-109	105.75	25.8	11183.06	2723.06	24.4
4.	110-119	116.33	26.3	13532.66	3061.80	26.6
5.	120-129	126.33	27.6	15959.26	3490.49	28.6
6.	130-139	135.00	28.2	18225.00	3800.00	30.39
7.	140-149	145.66	29.2	21216.83	4247.44	32.6
8.	150-159	156.00	32.40	24336.00	5054.40	34.72
9.	160-169	165.66	35.60	27443.23	5897.49	36.70
10.	170-179	177.50	39.7	31506.25	7039.65	39.14
11.	180-189	185.00	41.70	34225.00	7714.50	40.69
12.	190-199	195.00	44	38025.00	8585.05	42.75
13.	200-209	200.66	46.3	40264.43	9282.53	43.91
		$\Sigma X=1890.21$	$\Sigma Y=422.99$	$X^2=292390.97$	$\Sigma XY=65103.36$	

$X=145.40, Y=32.53, Y=2.6+0.2059\ X$

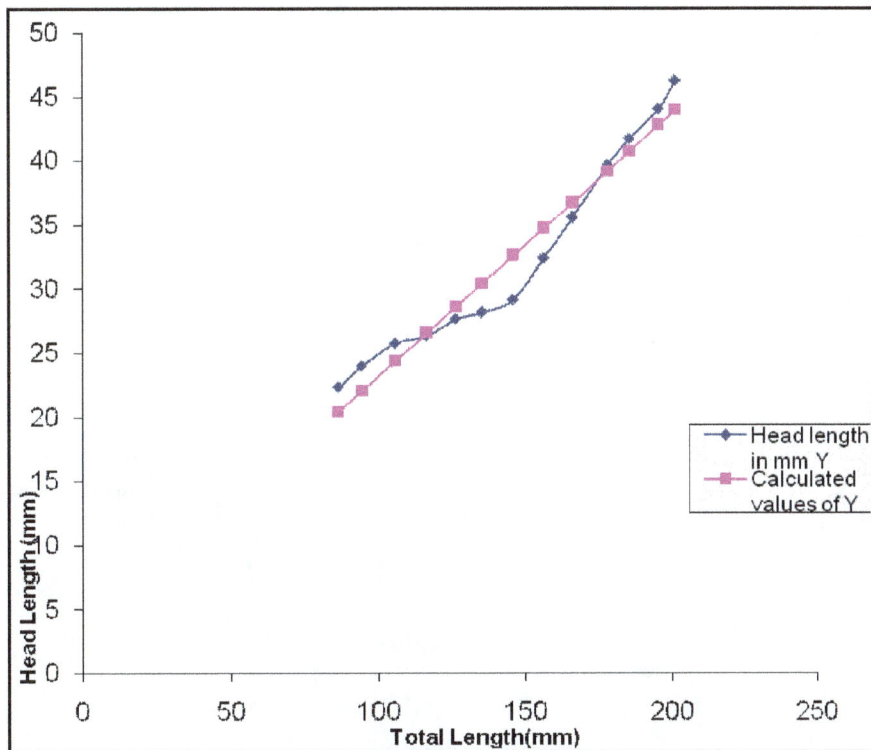

Figure 21.2

Table 21.3: Relationship Between Total Length and Orbit Diameter in *G. biocellatus*

Sl.No.	Length Group in mm	Total Length X in mm	Orbit Diameter in mm Y	X2	XY	Calculated Values of Y
1.	80-89	86.66	3.06	7509.95	282.51	2.90
2.	90-99	94.66	3.33	8964.3000	315.21	3.12
3.	100-109	105.75	3.35	11183.06	354.26	3.42
4.	110-119	116.33	3.46	13532.66	403.66	3.72
5.	120-129	126.33	3.96	15959.26	500.26	4.00
6.	130-139	135.00	4.50	18225.00	607.50	4.24
7.	140-149	145.66	4.62	21216.83	672.94	4.53
8.	150-159	156.00	4.66	24336.00	726.96	4.82
9.	160-169	165.66	4.70	27443.23	778.60	5.08
10.	170-179	177.50	5.73	31506.25	1017.07	5.41
11.	180-189	185.00	5.80	34225.00	1073.00	5.62
12.	190-199	195.00	5.83	38025.00	1136.85	5.90
13.	200-209	200.66	5.86	40264.43	1175.86	6.05
		$\Sigma X=1890.21$	$\Sigma Y=58.86$	$X^2=292390.97$	$\Sigma XY=9044.68$	

X=145.40, Y=4.5276, Y=0.5007+0.0277 X

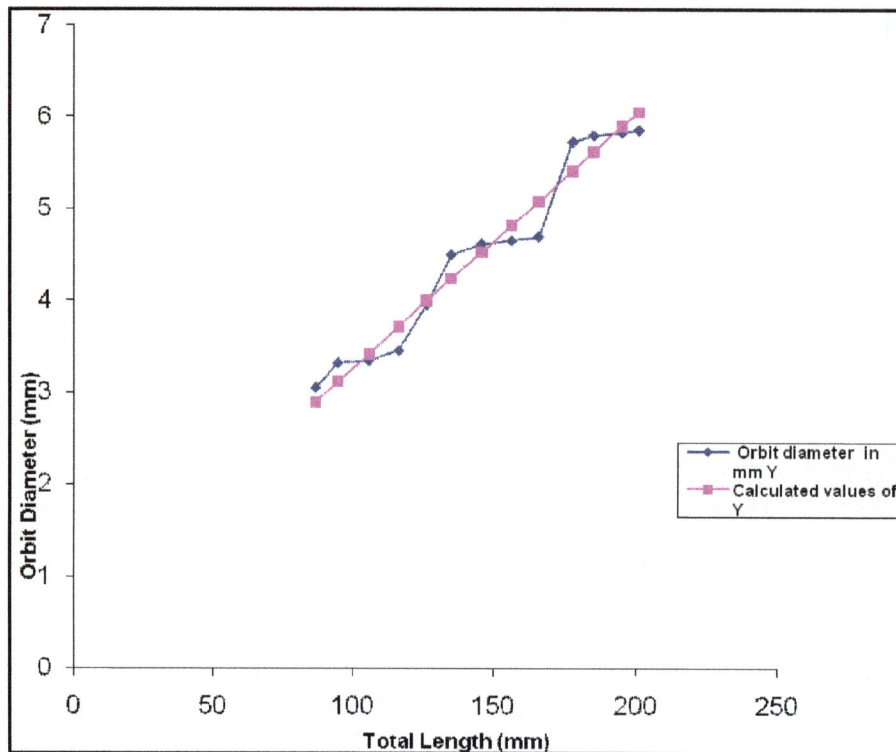

Figure 21.3

Table 21.4: Relationship Between Total Length and Interorbital Space in *G. biocellatus*

Sl.No.	Length Group in mm	Total Length X in mm	Interorbital Space in mm Y	X2	XY	Calculated values of Y
1.	80-89	86.66	3.16	7509.95	273.84	3.07
2.	90-99	94.66	3.30	8964.3000	312.37	3.35
3.	100-109	105.75	3.92	11183.06	414.54	3.74
4.	110-119	116.33	4.06	13532.66	472.29	4.11
5.	120-129	126.33	4.20	15959.26	530.58	4.47
6.	130-139	135.00	4.23	18225.00	571.05	4.77
7.	140-149	145.66	5.33	21216.83	776.36	5.15
8.	150-159	156.00	5.52	24336.00	861.12	5.51
9.	160-169	165.66	5.56	27443.23	921.06	5.85
10.	170-179	177.50	6.60	31506.25	1171.50	6.27
11.	180-189	185.00	6.73	34225.00	1245.05	6.54
12.	190-199	195.00	6.80	38025.00	1326.00	6.89
13.	200-209	200.66	6.90	40264.43	1384.55	7.09
		$\Sigma X=1890.21$	$\Sigma Y=66.31$	$X^2=292390.97$	$\Sigma XY=10260.31$	

$$X = 145.40, Y = 5.10, Y=0.012+0.0353 X$$

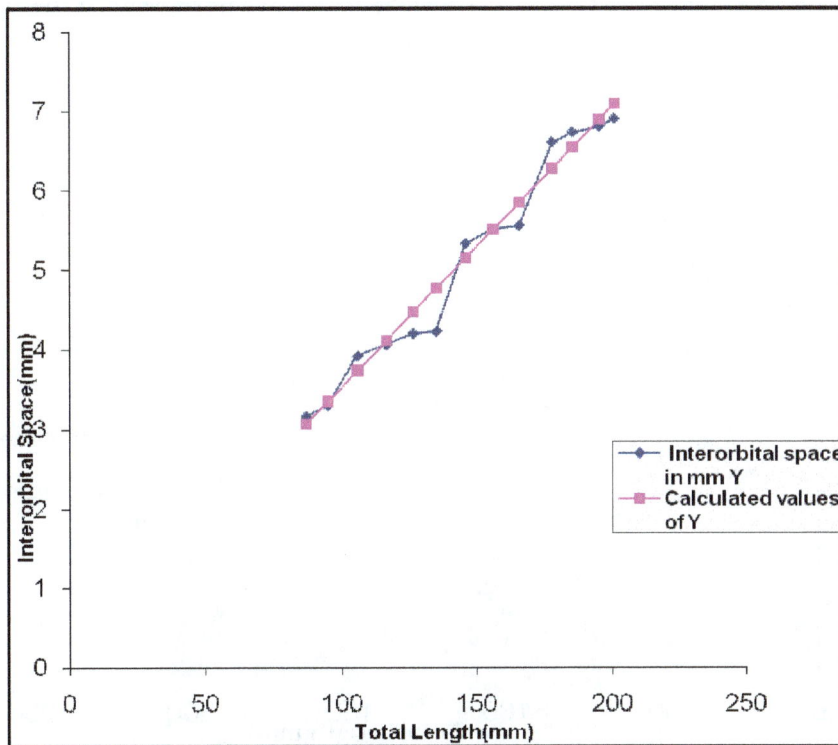

Figure 21.4

Table 21.5: Relationship Between Total Length and Length of the Snout in *G. biocellatus*

Sl.No.	Length Group in mm	Total Length in mm X	Length of the Snout in mm Y	X2	XY	Calculated Values of Y
1.	80-89	86.66	7.33	7509.95	635.21	8.09
2.	90-99	94.66	8.16	8964.3000	772.42	8.77
3.	100-109	105.75	9.87	11183.06	1043.75	9.65
4.	110-119	116.33	10.16	13532.66	1048.71	10.6
5.	120-129	126.33	11.46	15959.26	1505.04	11.5
6.	130-139	135.00	12.53	18225.00	1691.55	12.19
7.	140-149	145.66	13.56	21216.83	1975.14	13.10
8.	150-159	156.00	14.57	24336.00	2272.92	13.97
9.	160-169	165.66	15.65	27443.23	2592.57	14.79
10.	170-179	177.50	15.80	31506.25	2804.50	15.80
11.	180-189	185.00	16.83	34225.00	3113.55	16.43
12.	190-199	195.00	17.06	38025.00	3326.70	17.28
13.	200-209	200.66	17.10	40264.43	3431.78	17.76
		$\Sigma X=1890.21$	$\Sigma Y=170.08$	$X^2=292390.97$	$\Sigma XY=26213.84$	

$X=145.40, Y=13.08, Y=0.75+0.0848\ X$

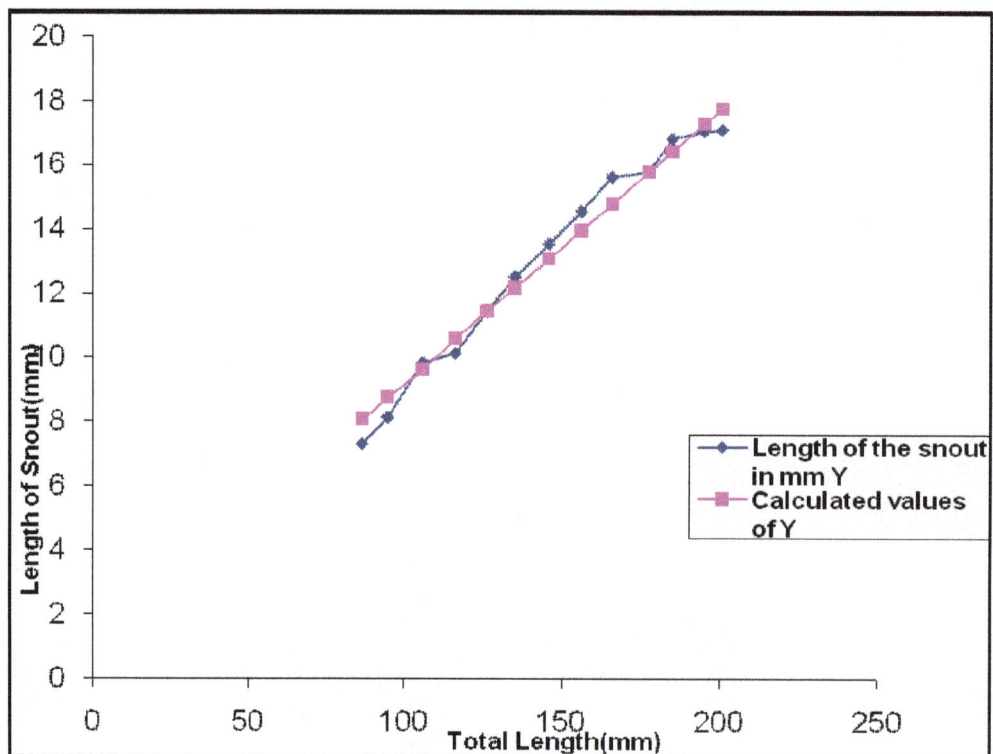

Figure 21.5

Table 21.6: Relationship Between Total Length and Depth at the Origin of Fin in *G. biocellatus*

Sl.No.	Length Group in mm	Total Length in mm X	Depth at the Origin of Dorsal Fin in mm Y	X2	XY	Calculated Values of Y
1.	80-89	86.66	11.50	7509.95	996.59	10.7
2.	90-99	94.66	12.2	8964.3000	1151.06	11.70
3.	100-109	105.75	13.2	11183.06	1390.61	13.1
4.	110-119	116.33	13.60	13532.66	1582.08	14.4
5.	120-129	126.33	14.1	15959.26	1783.77	15.60
6.	130-139	135.00	15.3	18225.00	2069.55	16.67
7.	140-149	145.66	16.60	21216.83	2417.95	17.98
8.	150-159	156.00	17.63	24336.00	2750.28	19.26
9.	160-169	165.66	19.7	27443.23	3256.87	20.45
10.	170-179	177.50	20.80	31506.25	3692.00	21.91
11.	180-189	185.00	22.25	34225.00	4116.25	22.84
12.	190-199	195.00	24.66	38025.00	4808.70	24.07
13.	200-209	200.66	25.87	40264.43	5191.07	24.76
		$\Sigma X=1890.21$	$\Sigma Y=227.3$	$X^2=292390.97$	$\Sigma XY=35206.78$	

$$X=145.40, \quad Y=17.48, \quad Y=0.03+0.1233\, X$$

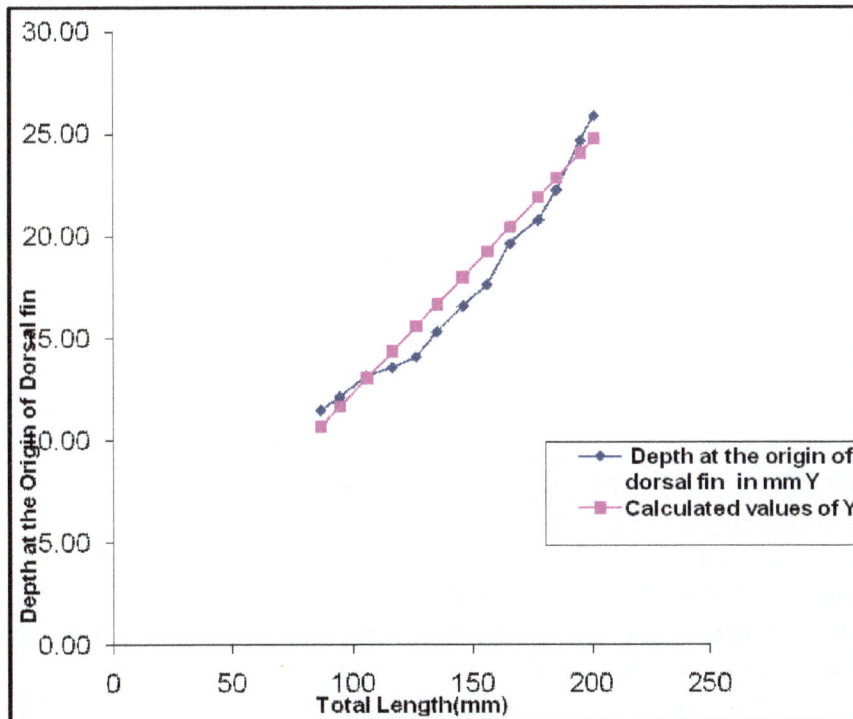

Figure 21.6

☆ *Inter-orbital space*: The space between the two orbits.

☆ *Length of the snout*: From the tip of the snout to the anterior margin of the orbit.

☆ *Depth at the origin of the dorsal fin*: Height of the fish from the origin of the dorsal fin, on a line perpendicular to the long axis of the fish.

To express the relationship between two morphometric measurements, the equation for the regression line Y = a + b X was used where, 'a' and 'b' are constants 'X' represents the total length and 'Y' the variable, such as the standard length, head length, orbit diameter, inter-orbital space, length of the snout and depth at the origin of the dorsal fin.

Results and Discussion

On plotting the calculated values for total length against these variable, the points representing the observed values for total length were found to lie closely round the regression line (Figures 21.1–21.6).

The relationship between the total length and the different variables are shown (Tables 21.1–21.6) by the following equations.

Total length (X) and Standard length (Y)

$$Y = 14.72 + 0.6977 X$$

Total length (X) and Head length (Y)

$$Y = 2.6 + 0.2059 X$$

Total length (X) and Orbit diameter (Y)

$$Y = 0.5007 + 0.0277 X$$

Total length (X) and Inter-orbital space (Y)

$$Y = 0.012 + 0.0353 X$$

Total length (X) and Length of the snout (Y)

$$Y = 0.75 + 0.0848 X$$

Total length (X) and Depth at the origin of dorsal fin (Y)

$$Y = 0.03 + 0.1233 X$$

References

Ganguly, D.N., Mitra, B. and Bhattacharya, N., 1959. On the inter-relationship between total length, standard length, depth and weight of Lates calcurifer. *Proc. Nat. Inst. Sci., India,* 25: 175-187.

Jadhav, P.G., 1974. Biology of *Nemachilus botia* (Ham). *Ph.D. Thesis.* Marathwada University, Maharashtra, India.

Madalapure,V.R., 1973. Biology of *Barbus* (*Puntitus*) *ticto* (Hamilton). *Ph.D. Thesis,* Marathwada University, Aurangabad, Maharashtra, India.

Somvanshi, 1976. Biology of Gorra mullya (Sykes) from Marathwada. *Ph.D. Thesis,* Marathwada University, Aurangabad, Maharashtra.

Sugunan, V.V. and Shankaran, J.M., 1972. Analysis of morphometric characters of *Selar kalla* (Cuv.) *Indian J. Mar. Sci.,* 1(1): 92–93.

Chapter 22

Gonadosomatic and Hepatosomatic Indexes of the Freshwater Male Fish *Mystus cavasius* (Ham.) during Various Reproductive Phases

☆ *Ashashree Hassan Mallikarjuna*

Introduction

Maturity and spawning is an important biological aspects to be studied in fishes as it is useful in its management (Anju and Anoop, 2003). Fish is the major food resource for human being. The nutritive value of various species and types of fishes compares more favorable than muscles of beef, pork and mutton (Parulekar and Bal, 1969). Since fish represents a much larger source of potential proteins for human consumption than has been utilized, it seems more desirable to bring the most important facts relating to the food value of economically important species, on seasonal basis and in relation to important metabolic process of feeding, growth and breeding. (Bumb and Parulekar, 2002).

The biological indexes of reproduction show the way in which fish use environmental and energetic resources. The Gonadosomatic index is a good indicator of reproductive activity, than being used in determining fish reproductive cycle stages (Devalming *et al.*, 1982). Variations in the hepatosomatic index of teleost are related to the liver capacity to store glycogen, physiological conditions, reproductive activity, feeding habits and food availability in their habitat (Dias *et al.*, 2000).

Mystus cavasius (Ham.) is a commercially important, having good price per kg in the market. The culture practice of this fish is not reported from this area. The gonado somatic index and hepato somatic index is one of the important parameter of the fish biology of Bhadra reservoir, which gives the detail idea about the fish production. Hence, the present study was under taken. Since gonado somatic index is an important tool in establishing the breeding period in animal and in fish (Saxena, 1986) and the liver has a role in the testis development therefore, the HSI was correlated with GSI.

In India some of the workers have been investigated the biology of freshwater fishes such as Ahirrao (2002) in *Mastacembelus armatus*; Sindhe and Kulkarni (2004) in *Notopterus notopterus*; Shankar and Kulkarni (2005) in *Notopterus notopterus*; Kaur and Kaur (2006) in *Channa punctatus* and Shendge and Mane (2006) in *Cirrhina reba*.

Material and Methods

The study was conducted during the period of December 2004 to December, 2005. During the present investigation fishes were collected from Bhadra reservoir (13°–42'-00"N latitude and 75°-38°-20"E longitude).The fishes were collected and brought to the laboratory and they were measured accurately for their total length, weight up to the nearest unit. Further fishes were sacrificed after decapitation, length and weight of the fish and weight of the testes and weight of the liver were recorded. The morphological changes of the testes and liver were observed during different stages.

GSI

The gonado somatic index was determined by using the formula of Giese (1959).

$$GSI = \frac{\text{Weight of the gonad}}{\text{Weight of the fish}} \times 100$$

HSI

The hepato somatic index was determined by following the formula of Gingerich (1982).

$$HSI = \frac{\text{Weight of the liver}}{\text{Weight of the fish}} \times 100$$

Statistical analysis: A one-way analysis of variance (ANOVA) followed by Duncan's test ($P<0.05$) was performed to compare the means of the biological indexes for monthly reproductive cycle.

Results

The percentage of the gonad in the total weight of the fish is known as gonado somatic index and these values are an index of maturation of fishes (Varghese 1971). In the present study, it has been observed that the values of gonado somatic index (GSI) and hepato somatic index (HSI) in the fish *Mystus cavasius* follow regular changes in the different months of four reproductive phases such as preparatory, pre-spawning, spawning and post-spawning phases. The GSI values vary from 0.2187±0.0186 to1.5836±0.0896. Whereas the HSI values vary from 1.84±0.202 to 0.90±0.14.

Gonadosomatic Index (GSI–TSI)

The Gonadosomatic index of the *Mystus cavasius* was recorded during 4 different reproductive phases is presented in the Table 22.1. The GSI increases during preparatory phase reaching maximum during pre-spawning phase, thereafter it reduced and reaching minimum during post-spawning.

Hepato-somatic Index (HSI)

The HSI of fish studied during the 4 phases of reproductive cycle is presented in Table 22.1. It indicates that maximum HSI was observed during post-spawning phase and the lowest HSI recorded during pre-spawning phase.

A comparison was made between GSI and HSI, shows that increase of GSI and decrease of HSI during pre-spawning phase indicates utilization of hepatic content for growth of testis suggesting an their relationship. Similar response of reduction in HSI and comparable increase of GSI during spawning and reduction of GSIand increase of HSI during post spawning phase also indicated liver involvement in testicular growth and spermatogenesis.

Table 22.1: Month-wise Values of GSI and HSI of Male Fish
Mystus cavasius

Months	GSI	HSI
January	0.38±0.05	3.62±0.01
February	0.59±0.04	2.45±0.09
March	0.77±0.01	2.09±0.09
April	1.15±0.02	1.26±0.05
May	1.39±0.01	1.24±0.03
June	1.45±0.09	1.10±0.01
July	1.64±0.02	0.80±0.11
August	1.44±0.05	1.05±0.01
September	1.36±0.01	1.69±0.01
October	1.29±0.11	3.00±0.02
November	0.61±0.03	3.99±0.03
December	0.34±0.01	5.95±0.08

Table 22.2: Changes in GSI and HSI in the Male Fish *Mystus cavasius*
during Four Reproductive Phases

Reproductive Phases	GSI	HSI
Preparatory phase	0.58±0.08	2.72±0.005
Pre-spawning phase	1.39±0.03	1.1±0.09
Spawning phase	1.36±0.07	1.9±0.04
Post-spawning phase	0.47±0.01	4.97±0.11

Discussion

The percentage of gonad in the total weight of fish is known as gonado somatic index and this value is an index of maturation of fishes (Varghese 1971 and Ahirrao, 2002). Nautiyal (1984) also stated that the high values of GSI of fishes are the indicator of peak activity of gonads. The changes in the GSI have been considered as a sensitive parameter to monitor gonadal growth (Hong yand and Jong-man yon, 1992).

The gonadal cycle of *Mystus cavasius* can be divided into four phases during one year period as preparatory phase during January to March, pre spawning phase during April to July, spawning phase during August to October and post spawning phase during November to December. The developing gonad is in the preparatory phase, developed and ripe gonad during pre spawning and regressed gonad during is during post-spawning has been noticed based on changes in the GSI and such studies have been reported in other Indian teleosts (Malhotra *et al.*, 1989 and Shankar and Kulkarni, 2005).

It is known that liver involve in the regulation of testicular growth by contributing the biochemical components and providing energy for spermatogenesis. The correlative changes between liver weight and gonadal activity has been shown to be associated with energy requirement of the gonads for the development of spermatogenesis (Htun Han, 1978). It has been also shown that GSI and HSI have relationship and this relationship is directly related to gonadal activity (Patil and Kulkarni, 1996). The results on comparison between GSI and HSI during different phases in *Mystus cavasius* in the present study shows that when GSI is increased the HSI is decreased during breeding phase of a fish and when GSI is decreased, HSI is increased during non-breeding phase.

According to Agostinho *et al.* (1992) increasing GSI values are associated with maturation, whereas decreasing values are related to gamete extrusion. GSI was increased from immature to mature stage but declined at spent stage, whereas The HSI was increased during spent phase and preparatory phase, this high value of HSI indicates heavier liver. It indicates that, GSI and HSI have inter relationship and it may be because of the involvement of the liver in the seasonal growth of the testes. HSI is inversely proportional to GSI. The evaluation of the HSI must take role of both endogenous and exogenous factors. The HSI varies with seasonal cycles (Delaronty 1953; Baborowski *et al.*, 1996 and Shankar and Kulkarni, 2005) because of the role in storage and metabolic activities, nutritional quality also affect relative liver size. Daniels and Robinson (1986), Cambray and Bruton (1984) have suggested the HSI of the fish *Babus anoplas* the lowering of Somatic condition factor following spawning was due to spawning activity and cessation of feeding during spawning. Singh and Singh (1979) found that the high values of HSI during preparatory and post-spawning and low levels during pre-spawning and spawning in the *Heteropneustis fossils*.

The GSI values corresponding to spermatogenetic activity in the testes of different teleosts have been observed Jaspers (1972), Bhatti and Al-Daham, Thakur (1978), Llewellyn (1979) and Yoneda *et al.* (2001). Shankar and Kulkarni (2005) suggests that, the reduction in the condition of HSI during spawning and post spawning indicates exhaustive condition after breeding activity in fish *Notopterus notopterus* and cessation of feeding during spawning. So a majority of teleost fishes all over the world are seasonal breeders and in the Indian sub continent a vast majority of the freshwater fishes breed during the heavy rain fall of monsoon months. (Jhingran, 1982). Environmental factors are well known tools, influence reproductive maturity (Cambray, 1994; Chadhuri, 1997).

In the fish *Mystus cavasius* spawning occurs during monsoon periods. Hence this fish can be categorized as monsoon spawners (Malhotra *et al.*, 1989). Similar observations were also made in *Mystus cavasius* (Bhatt, 1971); in *Mystus aor* (Ramkrishniah, 1982); in *Heteropneustis fossils* (Verma *et al.* (1985); in *P. sarana sarana* (Patil and Kulkarni, 1996); in *Macrozoarces americanus* (Wang and Crim, 1997); in *Padogobius martensi* (Cinquetti and Dramis, 2003); in *Iheringichthys labrosus* (Santos *et al.*, 2004) and in *Mystus montanus* (Arockiaraj *et al.*, 2004). The reports are also available on many other fishes from Indian waters showing that the GSI increased with the progressive development of the gonads until the gonads become ripe and the index declined sharply in the spawning and the spent fishes. There fore it can be concluded that in the present fish *Mystus cavasius* both GSI and HSI have inter-relationship and it may be due to the involvement of liver in the seasonal growth of the testes.

References

Agostinho, A.A., Julio, H.F. and Borghetti, J.R., 1992. Consideragoes sobre os impactos dos represamentos na ictiofauna e medidsas para sua atenuagao. Um estudo de caso: resseravatio de Itaipu. Revista Unimar, Maringa,14 (Suppl): 89–107.

Ahirro, S.D., 2002. Status of gonads in relation to total length and GSI in freshwater spiney eel, (pisces) *Mastacembelus armaatus* (Lacepede) from Marathwada Region, Maharashtra. *J. Aqua. Biol.*, 17(2): 55–57.

Antonysiluvai, J., Arockiaraj, Mohammed, Haniffa, A., Sankaran, Seetharaman and Singh, Shriprakash, 2004. Cyclic changes in gonadal maturation and histological observations of threatened freshwater catfish "Narikeliru" *Mystus montanus* (Jerdon, 1849), *Acta Ichthyo-logica et piscatorial*, 34(2): 253–226.

Bhatt, V.S., 1971. Studies on the biology of some freshwater fishes part VI. *Mystus cavasius* (Ham).

Bhatti, M.N. and Al-Daham, N.K., 1978. Annual cyclical changes in the testicular activity of a freshwater teleost, *Barbus luteus* (Heckel) from shaft Al-Arab. *Iraq. J. Biol.*, 13: 321–326.

Cambray, J.A. and Bruton, M.N., 1984. The reproductive strategy of a barb, *Barbus annnopllus* (Pisces: Cyprinide) colonizing a man made lake in SouthAfrica. *J. Zool. Lond.*, 204: 143–168.

Cambray, J.A., 1994. The comparative reproductive cycles of two closely related African minnows inhabiting two; different sections of Gamtoos river system. In: *Women Ichthyology: An Anthology in Hopnour of ET, R and Genie,* (Eds.) E.K. Balon, M.N. Bruton and D.L.G. Noakes. Kluwer Academic Publishers, Dordrecht, p. 247–268.

Chaudhuri, H., 1997. Environmental regulation of gonadal maturation and spawning in fishes. In: *Frontiers in Environmental and Metabolic Endocrinology,* (Ed.) S.K. Maitra. The University of Burdwan, Burdwan, West Bengal, p. 91–100.

Cinquetti, R. and Dramis, L., 2003. Histological, histochemical, enzyme histochemical and ultra structural investigation of the testis of *Padogobius martensi* between annual breeding seasons. *J. of Fish Biol.*, 63: 1402–1428.

Daniels, W.H. and Robinson, E.H., 1986. Protein and energy requirements of juvenile red drum (*Sciaenops ocellatus*). *Aquaculture,* 53: 243–252.

De Vlaming, V.L., Santos, G.B., and Chapman, F., 1982. On the use of the gonosomatic index. *Comparative Biochemistry and Physiology,* Vancouver, 73A(1): 31–39.

Giese, A.C., 1959. Comparative physiolgy: Annual reproductive cycle of marine invertebrates. *Ann. Rev. Physiolgy,* 21: 547–560.

Gingerich, W.H., 1982. Hepato toxiology of fishes. In: *Aquatic Toxixcology,* (Ed.) L.J. Weber. Raven Press, New York, p. 55–105.

Jaspers, E.J.M., 1972. Some spermatological aspects of channel catfish, *Lctalurus punctatus* (Rafinnesque). *Ph.D. Thesis,* Louisiana State Univ. Baton Rouge (La).

Jhingran, V.G., 1982. *Fish and Fisheries of India.* Hindustan Publishing Corporation (India), Delhi.

Kaur and Kaur, 2006. Impact of nickel chrome electroplating effluent on the protein and cholesterol contents of blood plasma of *Channa punctatus* (Bl.) during different phases of the reproductive cycle. *J. Environ. Biol.*, 27(2): 241–245.

Llewellyn, L.C., 1979. Some observation on the spawning and development of the Mitchellian freshwater hardyhead, *Craterocephalus fluviatilis,* Mc Culloch. *Austral. Zool.*, 20: 269–288.

Malhotra, Y.R., Youth, M.K. and Gupta, 1989. Reproductive cycles of freshwater fishes. In: *Reproductive Cycles of Indian,* (Ed.) S.K. Saidapur. Allied Publishers Ltd., New Delhi, pp. 58–105.

Patil, Manohar and Kulkarni, R.S., 1994. Ovarian and hepatic biochemical correlative changes in the freshwater fish, *Labeo boggut* (Sykes). *Uttar Pradesh J. Zool.*, 14(1): 70–74.

Parulekar, A.H. and Bal, D.V., 1969. Observations on the seasonal changes in the chemical composition of *Bregraceros mc Cleelandi* (Tnomoson). *J. Univ. Bombay*, 37(65): 88–92.

Ramkrishniah, M., 1992. Studies on the breeding and feeding biology of *Mystus aor* (Hamilton) of Nagarjunasagaar Reservoir. *Proc. Nat. Acad. Sci., India*, 63(B): 3.

Santos, J.E.D., Bazzoli, N., Rizzo, E. and Santos, G.B., 2004. Reproduction of the catfish *Iheringichthys labrosus* (Lu tken) (Pisces, Siluriformes) in furnas reservoir, Minas gerais, Brazil. *Rev. Bras. Zool.*, 21(2) curitiba.

Shankar, D.S. and Kulkarni, R.S., 2005. Somatic condition of the fish, *Notopterus notopterus* (Pallas) durilng different phases of the reproductive cycle. *J. Environmental Biology*, 26(1): 45–47.

Shendge, A.N. and Mane, U.H., 2006. Gonadosomatic index and spawning season of Cyprinid fish *Cirrhina reba* (Hamilton). *J.Aqua. Biol.*, 21(1): 127–129.

Singh, A.P. and Singh, T.P., 1979. Season fluctuation in lipid and cholesterol content of ovary, liver and blood serum in relation to annual sexual cycle in *Heteropneustes fossilis* (Bloch). *Endocrinology*, 73: 47–54.

Sindhe, V.R. and Kulkarni, R.S., 2004. Gonadosomatic and hepatosomatic indices of the freshwater fish *Notopterus notopterus* (Pallas) in response to some heavy metal exposure. *Journal of Environmental Biology*, 25(3): 365–368.

Slotte, A., 1999. Differential utilization of energy during wintering and spawning migration in Norwegian spring-spawning herring. *Journal of Fish Biology*, London, 54: 338–335.

Saxena, R., 1986. Comparative studies on the in two bottom feeder teleosts. *Indian J. Phy. Nat. Sci.*, 6: 39–97.

Sushma, Bumb and Parulekar, 2002. Annual cycle of metabolites in relation to breeding and feeding in *Ambassis commersoni* in Goa waters. *Ad. Bios*, (2191): 67–70.

Tacares-dias, M., Martins, M.L. and Moraes, F.R., 2000. Relacao hepatosomatica e esplenosmatica em peixes teleosteos de cultivo intensive. *Revista Brasileira de zoologia, Curitiba*, 17(1): 273–281.

Thakur, N.K., 1978. On the maturity and spawning of an air-breathing catfish *Clarias batrachus* (Linn). *Matsya*, 4: 59–66.

Thapliyal, Anju and Dobriyal, Anoop K., 2003. Maturation and spawning biology of a hill stream catfish *Pseudecheneis* (Mc clelland) from the Garhwal Himalaya, Uttarakhand. *Ad. Bios.*, 22 (1): 9–22.

Varghese, T.J., 1971. Studies on maturation, spawning and sex ratio of *Coilia ramcarati* Gunther. *J. Indian Fish. Asso.*, 1(1): 58–67.

Wang, Z. and Crim, L.W., 1997. Seasonal changes in the biochemistry of seminal plasma and sperm motility in the ocean pout, *Macrozoarces americanus*. *J. of Fish Physiology and Biochemistry*, 16: 77–83.

Verma, G.P., Sahu, K.C., Mohapatra, S., Mohapatra, S. and Das, C.C., 1985. A Comparative histobiochemical study of vitellogenesis in teleost *Channa punctatus and Heteropneustis fossilis*. In: *Recent Advances in Zoology*, (Eds.) C.B.L. Srivastava and Suresh Goel. Rastogi Company, Meerut, India, 45–60.

Chapter 23

Effect of Ascorbic Acid on Lead Toxicity in Testis of Freshwater Bivalve

☆ *R.S. Jawale, S.P. Zambre, D.H. Thorat and R.J. Chaudhari*

Introduction

In the present era of industrial development, enormous synthetic chemicals are produced and release everyday. Further the industrial out put and effluent discharges are also increasing rapidly. The composition of these discharges is very complex and the cumulative action of the toxicants from complex mixture causes enormous amount of stress in the recipient ecosystem. Aquatic ecosystems are used as dumping grounds for industrial discharge and it has become very essential to take sufficient interest and precaution to safeguard our aquatic resources. Because of transboundary contamination and consequential risks and hazards, this has become a global issue.

Lead is a relatively abundant metal in nature. A biologically non-essential element, the metal finds extensive use in modern times (Landrigan *et al.*, 1990). Jurbiewicz *et al.* (1998) studied the response of aquatic molluscs to heavy metal pollution and their application a bioindicators.

The ability of aquatic molluscs to accumulate heavy metals is well documented. In particular, filter-feeding bivalves are able to concentrate heavy metals such as lead, mercury, copper, zinc and cadmium in their tissues (Kock, 1983; Abaychi, 1988; Phillips and Rainbow, 1988 and Cossa, 1989). These bioaccumulated molecules cause several serious problems in the organisms.

Ascorbic acid is an antioxidant vitamin. By this function, it helps prevent oxidation of water-soluble molecules that could otherwise create free radicals, which may generate cellular injury and disease. Ascorbic acid acts as a detoxifier and may reduce the side effects of drugs such as cortisone, aspirin and insulin; it may also reduce the toxicity of the heavy metals lead, mercury and arsenic. Rao *et. al,.* (1994) studied the ascorbate effect on methyl mercury toxicity in reproductive organs of Guinea pigs and found the recovery in the metabolic functions of the reproductive organs and beneficial effect of ascorbic acid against methyl mercury toxicity. In the present work-studies on the detoxification of heavy metals, lead and zinc is carried out on the experimental animal, the *Lamellidens corrianus*.

Materials and Methods

Freshwater bivalves *Lamellidens corrianus* used in the present study were collected from Jayakwadi dam area. They were brought in the laboratory, cleaned by soft brush and maintained in trough containing dechlorinated water for acclimatization for 5 days. Lethal toxicity tests were conducted over 24 and 96 hours. Stock solution of the toxicants (PbNO$_3$; 6.810ppm) was prepared in the double glass distilled water and added to the test medium to get the desired concentration.

The acclimatized active *Lamellidens corrianus* were divided into three groups, A, B and C. The 'A' group of *Lamellidens corrianus* were kept as control, while 'B' group(PbNO$_3$; 6.810ppm) up to 20 days while *Lamellidens* corrianus from group 'C' were exposed to chronic concentration of heavy metal with 50 mg/L ascorbic acid up to 20 days. After 20 days exposure to heavy metals, *Lamellidens corrianus* from group B were divided into two subgroup, D and E. The Lamellidens corrianus from group D were allowed to self cure naturally in normal water while those group 'E' were exposed to 50 mg/L ascorbic acid up to 9 days.

The *Lamellidens corrianus* from A, B, and C groups were dissected after 20 days, while those of recovery were dissected after 9 days, their testis were removed and fixed in Bouin's fluid for 24 hours, washed and dehydrated in alcohol grades, cleared in toluene and embedded in paraffin wax (58–60°C). Serial sections of 6 µ thickness were cut and stained with Mallory's triple stain. The stained sections were examined under light microscope for histopathological impact of heavy metal salts.

Observations and Results

Testis of the *L. corrianus* consists of testicular follicles which have densely populated spermatogenic stages. The follicles are surrounded by the intertubular connective tissue with some muscles and Collagen fibers. The germinal epithelium of the follicle rests on a thin basement membrane (Figure 23.1). The testicular follicles are scattered in the fibrous connective tissue through which the support and the nutrients are provided to the growing spermatozoa. The connective tissue also contains some bundles of the muscle fibers which when contract causes the release of the sperms.

The germinal epithelial cells present at the periphery of the testicular follicles have larger nuclei which divide to form the sperms. Thus there appears the array of different spermatogenic stages from the periphery towards the centre consisting of spermatogonia, primary spermatocyes, secondary spermatocyes, spermatids and the sperms. The sizes of these stages reduce and also the nuclei get constricted due to the loss of the cytoplasm and the nucleoplasm respectively. The fine nuclei of these stages appears densely stained granular structures. The tails of the sperms being very narrow beyond the limit of the resolving power of the microscope used, are not visible.

As compared to the testis of control bivalve, *Lamellidens corrianus* after chronic exposure to lead (6.810 ppm) marked histopathological changes were induced. This exhibited an initial reaction on germinal epithelial damage, together with necrotic changes. The basement membrane and interfollicular connective tissue are affected and the delamination of the germinal epithelium and its necrosis is clearly visible. Different spermatogenic stages are compactly aggregated and thus the vacuoles are developed after 20 days exposure to lead (Figure 23.2) with the necrosis and damage of the connective tissue and the germinal epithelial layer. The cellular identity is lost at most of the places. (Figure 23.3) shows the effect of ascorbic acid protection against the simultaneous exposure to lead. The damage

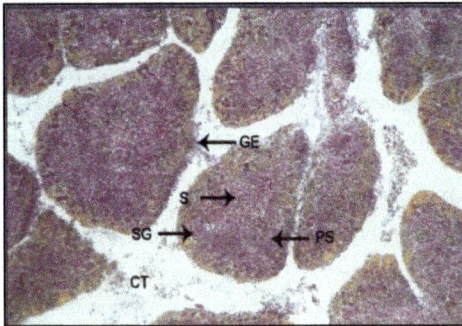

Figure 23.1: Normal Testis

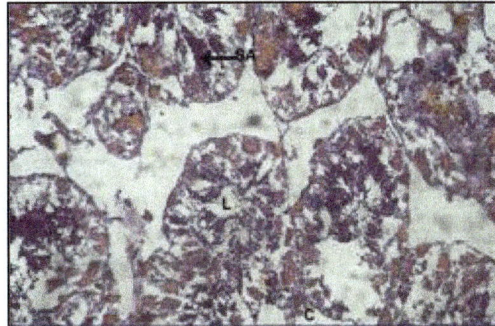

Figure 23.2: Testis Exposed to Lead for 20 Days

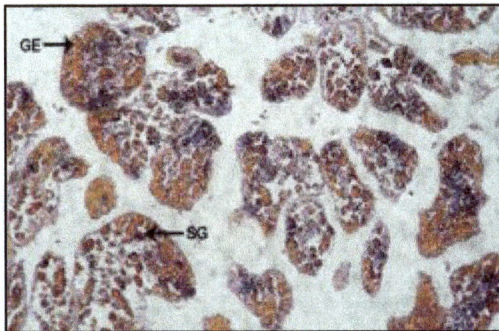

Figure 23.3: Testis Exposed to Lead with Ascorbic Acid for 20 Days

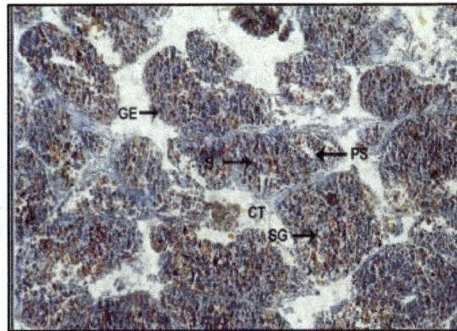

Figure 23.4: Testis on Exposure to Normal Water for Recovery on 9th Day After Exposure to Lead for 20 Days

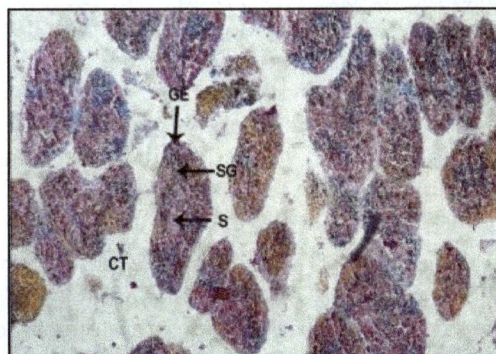

Figure 23.5: Testis on Exposure to Ascorbic Acid for Recovery on 9th Day After Exposure to Lead for 20 Days

induced in the structure of testicular follicle is less and tissue structure is maintained to somewhat normal level. In comparison with the structure of lead exposed testis, there seems to be the slow and progressive improvement in the recovery of the structures with respect to cellular integrity, basement membrane and restoration of the germinal epithelium. Figures 23.4 and 23.5 shows the recovery after

20 days exposure to lead in the histological structure in 50 mg ascorbic acid per liter dechlorinated water after three and nine days. In comparison with the structure of lead exposed testis, there seems to be fast and progressive improvement in the recovery of the structures with respect to cellular integrity, basement membrane, restoration of germinal epithelial layer and distribution of the spermatogenic stages. As compared to the structure of testicular follicle during the recovery in normal water the, improvement is rapid and the structure is fully recovered on the ninth day with normal cellular integrity, basement membrane, germinal epithelial cells and distribution of the spermatogenic stages etc. The differential staining of the testicular follicle components in (Figure 23.5) indicates the restoration of the functioning of the testis.

Discussion

It is well known that histological approach is the most valuable tool for assessing the action of toxicant at tissue level for providing data concerning tissue damage and manifestation of structural and functional changes in tissues and organs (Sprague, 1971). Histopathological abnormalities caused due to toxicity of heavy metals in animals have been reported earlier (Khalid Shareef *et al.*, 1986, Srivastava *et al.*, 1982; Srivastava and Maurya, 1991 and Saxena and Saxena, 1997). Histopathological changes in various tissues affected of various compounds of the heavy metals is mainly studied in invertebrates, amphibians and mammals (Laborda *et al.*, 1986; and Joshi and Patil, 1995). The toxicity of copper lead and zinc on freshwater fishes is still little explored (Sastri and Tyagi, 1982; Bengeri and Patil, 1995). Heavy metals being non-degradable causes physiological stress even at very low concentrations because of bio-accumulation. Heavy metals decreases the growth rate (Galtsoft *et. al.*1964) and exhausts the biochemical reserves (Bayne and Thompson, 1970) in bivalves.

Visible histological abnormalities caused due to toxicity of heavy metals in animals have been reported earlier (Srivatsava *et al.*, 1982, Rameshkumar *et. al.* 1988; Srivatsava and Maurya, 1991; Dhanapakiam *et al.*, 1998; and Anusaya and Christy, 1999).

Arrested ovarian recrudescence may be attributed to inhibition of gonadotrophin levels due to treatment with pesticide (Singh and Singh, 1981). Several investigators observed the similar results (Harilal and Sahai, 1986; Ansari *et al.*, 1987; Umarani and Rajendranath, 1988; Rastogi and Kulshrestha, 1990). Cessation of spermatogenesis observed can be due to necrosis of germinal epithelium, these changes suggest the possible inhibition of testicular steroidogenic enzyme activities and androgen production (Bagchi *et al.*, 1990). Deleterious effects of pesticides on testis of fish had been reported earlier (Kapur *et al.*, 1978; Pandey and Shukla, 1980 and Khillare and Wagh, 1989). Victor *et. al.*(1980) on exposure to mercury and cadmium observed degenerative lesions in the ovary, the impairment of vitellogenesis in *Lepidocephalichthys thermalis*.

Reduced productivity of sperm and ova was observed in the gonad of fish, *Carassius auratus* injected with cadmium (Tafanelli and Felt, 1975). Sangalang and O'Halloran (1972) and Ahsan and Ahsan (1974) reported the interference of cadmium in the spermatogenic activity of *Salvelinus fontinalis* and *Clarius batrachus* respectively.

Ascorbic acid is essential for the normal development and in most species it is produced endogenously, thus resembling a hormone. Glucose and other hexoses convertible to glucose serve as the starting material for biosynthesis of ascorbic acid. Halver (1972) states that the ascorbic acid plays a major role in tissue synthesis and growth processes and obviously mediates rapid tissue repair in trauma or disease condition.

Ascorbic acid is an excellent reducing agent, which is able to serve as a donor antioxidant in free radical mediated oxidation processes and is also able to reduce metal such as Cu and Fe (Beuttner and Jurkiewicz 1996). Thus the effect of the other toxic metal can be protected by ascorbic acid. Ascorbate protected the specific binding sites of the receptors by virtue, of its ability to reduce Hg++ to insoluble Hg+ (Scheuhanner and Cherian, 1985). Ascorbic acid readily forms salts of several metals and reduces their activity.

The protection provided by the ascorbic acid against the lead in the present study may be due to its antioxidant activity that has scavenged the ROS generated due to heavy metals, its role in collagen synthesis which has rapidly replaced the basement membrane and due to its binding to metal ions.

References

Abaychi, J.K. and Mustafa, Y.Z., 1988. The Asiatic clam, *Corbicula fluminea* an indicator of trace metal pollution in the shattal Arab river, Iraq. *Environ. Pollut.,* 54: 109–122.

Ahsan, S.N. and Ahsan, J., 1974. Degenerative changes in the testis of *Clarias batrachus* (Linn.) caused by cadmium chloride. *Indian J. Zool.,* 15: 39–40.

Anusuya, D. and Christy, I., 1999. Histological alterations in Tadpole of *Bufo melanostcus* exposed to a sub-lethal concentration of cadmium. *Ecol. Env. and Cont.,* 5(2): 165–169.

Bengeri, K.V. and Patil, H.S., 1995. Uptake of zinc by two freshwater fishes *Labeo guntea* and *Labeo rohita*. In: *Third Congress on Toxicology in Developing Countries,* Cario, Egypt.

Buttner and Jurkiewicz, B.A., 1996. Catalytic metal ascorbates and free redicals, combination to avoid. *Radiate Res.,* 145(5): 532–541.

Cossa, D., 1989. A review of the use of *Mytilus* spp. as quantitative indicators of cadmium and mercury contamination in costal waters. *Oceano Acta,* 12: 417–423.

Dhanapakim, P., Ramaswamy, V.K. and Sampooyani, 1998. A study on the Histopathological changes in gills of *Channa punctatus* in Cauery river water. *J. Environ. Biol.,* 19(3): 265–268.

Galtsoft, P.S., 1964. The American oyster, *crasotrea verginica.*

Halver, J.E., 1972. The role of ascorbic acid in fish diseases and tissue repair. *Bull. Jap. Soc. Sci. Fisheries,* 35(1): 79–92.

Harilal, K.S. and Sahi, Y.N., 1986. Effect of carbaryl on the gonands of *Heteropneustes fossilis* (Bloch). In: *Environment and Ecotoxicology,* (Eds.) R.C. Dalela *et al.*). Academy of Environmental Biology (India), p. 277–281.

Jurkiewicz, K.E., 1998. Responses of aquatic molluscs to heavy metals pollution and their applications in bionidication. *Wiadomosci Ekologiczne,* 44(3): 217–234.

Landrigan, P.J., Silvergeld, E.K., Fronies, J.R. and Pfeffer, R.M., 1990. Lead in modern work place. *American J. Public Health,* 80: 907–908.

Phillips, D.J.H. and Rainbow, P.S., 1988. Barnacles and mussels as biomonitors of trace elements: A comparative study. *Mar. Ecol. Pro. Ser.,* 49: 83–93.

Rao, M.V., Mehta, A.R. and Patil, J.S., 1994. Ascorbate effect on methyl mercury toxicity in reproductive organs of male Gunea pigs. *Indian Journal of Environmental and Toxicology,* 4 2): 53–58.

Rastugi, A. and Kulshrestha, S.K., 1990. Effect of sublethal dose of three pesticide on the ovary of *Carp minnos, Rasbora daniconius*. *Bull. Env. Cont. Toxicol.,* 45: 742–747.

Sangalang, G.B. and Halloran, M.K.O., 1974. Adverse effect of cadmium on brook trout testis and *in vitro* testicular androgen synthesis. *Bio. Repro.,* 9: 347–403.

Sastry, K.V. and Tyagi, B.K., 1982. *Toxicol. Lett. (Amst),* 11: 17–22.

Saxena, Saraj and Saxena, Seema, 1997. Effect of zinc on the development o gonands in the tadpoles of toad bufo andersonli. *J. Ecotoxicol. Environ. Moni.,* 7(1): 11–15.

Scheuhanner, A.M. and Cherian, 1985. Effect of heavy metal cations sulfhydryl reagents and other chemichal agents on striatal O_2 dopamine receptors. *Biochem. Pharmacol.,* 34(19): 3405–3413.

Shareef, Khalid, Shareef, Shakeela and Wagh, S.B., 1986. Effect of sublethal dose of squin (quinophos 25 per cent W/W) on liver and kidney of freshwater Cyprinid fishes *Barbus ticto* and *Rasbora daniconius* (Ham.). *Env. Biol. Coast. Ecosystem,* p. 93–100.

Sprague, J.B., 1971. The measurement of pollutant toxicity to fish III, sublethal effects and safe concentration. *Water Res.,* 5: 245–266.

Srivastava, V.M.S. and Maurya, R.S., 1991. Effect of chromium stress on gill and intestine of *Mystus vittatus* (Block) Scanning E.M. *Study. J. Ecobiol.,* 3(1): 69–71.

Srivastava, V.M.S., Tripathi, R.S. and Saxena, A.K., 1982. Chromium included histopathological changes in some tissues of *Puntius sphore* (Ham). *J. Biol. Res.,* 2: 67–68.

Tanfanelli, R. and Summerfelt, R.C., 1975. Cadmium induced histopathological changes in gold fish. In: *Pathology of Fishes,* (Ed.)W.E. Ribelin and G. Migaki. University of Wysconsin Presss, Medican Wis., p. 613–645

Umarni, S. and Rajendranath, T., 1988. Toxicity of kitazin on the oocytes of the fish *Channa punctatus* (Bloch). *Indian J. Comp. Anim. Phys.,* 6(2): 140–143.

Victor, B.S., Mahalingam and Sarogini, R., 1980. Toxicity of mercury and cadmium on oocyte differentiations and vitellogensis of teleost *Lepidocepalichtyhs thermalis* (Bleeker). *J. Environ. Biol.,* 7(4): 209–214.

Chapter 24

Studies on Fluoride Content in Borewell Water in Malkapur Town in Buldhana District, Maharashtra

☆ *C.M. Bharambe, J.H. Pawar and Alfia Tabassum*

Introduction

Fluorine is the most electronegative element, distributed ubiquitously as fluorides in nature. Water is the major medium of fluoride intake by humans (WHO, 1984). Fluoride can rapidly cross the cell membrane and is distributed in skeletal and cardiac muscle, liver, skin (Carlson *et al.*, 1960) and erythrocytes (Jacyszyn *et al.*, 1986). Fluorosis is a major public health problem resulting from long-term consumption of food grains products with high fluoride levels. It is characterized by dental mottling and skeletal manifestations such as crippling deformities, osteoporosis, and osteosclerosis. In India, as many as 15 states are affected by endemic fluorosis, and an extensive belt of high fluoride in food and soil is reported in South India (Krishnamachari, 1976; Pandit *et al.*, 1940; Satyanarayana *et al.*, 1953 and Siddiqui, 1955). Many district of Maharashtra was found to be a fluorosis endemic area by (Nawlakhe *et al.*, 1993; Susheela, 1999 and Bellack *et al.*, 1958). However, a detailed survey for clinical manifestations attributable to fluoride toxicity has not been recorded.

According to Teotia and Teotia (1984) more than 0.5 to 1.0 mg/lit fluoride in drinking water caused "Yellowing of teeth" in infants and children. Short *et al.* (1937); have reported skeletal fluorosis from India as early as 1937. Teotia *et al.* (1971) reported that skeletal fluorosis is not confined to adult but also afflicts infants and children in the age range of 6 months to 14 years. Further, more than 1 million people in India are affected with skeletal fluorosis (Teotia and Teotia, 1984). It is also estimated that nearly 25 million people are affected with fluoride poisoning due to the environment pollution (Susheela, 1990).

With the increasing population and rapid expansion of the town adequate supply of drinking water has assumed a great significance in the town planning. In the recent all mega cities, there is great reduction in water supply and the gap between demand and supply of drinking water is growing more. As such alternate day drinking water supply has become a common phenomenon in city.

The fluoride is most toxic compound widely distributed in nature and universally present in soil, water and atmosphere. Concentrations of fluoride in bore well drinking water in Malkapur Town vary between 0.2 mg/lit to 2.42 mg/lit. In the present paper the fluoride content in different bore well drinking water of Malkapur Town and their effect on school going children and different age group of human are reported.

Materials and Methods

The water samples were collected from 55 bore wells from different colonies and most of them are government hand pumps, for the detail study of fluoride content in Malkapur town. Malkapur town is divided in four zones namely North, East, West and South zone. The water samples were collected in wide mouthed screw caped airtight and opaque polythin containers. Each sample was comprised of 1 lit of waters. Fluoride from Borewell water samples were estimated by standard method suggested by APHA (1989).

Results and Discussion

In the present work all 55 borewell water samples analyzed from different areas of different zones of Malkapur Town (Table 24.1). Out of 55 water samples in 23 samples were found fluoride percentage lower than permissible limit (0.5 mg/lit) and while 10 samples were found fluoride percentage more than permissible limit (1.0 mg/lit) (Tables 24.2–24.5)

Table 24.1: Showing Number of Collection of Water Samples from the Borewells from Different Zones of Malkapur Town

Name of the Zone	Number of Samples
West Zone	10
North Zone	17
East Zone	08
South Zone	20

Table 24.2: Fluoride Content in Drinking Water from the Wells from West Zone of Malkapur Town and their Respective Colonies

Sl.No.	Name of the Sampling Site West Zone of Malkapur Town	Percentage of Fluoride (mg F/L)
1.	Adarsha Nagar	0.30*
2.	Gadegaon	0.36*
3.	Irrigation Colony	0.20*
4.	Gandhi Nagar	0.82
5.	Shindhi Colony	0.38*
6.	Matamahakali Nagar	0.40*
7.	Muktai Nagar	0.58
8.	Salipura	1.58**
9.	Cycle Pura	0.58
10.	Mali Pura	1.38**

Table 24.3: Fluoride Content in Drinking Water from the Wells from North Zone of Malkapur Town and their Respective Colonies

Sl.No.	Name of the Sampling Site North Zone of Malkapur Town	Percentage of Fluoride (mg F/L)
1.	Taj Nagar	0.58
2.	Peelu Takiya	0.78
3.	Quresh Nagar	0.80
4.	Ahemad Shah Pura	0.60
5.	Raj Mohalla	0.32*
6.	Faqeer Pura	0.30*
7.	Jameel Shah Pura	0.40*
8.	Dhangar Pura	0.78
9.	Eid Gah Plot	0.46*
10.	Mushtaque Ali Nagar	1.28**
11.	Madar Tekadi	0.60
12.	Momin Pura	1.38**
13.	Bada Bajar	0.52
14.	Chota Bajar	0.40*
15.	Baradari	0.20*
16.	Mohan Pura	0.20*
17	Bramhan Wadi	0.30*

Percentage of F.L. 0.5 to 0.1; 0.2 to 0.5*, 1.0 to 2.0 ** above

Table 24.4: Fluoride Content in Drinking Water from the Borewells from East Zone of Malkapur Town and their Respective Colonies

Sl.No.	Name of the Sampling Site East Zone of Malkapur Town	Percentage of Fluoride (mg F/L)
1.	Pant Nagar	0.80
2.	Vishnu Wadi	0.48*
3.	Ram Wadi	0.46*
4.	Laxmi Nagar	0.72
5.	Purohit Colony	1.80**
6.	Saoji Fhaeel	0.20*
7.	Shivaji Nagar	0.56
8.	Birala Temple	0.78

Fluoride levels in the water from the three bore wells ranged from 0.60 to 13.4 ppm. Earlier studies have indicated that the incidence and severity of chronic fluoride intoxication are greatly influenced by socio-economic, climatic, and nutritional status (Sarala and Ramakrishna, 1993; Krishnamachari and Krishnaswamy, 1973 and Teotia *et al.,* 1988).

Fluoride toxicity affects children more severely and after shorter exposure to fluoride than adults, owing to the greater and faster accumulation of fluoride in the metabolically more active growing bones of children (Teotia *et al.*, 1983).

Several workers have reported skeletal and crippling fluorosis at fluoride levels above 1.4 ppm and 3.0 ppm respectively (Sarala and Ramakrishna, 1993; Leone *et al.*, 1970 and Choubisa *et al.*, 1997). Elevated plasma alkaline phosphatase and parathyroid hormone levels support the diagnosis of associated metabolic bone disease and secondary hyperparathyroidism in fluorosis (Teotia and Teotia 1988).

Table 24.5: Fluoride Content in Drinking Water from the Wells from South Zone of Malkapur Town and their Respective Colonies

Sl.No.	Name of the Sampling Site South Zone of Malkapur Town	Percentage of Fluoride (mg F/L)
1.	Chalis Bigha	0.40*
2.	Vittal Nagar	0.48*
3.	Prashant Nagar	0.40*
4.	Telephone Colony	0.20*
5.	Ganesh Nagar	0.42*
6.	Deshmukh Nagar	0.20*
7.	Dwarka Nagar	0.58
8.	Anand Society	0.80
9.	Harikishan Society	0.88
10.	Cheetanya Wadi	1.68**
11.	Hanuman Nagar	2.18**
12.	Ganpati Nagar	1.98**
13.	Hegadewar Nagar	0.58
14.	Professor Colony	0.40*
15.	Dipak Nagar	0.58
16.	Gopal Krishna Nagar	0.62
17.	Wrundhawan Nagar	1.56**
18.	Yeshodham	1.72**
19.	Bansilal Nagar	0.88
20.	Gokul Nagar	0.58

A marked increase in serum fluoride levels was also observed, which is consistent with the observations of in juvenile fluorosis (696 IU/L) (Teotia *et al.*, 1971). Similar results have been reported in earlier studies on fluorotic individuals and experimental animals (Tsunoda *et al.*, 1985 and Chinoy *et al.*, 1991). Fluoride is known to inhibit protein synthesis mainly due to impairment of peptide chain initiation and by interfering with peptide chains on ribosomes (Chinoy *et al.*, 1992 and Mathews *et al.*, 1996).

Conclusions

Fluoride concentration of 0.5 mg/L in drinking water helps in preventing dental carries in children and promotes the formation of hard, strong and decay-resistant teeth. But fluoride concentration crosses 1.0 mg/L, it becomes toxic. It causes discoloration of teeth, dental damage, bone fluorosis and skeletal abnormalities *i.e.* steepness of the vertebral column and bulging in joints in children.

Considering the harmful effects of fluoride on human being, drinking water content fluoride significantly below the optimal level, fluoride supplementation may be necessary. Due to high fluoride content of dental and skeletal fluorosis as an expression of environmental fluoride toxicity.

References

APHA, 1989. *Standard Methods for the Examination of Water and Wastewater,* 17th edn. American Public Health Association, Washington, DC.

Bellack, E. and Schouboe, P.J., 1958. Rapid photometric determination of fluoride in water. *Anal. Chem.,* 30(12): 2032–2034.

Carlson, C.H., Armstrong W.D. and Singer, L., 1960. Distribution and excretion of radiofluoride in the human. *Proc. Soc. Exp. Biol. (NY),* 104: 235–239.

Chinoy, N.J., 1991. Effects of fluoride on physiology of animals and human beings: A review. *J. Environ. Toxicol.,* 1: 17–32.

Chinoy, N.J., Narayana, M.V., Sequeira, E., Joshi, S.M., Barot, J.M. and Purohit, R.M., 1992. Studies on effects of fluoride in 36 villages of Mehsana district, North Gujarat. *Fluoride,* 25(3): 101–110.

Choubisa, S.L., Choubisa, D.K. and Joshi, S.C., 1997. Fluorosis in some tribal villages of Dungarpur district of Rajasthan, India. *Fluoride,* 30(4): 223–228.

Jacyszyn, K. and Marut, A., 1986. Fluoride in blood and urine in humans administered fluoride and exposed to fluoride-polluted air. *Fluoride,* 19(1): 26–32.

Krishnamachari, K.A.V.R. and Krishnaswamy, K., 1973. *Genu valgum* and osteoporosis in an area of endemic fluorosis. *Lancet.,* 2: 887–889.

Krishnamachari, K.A., 1976. Further observations on the syndrome of genu valgum of South India. *Indian J. Med. Res.,* 64(2): 284–291.

Leone, N.C., Martin, A.E., Minoguchi, G., Schlesinger, E.R. and Siddiqui, A.H., 1970. Fluorides and general health. In: *Fluorides and Human Health.* WHO, Geneva, p. 273–320.

Mathews, M., Barot, V.V. and Chinoy, N.J., 1996. Investigations of soft tissue functions in fluorotic individuals of North Gujarat. *Fluoride,* 29(2): 63–71.

Pandit, C.G., Raghavachari, T.N., Rao, D.S. and Krishnamurthi, V., 1940. Endemic fluorosis in South India: A study of factors involved in the production of mottled enamel in children and severe bone manifestations in adults. *Indian J. Med. Res.,* 28: 533–535.

Sarala, Kumari, D. and Ramakrishna Rao, P., 1993. Endemic fluorosis in the village Ralla Ananthapuram in Andhra Pradesh: An epidemiological study. *Fluoride,* 26(3): 177–180.

Susheela, A.K., 1990. Too much can cripple you. *Health for Millions,* 1695: 48–52.

Susheela, A.K., 1999. Fluorosis management programme in India. *Curr. Sci.,* 77(10): 1250–1256.

Teotia, S.P. and Teotia, M., 1984. Endemic flurosis in India. A challenging National Health Problem. *Japi.*, 32(4): 347–352.

Teotia, S.P.S., Teotia, M. and Singh, D.P., 1983. Fluoride and calcium interactions, syndromes of bone disease and deformities (Human studies). In: *Proceedings of the International Symposium: Disorders of Bone and Mineral Metabolism*. Henry Ford Hospital, Michigan, USA, p. 502–503.

Teotia, S.P.S. and Teotia, M., 1988. Endemic skeletal fluorosis: clinical and radiological variants (review of 25 years of personal research). *Fluoride*, 21(1): 39–44.

Teotia, M., Teotia, S.P. and Kanwar, K., 1971. Endemic skeletal fluorosis. *Arch. Dis. Child*, 46: 686–691.

Tsunoda, H., Makaya, S., Sakurai, S., Itai, K., Yazaki, K. and Tatsumi, M., 1985. Studies of the effects of environmental fluoride on goats. In: A.K. Susheela (Ed.) p. 13–17.

WHO, 1984. *Environmental Health Criteria for Fluorine and Fluorides*. Geneva, p. 1–136.

Chapter 25

Study on Fluoride Content in Food Grains in Malkapur Tahasil of Buldhana District, Maharashtra

☆ *C.M. Bharambe, Alfia Tabassum and J.H. Pawar*

Introduction

Malkapur is also known as the entrance of Vidarbha, is located at 20°31?N 76°07?E?/20.52°N 76.12°E. It has an average elevation of 255 meters (839 feet). It is a major industrial and educational center in Buldhana district. Malkapur is known for producing cotton. In the British era it was known as the "white gold of Vidarbha." Malkapur has a vast paper industry and the primary paper suppliers to Mumbai are in Malkapur. Malkapur also has many dal mills, and a major agricultural industries specializing in Mahyco and Ankur seeds. Chemical plants are also situated at MIDC. Malkapur is a well-known food grains and cloth market due to its proximity to Madhya Pradesh.

Dental fluorosis has increased in the last decade in both optimally fluoridated and no fluoridated areas in many countries, as well as in India. Thus, the study of the sources of fluoride intake deserves special attention. Among the main sources of fluoride intake are fluoridated food grains, water, powdered milk reconstituted with fluoridated water, inadvertent ingestion of fluoridated toothpaste, inappropriate use of dietary supplements, as well as beverages processed with fluoridated water. The dietary habits during have changed substantially in the last decades and a large increase in the consumption of hybrid food grains and manufactured products has been observed. Monitoring of the fluoride intake requires knowledge on the fluoride content of the food grains commonly consumed by children at the age of risk for dental fluorosis.

Food grains showed that high levels (0.5-7.3 mg/g) major effect on brain function. Other studies have indicated that high fluoride intake has a negative effect on the brains of animals and humans, and that a brain system in the process of development is one of the most susceptible targets for the toxic effects of fluoride. At present, however, there is very little research on the mechanisms involved in brain damage caused by fluoride poisoning.

The present work showed that some food grains analyzed should be important contributors to total fluoride content and daily fluoride consumed by Malkapurian peoples.

Materials and Methods

The food grains were collected from the 70 villages belongs to the Malkapur Tahsil area; the food grains include Jawar, Wheat, Bajara, Maize, Soya been and cereals. The samples of water for the analysis were collected during the February 2008 and March 2009.

To study the fluoride content of food grains 50 gm of each grain sample was collected from each village and 10 gm of each grain sample was cleaned for sand and dust. Fluoride content of food grains were estimated by standard method suggested by APHA (1989) with Orion Ion analyzer, model EA 940, Fluoride electrode.

Results and Discussion

The food grains samples were collected from the 70 villages belongs to the different zones of Malkapur Tahsil area (Table 25.1).The food grains include Jawar, Wheat, Bajara, Maize, Soya been etc.

Altogether 70 food grain samples from different zones were analyzed. 44 samples were found fluoride percentage lower than permissible limit 0.5 mg/kg. and 6 samples were with more than permissible limit *i.e.* 1.0 mg/kg (Tables 25.2–25.5).

Table 25.1: Number of Food Grain Samples Collected from the Villages from Different Zone of Malkapur Tahsil

Name of the Zone	Number of Samples
West Zone	17
North Zone	30
East Zone	10
South Zone	13

Table 25.2: Fluoride Content of Food Grains from the Villages from West Zone of Malkapur Tahsil

Sl.No.	Name of the Sampling Site West Zone of Malkapur Town	Percentage of Fluoride (mg/kg)
1.	Malkapur	0.40*
2.	Jalaland	0.70
3.	Lonawadi	0.72
4.	Wakodi	0.20*
5.	Nimbhari	0.20*
6.	Datala	0.22*
7.	Rastapur	0.20*
8.	Shirdhan	0.20*
9.	Morkhed	0.20*
10.	Dudhalgaon Bk.	0.42*
11.	Aland	0.82
12.	Wadji	0.72
13.	Harankhed	0.84

Contd...

Table 25.2–Contd...

Sl.No.	Name of the Sampling Site West Zone of Malkapur Town	Percentage of Fluoride (mg/kg)
14.	Chinchkhed	0.68
15	Jambuldhaba	1.58**
16	Jalaland	0.36*
17	Goolkhed	0.32*

Table 25.3: Fluoride Content of Food Grains from the Villages from North Zone of Malkapur Tahsil

Sl.No.	Name of the Sampling Site North Zone of Malkapur Town	Percentage of Fluoride (mg/kg)
1.	Vajirabad	0.20*
2.	Khamkhed	0.40*
3.	PimpalKhuta Mahadev	0.30*
4.	Bhalegaon	0.72
5.	Hinganakaji	0.20*
6.	Devdhabha	0.20*
7.	Khadki	0.20*
8.	Gorad	0.20*
9.	Kamardipur	0.20*
10.	Rangaon	0.52
11.	Nimboli	0.58
12.	Shivani	0.62
13.	Ghedi	0.44*
14.	Hingane Dharangaon	0.66
15.	Kund	0.26*
16.	Leha Khupa	0.26*
17.	Dharangaon	0.48*
18.	Tandulwadi	0.20*
19.	Talaswada	0.30*
20.	Maiswadi	0.40*
21.	Korwad	1.16**
22.	Chinchol	1.42**
23.	Waghol	0.52*
24.	Dasarkhed	0.30*
25.	Vivara	0.40*
26.	Tighara	0.66
27.	Bhanguara	0.60
28.	Ranthumb	0.72
29.	Chikhli	0.80
30.	Dudhalgaon Kd.	0.20*

Table 25.4: Fluoride Content of Food Grains from the Villages from East Zone of Malkapur Tahsil

Sl.No.	Name of the Sampling Site East Zone of Malkapur Town	Percentage of Fluoride (mg/kg)
1.	Panhera	0.40*
2.	Bhahapura	0.36*
3.	Wadoda	0.72
4.	Anurabad	0.80
5.	Zodaga	0.30*
6.	Harsoda	0.40*
7.	Narvel	1.20**
8.	Hingane Nagapur	0.36*
9.	Telkhed	0.20*
10.	Kalegaon	0.32*

Table 25.5: Fluoride Content of Food Grains from the Villages from South Zone of Malkapur Tahsil

Sl.No.	Name of the Sampling Site South Zone of Malkapur Town	Percentage of Fluoride (mg/kg)
1.	Khaparkhed	0.62
2.	Waghul	0.20*
3.	Belad	0.30*
4.	Ghirni	0.52
5.	Gahukhed	0.42*
6.	Makner	0.30*
7.	Umali	1.18**
8.	Varkhed	0.20*
9.	Bhadagani	1.52**
10.	Lasura	0.32*
11.	Samali	0.22*
12.	Pimpalkhuta Bk	0.77
13.	Nimkhed	0.44*

Dental fluorosis has increased in the last decade in both optimally fluoridated and no fluoridated areas in many countries, as well as in India has been demonstrated (Adair *et al.*, 1978; Fomon *et al.*, 2000; Levy *et al.*, 2001 and Pereira *et al.*, 2000).

Cheng (1989) studied of children from an area with fluoride contaminated food grains showed that high levels (0.5-7.3 mg/g) major effect on brain function.

The peoples are not enough known about the fluoride content of their food grains; it is possible to estimate the fluoride intake only through fluoride assay. Several previous studies determined the fluoride content of food grains, such as Jawar (Howat and Nunn, 1981 and Van Winkle *et al.*, 1995), wheat and say been (Dabeka *et al.*, 1982; Heilman *et al.*, 1997; Ophaug *et al.*, 1980, 1985; Rojas-Sanches *et al.*, 1995; Vlachou *et al.*, 1992) and other food grains (Adair *et al.*, 1978; Buzalaf *et al.*, 2002; Kiritsy *et al.*, 1996).

Main sources of fluoride intake are fluoridated food grains and also water, powdered milk reconstituted with fluoridated water, inadvertent ingestion of fluoridated toothpaste, inappropriate use of dietary supplements, as well as beverages processed with fluoridated water (Buzalaf *et al.*, 2001 and Mascarenhas, 2000).

Four food grains analyzed (Jawar, Wheat, Bajara and Maize) were considerable fluoride concentration of 1.54µg/g and additional risk factors for dental fluorosis, especially when associated to other fluoride sources (Adair *et al.*, 1978; Buzalaf *et al.*, 2002; Kiritsy *et al.*, 1996; Trautner *et al.*, 1986 and Larsen *et al.*, 1988).

Fomon *et al.* (2000) reported that dietary habits changed substantially during the last decades and a large increase in the consumption of hybrid food grains and manufactured products has been observed.

The permissible limit of fluoride content in food grains recommended by ISI and WHO is 0.5 to 1.0 mg/kg. However, Adair and Wei (1978), on the basis of their studies, they have suggested the maximum permissible limit of fluoride below 0.5 mg/kg.

Conclusions

It is an age of agriculturalization; rapidly growing hybrid food has unfortunately resulted in a complex range of health problems due to ingestion of hybrid food grains. One of the most important health hazards resulting from hybrid food grains is fluorosis. The low/high level of fluoride affects bones, teeth and soft tissues of the human body. It is caused by cumulative action of fluoride content in food ingested over prolonged periods

References

APHA. 1989.Standard methods for the examination of water and wastewater 17th edn. American public health Association, Washington, DC.

Adair, S.M. and Wei, H.Y., 1978. Supplemental fluoride recommendations for infants based on dietary fluoride intake. *Caries Res.*, 12(2): 76–82.

Buzalaf, M.A., Granjeiro, J.M., Damante, C.A. and Ornelas, F., 2002. Fluctuations in public water fluoride level en Bauru, Brazil. *J. Public Health Dent.*, 62: 173–176.

Buzalaf, M.A., Granjeiro, J.M., Damante, C.A. and Ornelas, F., 2001. Fluoride content of infant formulas prepared with deionized, bottled mineral and fluoridated drink water. *J. Dent Child.*, 68(1): 37–41.

Buzalaf, M.A., Granjeiro, J.M., Duarte, J.L. and Taga, M.L., 2002. Fluoride content of infant foods in Brazil and risk of dental fluorosis. *ASDC J. Dent. Child.*, 69(2): 196–200.

Buzalaf, M.A., Whitford, G.M. and Cury, J.A., 2001. Fluoride exposures and dental fluorosis: A literature review. *Rev. Fac. Odontol. Bauru*, 9(1): 1–10.

Cheng, Z., 1989. An epidemiological study to determine amounts for endemic fluoride poisoning. *Chinese Journal of Control of Endemic Diseases*, 4(1): 18.

Dabeka, R.W., McKenzie A.D., Conacher, H.B. and Kirkpatrick, D.C., 1982. Determination of fluoride intakes by infants. *Can. J. Public Health*, 73: 188–191.

Fomon, S.J., Ekstrand, J. and Ziegler, E.E., 2000. Fluoride intake and prevalence of dental fluorosis: Trends in fluoride intake with special attention to infants. *J. Public Health Dent.*, 60(3): 131–139.

Howat, A.P. and Nunn, J.N., 1981. Fluoride levels in milk formulations. *Brit. Dent. J.*, 150: 276–278.

Kiritsy, M.C., Levy, S.M., Warren, J.J., Guha-Chowdhury, N., Heilman, J.R. and Marshall, T., 1996. Assessing fluoride concentrations of juices and juice-flavored drinks. *J. Am. Dent. Ass.*, 127: 895–902.

Larsen, M.J., Senderovitz, F., Kirkegaard, E., Poulsen, S. and Fejerskov, O., 1988. Dental fluorosis in the primary and permanent dentition in fluoridated areas with consumption of either powdered milk or natural cow's milk. *J. Dent. Res.*, 67: 822–825.

Levy, S.M., Warren, J.J., Davis, C.S., Kirchner, H.L., Kanellis, M.J. and Wefel, J.S., 2001. Patterns of fluoride intake from birth to 36 months. *J. Public Health Dent.*, 61: 70–77.

Mascarenhas, A.K., 2000. Risk factors for dental fluorosis: A review of the recent literature. *Pediatr. Dent.*, 22: 269–277.

Rojas, S.F., Kelly, S.A., Drake, K.M., Eckert, G.J., Stookey, G.K. and Dunipace, A.J., 1999. Fluoride intake from foods, beverages and dentifrice by young children in communities with negligibly and optimally fluoridated water: A pilot study. *Comm. Dent. Oral. Epidemiol.*, 27: 288–297.

Trautner, K. and Siebert, G., 1986. An experimental study of bioavailability of fluoride from dietary sources in man. *Arch. Oral. Biol.*, 31: 223–228.

Van, W.S., Levy, S.M. and Kiritsy, M.C., 1995. Water and formula fluoride concentrations: significance for infants fed formula. *Pediatr. Dent.*, 17: 305–310.

Vlachou, A., Drummond, B.K. and Curzon, M.E., 1992. Fluoride concentrations of infant foods and drinks in the United Kingdom. *Caries. Res.*, 26: 29–32.

Chapter 26

Effect of Temperature and Light on Metabolism of Snail *Bellamya bengalensis* from Nalganga Reservoir

☆ *C.M. Bharambe, C.D. Morey and P.M. Salok*

Introduction

Temperature is one of the important environmental factor governing distribution and survival of various forms of life on the earth. The range of temperature is much greater however for a active life, it is normally between narrow range of between 15° and 40° C. Nevertheless, organisms can grow and can be metabolically active at the temperature range of 2 to 10° C, a range some time referred to as the "Biokinetic Zone". Because temperature acts as rate controlling factor for various biochemical reactions engaged in maintenance of life, can greatly influence the distribution of biomass on the earth.

Molluscs have been the favourites of all as they have invaded man's dominance in every field. The metabolic rate of aquatic invertebrates decreases with decreased ambient temperature and increase with increase. Sokolova *et al.* (2003) investigated that the effect of temperature and light as an important environmental factor, influencing all life functions of an organism through changes in the rates of biochemical and physiological processes.

The importance of continuous measurements of respiratory rate to account for diurnal rhythms in respiration was stressed, indicates that the oxygen consumption in both the sexes of *Bellamya bengalensis* is minimum at noon and that it progressively increase to a maximum value at midnight and there after it decreases. This may be due to the differences in the levels of activity of the snails concerned at different time of day and night. The weight specific oxygen consumption is dependent on the level of activity during the test period and the greater levels of activity results is increased metabolic rate The main aim of the present investigation is to find out the rate of oxygen consumption of the snail *Bellamya bengalensis* at different temperatures and light intensities during time of day and night.

Materials and Methods

The snail *Bellamya bengalensis* were collected from Nalganga reservoir, Nalgangapur, Buldhana District which is 24 km away from Malkapur, and were maintained in the plastic troughs containing sufficient dechlorinated tap water. Healthy adult snail of uniform weight (0.8±0.18 males and 1.4±0.18 females) were sorted, cleaned, maintained separately sex wise in the groups of ten in plastic troughs. They were acclimated to the laboratory conditions for 5 days prior to experiment.

Throughout the experiment the water used was dechlorinated or by aged tap water. The physico-chemical parameters of aged tap water were determined periodically as per standard methods for the examination of water and wastewater. Standard Winkler's Titration Method was adapted for measuring oxygen content of the water samples (Welsh and Smith 1960; APHA 1989). The same water also served as control.

All the experiments were carried out at the time of minimal activity *i.e.* between 11am to 1pm. *Bellamya bengalensis* were transferred from the control through to the respiratory chambers maintained at four different temperatures *i.e.* 10°, 20°, 30° and 40°C with a minimum variation of ±1°C. Higher temperature (40°C) was obtained by maintaining the respiratory chambers in hot water bath. At each temperature ten individuals of each sex were used. The temperature of control water was 26°C. Measurement of oxygen consumption were also made at various times of the day and night (8, 12, 16, 24, 04 hrs).

Results and Discussion

Oxygen consumption at different temperature: Oxygen consumption measurements (mg/g wet body weight/h) of *B. bengalensis* exposed to different temperatures (10°, 20°, 30° and 40°C) increased with increasing temperatures and with decreased in temperatures. The oxygen consumption was taken for one hour and constant for each set of experiment (Table 26.1) but not uniform throughout the temperature range. It is clear that the oxygen consumption increased in both the sexes of *B. bengalensis* with increasing temperatures as reported in other molluscs (Kinne 1970; Debbagh 1972; Akerlund 1974; Ivleva 1980; McMahon *et al.*, 1981; Hawkins 1995; Gehrke *et al.*, 2004; George *et al.*, 2004). The respiratory rates of the tropical gastropods *Nerita lessellata;* was depressed at 20° C and were also not size dependent at that temperature. A comparism of the results with these gastropods show that *B. bengalensis* exhibits increased respiratory rate upto 35° C and decreased at low temperature 10° C, similar to the other tropical gastropods. Prosser (1991) and Gehrke *et al.* (2004) stated that the poikilothermic animals exhibit homeostasis over a certain range of the oxygen consumption between 26° to 30° C indicating that homeostatic mechanisms are probably in operation in both the sexes of *B. bengalensis* within this temperature range.

It is interesting to note that there is no significant difference in the oxygen consumption between the two sexes at ambient (26° C) and extreme temperatures selected (10° and 40° C) and the difference is significant in the intermediate range (20°–30° C) and the oxygen consumption is more in females. This may be due to basic physiological differences in the intermediate temperature.

**Table 26.1: Oxygen Consumption (mg/g soft body weight/h) of *B. bengalensis*
Exposed to Different Temperatures**

Sl.No.	Temperature (0° C)	Average Weight (g)		Oxygen Consumption (mg/g/h)	
		Male	Female	Male	Female
1.	10	0.937±0.12	1.382±0.18	0.318±0.08 (−44.01)	0.372±0.05 (−44.56)
2.	20	0.978±0.14	1.238±0.16	0.482±0.08 (−15.14)	0.568±0.03 (−15.35)
3.	26 (Control)	0.985±0.18	1.313±0.14	0.568±0.11	0.671±0.03
4.	30	0.908±0.23	1.298±0.16	0.558±0.08 (−1.76)	0.714±0.03 (6.40)
5.	40	0.836±0.28	1.282±0.18	0.718±0.14 (−1.76)	0.868±0.05 (6.40)

Values are mean of five replicates±SE; Figures in parentheses indicate percent change over control.

Relatively low oxygen consumption values recorded for both the sexes at temperature between 26° and 40° C which represents that the metabolism is temperature dependent over this range. This approximates to the habitat temperature of this species as observed by Clarke *et al.* (1998, 2000) in the *Limpet patella*. The very low oxygen consumption values at temperature (30°–40°C) observed and this may be an adaptive significance. The oxygen consumption of *B. bengalensis* is not uniform throughout the temperature range studied (Shirley *et al.*, 1978, 2000; Sokolova *et al.*, 2003).

**Table 26.2: Oxygen Consumption (mg/g soft body weight/h) of *B. bengalensis*
Exposed to Different Times of the Day and Night**

Sl.No.	Time of Day (hr)	Average Weight (g)		Oxygen Consumption (mg/g/h)	
		Male	Female	Male	Female
1.	8	0.0.923±0.06	1.218±0.09	0.618±0.05 (4.04)	0.712±0.07 (6.91)
2.	12 (Control)	0.918 0.09	1.186±0.07	0.594±0.06	0.666±0.08
3.	16	0.894±0.08	1.318±0.04	0.616±0.04 (3.70)	0.718±0.06 (7.80)
4.	20	0.928±0.05	1.238±0.07	0.712±0.02 (19.86)	0.828±0.06 (24.32)
5.	24	0.932±0.08	1.220±0.06	0.812±0.04 (36.70)	0.864±0.08 (34.74)
6.	4	0.928±0.04	1.194±0.05	0.630±0.02 (6.06)	0.714±0.02 (7.20)

Values are mean of five replicates±SE; Figures in parentheses indicate percent change over control.

Oxygen Consumption at Day Light

Oxygen consumption by the male and female *Bellamya bengalensis* at different light intensities of the day and night times. Indicated that the oxygen consumption was more in the evening periods till late night when they are active. Later on they become lethargic and consume less oxygen between 4 and 8 hrs. The increase in the oxygen consumption was 6.0 and 4.0 mg/g/h respectively by the male snails. However in female snail increase in the oxygen consumption was 7.2 mg/g/h at 4 hr and 6.9 mg/g/h at 8 hr. At 20 hr and 24 hr the increased oxygen consumption by males was 19.9 per cent and 36.7 per cent respectively. The female snails consume still more oxygen (Table 26.2); Gehrke *et al.* (2004) found that both the sexes of gastropods were least active between 21 and 23 hrs, moderately active between 19 and 21hrs and females were most active between 13 and 15 hrs while males did not show such a period of peak activity. The results of this study indicate that in the respiratory rate of *B. bengalensis* is minimum at noon (12 hr) and increased up to midnight (24 hrs). Kinne (1970) has observed that the rate of oxygen consumption of animals is the highest at 10 hr and lowest at 4 hr. and concluded that the respiratory pattern observed was an integral part of the daily respiratory curves and suggested as endogenous origin.

In diurnal studies, more oxygen consumption was recorded at low temperature at night and during day light less oxygen was consumed by the snail indicating that they are much active during night and during day time. They remain inactive or partial dormancy. These results are in accordance with the results obtained on other invertebrates (Brody 1945; Zeuthen 1953; Hemmingsen 1960; William 1973; Marsden 1979; Daoud 1984; Dabbagh and Marina, 1986).

References

Akerlund, G., 1974. Oxygen consumption in relation to environmental oxygen concentrations in the ampullariid snail, *Marisa cornuarietis* (L). *Comp. Biochem. Physiol.*, 47A: 1065–1975.

APHA, 1989. *Standard Methods for the Examination of Water and Wastewater*, 17th edn. American Public Health Association, Washington, DC.

Brett, J.R., 1941. Temperature versus acclimation in planting of speckled trout. *Trans. Amer. Fish. Soc.*, 70: 397–403.

Brody, S., 1945. *Bioenergetics and Growth with Special Reference to the Efficiency Compels in Domestic Animals*. Reinhold Publ. Corp., New York, p. 1023.

Clarke, A., 1998. Temperature and energetics: an introduction to cold Ocean physiology. In: *Cold Ocean Physiology*, p. 3–30.

Clarke, A.P., Mill, P.J. and Graham, J., 2000. The nature of heat coma in *Littorina littorea* (Mollusca: gastropoda). *Mar. Biol.*, 137: 447–451.

Daoud, Y.T., 1984. Ecology and bioenergetics of two species of *Asellus* in Ruttand water. *Ph.D. Thesis*. Univ. Leicester, England.

Debbagh, K.Y. and Marina, B.A., 1986. Relationship between oxygen uptake and temperature in the terrestrial isopod *Porcellionides pruinosus. J. Arid Environ.*, 11: 227–233.

Gehrke, P.C. and Donald, R.F., 2004. Effect of temperature and dissolved oxygen on heart rate, ventilation rate and oxygen consumption of spangled perch, *Leiopotherapon unicolor. J. Comp. Physiol. B. Biochem. Sys. Environ. Physiol.*, 157(6): 771–782.

George, M.B., Borchers, P., Brown, C.R. and Donnelly, D., 1988. Temperature and food as factors influencing oxygen consumption of intertidal organism, particularly Limpets. *American Zoologist,* 28(1): 137–146.

Hawkins, A.J., 1995. Effects of temperature change on ectotherm metabolism and evolution. Metabolic and physiological interrlations underlying the superiority of myltylocus heterozygotes in heterogeneous environments. *J. Therm. Biol.,* 20: 23–33.

Hemmingsen, A.M., 1960. Energy metabolism as related to body size and respiratory surfaces, and its evolution. *Rep. Steno. Mem. Hosp. Nordisk. Insulin Lab.,* 9: 1–110.

Meval, V., 1980. The dependence of crustacean respiration rate on body mass and habitual temperature. *Int. Rev. Gas. Hydrobiology,* 65: 1–47.

Kinne, O., 1970. Temperature, Invertebrates. In: *Marine Ecology,* (Ed.) O. Kinne. Wiley, London, 1: 407–514.

Marsden, I.D., 1979. Seasonal oxygen consumption of the salt marsh isopod, *Sphaeroma regicauda. Mar. Biol.,* 51: 329–337.

McMahon, F.R. and Russell, W.D., 1981. The effects of physical variables and acclimation on survival and oxygen consumption in the high Littoral Salt-Marsh snail, *Melampus bidentatus* Say. *Biol. Bull.,* 161: 246–269.

McMahon, R.F. and Russell, W.D., 1977. Temperature Relations of aerial and aquatic respiration in six littoral snails in relation to their vertical zonation. *Biol. Bull.,* 152: 182–198.

Prosser, C.L., 1991. Environmental and metabolic. *Animal Physiology.* Wiley-Liss, New York.

Shirley, T.C., Guy, D.J. and Stickle, W.B., 1978. Seasonal respiration in the marsh periwinkle, *Littoring irroata. Biol. Bull.,* 154: 322–334.

Sokolova, I.M. and Portner, H.O., 2003. Metabolic plasticity and critical temperatures for aerobic scope in a eurythermal marine invertebrate (*Littorina saxatilis,* Gastropoda: Littorinidae) from different latitudes. *J. Experi. Biol.,* 206: 195–207.

Welsh, J. and Smith, R.T., 1960. *Laboratory Exercises in Invertebrate Physiology.* Burgess Publishing Co., Minneapalis.

William, S.B., 1973. The reproductive physiology of the intertidal prosobranch *Thais lamellosa* (Gmelin). Seasonal changes in the rate of oxygen consumption and body component indexes. *Biol. Bull.,* 144: 511–524.

Zeuthen, E., 1953. Oxygen uptake as related to body size in organism. *Quart. Rev. Biol.,* 28: 1–12.

Zeuthen, E., 1970. Rate of living as related to body size in organism. *Pal. Arch. Hydrobiol.,* 17: 21–30.

Chapter 27

Impact of Copper Sulphate on the Oxygen Consumption of the Freshwater Fish *Labeo rohita*

☆ *C.M. Bharambe, P.M. Salok and C.D. Morey*

Introduction

Heavy metals in aquatic ecosystem show levels above the expected background. Extensive uses of various chemical contaminants are known to adversely affect growth, survival and physiology of freshwater fauna. The toxicity of the chemical contents in invertebrates is mainly attributed to the central nervous system. It is further are known to influence other physiological processes including respiration. Respiration is the sign of life and index of all biochemical activities that occur due in the effect of toxicants on the overall metabolism of the exposed animals. The changes in the oxygen consumption due to pollution stress create a physiological imbalance in the organisms.

The genus *Labeo rohita* is of common in the lakes and streams of Vidarbha region and presents itself as an excellent experimental animal for physiological and toxicological studies. *Labeo rohita*, the most commonly occurring species of *Labeo rohita* in this region.

The survey of literature reveals that considerable work has been done on the effect of pollutants on the respiration of marine animals but comparatively little attention has been focused on their freshwater counterparts.

Hence the present investigation was undertaken to study respiratory metabolism of the freshwater fish, *Labeo rohita* after exposure to copper sulphate for varying period such as 24, 48, 72 and 96 hours. The present work was thus also undertaken keeping in view the health, fishery and economic importance.

Materials and Methods

A freshwater fish *Labeo rohita* were collected from the Nalganga Reservoir, Nalgangapur, Buldhana district and were brought to the laboratory. Healthy adult fish of uniform weight 30-40gm were sorted, cleaned, and maintained separately in the groups of ten in plastic troughs. They were acclimated to the laboratory conditions for 5 days prior to experiment. The animals were fed with small pieces of goat muscle to avoid effects of starvation. The animals were subjected to sublethal concentration (2.0 ppm) of $CuSO_4$ and oxygen consumption was studied after 0, 24, 48, 72 and 96 hours of exposure.

The respiratory metabolism was studied by modified "Winkler's method" (Welsh and Smith, 1959, 1960). The set designed by Saroja, 1959 was used to determine the oxygen consumption. Specialized respiratory chamber was used which was back colored glass bottle facilitated by inlet, outlet and control openings. After taking all precautions about respiratory chambers, a medium sized animal was kept in airtight respiratory chamber and initial water sample was collected. The animal was allowed to stay in the chamber for one hour and at the end, the final sample was collected. By this method oxygen consumption in initial and final water samples was determined and the difference between the two readings was the amount of oxygen consumed by animal during one hour.

Results and Discussion

The freshwater fish *Labeo rohita* showed variations in total oxygen consumption and rate of oxygen consumption when exposed to copper sulphate up to 96 hours. In the present investigation it was showed that gradual-decreasing trend was seen in total oxygen consumption and rate of oxygen consumption upto 96 hours as compared to control. The results were tabulated in Table 27.1 for oxygen consumption.

Table 27.1: Effect of Copper Sulphate on Total Oxygen Consumption and Rate of Oxygen Consumption in *Labeo rohita*

Sl.No.	Duration of Exposure Period (hrs)	Average Weight of Fish (gm)	Total Oxygen Consumption in Fish (mg of O_2/animal/hr)	Rate of Oxygen Consumption in Fish (mg of O_2/gm/hr)
1.	00 (Control)	30	2.561±0.45	0.0854±0.02
2.	24	35	1.545±0.22	0.0442±0.01
3.	48	32	0.925±0.25	0.0289±0.07
4.	72	38	0.842±0.18	0.0222±0.08
5.	96	40	0.665±0.20	0.0166±0.04

Tilak and Satyavardhan (2002) showed that the amount of oxygen consumption was initially increased and then gradually decreased when the fish, *Channa punctatus* were treated with fenvalerate.

Newell (1973) reported that oxygen consumption represents the physiological state of metabolic activity and may be an indictor of metabolic stress. The pollutants may induce stress to exposed animals.

Malla Reddy (1987) observed significant drop in rate of oxygen consumption in *Cyprinus carpio* exposed to both fenvalerate and cypermethrin.

The results obtained indicate that the total oxygen consumption of *Labeo rohita* decreased gradually when exposed with sublethal concentration *i.e.*2 ppm of copper sulphate. The control set showed maximum respiratory metabolism, but when the animals were exposed to copper sulphate, they showed marked decline in total oxygen consumption and rate of oxygen consumption.

Newell, (1973) reported that oxygen consumption represents the physiological state of metabolic activity and may be an indictor of metabolic stress. The pollutants may induce stress to exposed animals.

Many workers have shown the harmful effects of heavy metals on histological structure of gills of crustaceans (Eller, 1971; Dixcon and Leduc, 1981; Sarojini and Indra, 1990; Nilknath and Sawant, 1993; Ramanna Rao and Ramamurthy 1996; Khan *et al.*, 2000).The decline in the rate of oxygen consumption may be the result of formation of coagulated mucous over the gills and body surface of the crab. Similar changes were reported (Chinnayya, 1971; Nagabhushanam and Diwan, 1972 and Nagbhushanam and Kulkarni, 1981).

A significant drop in rate of oxygen consumption in *Cyprinus carpio* exposed to both fenvalerate and cyper methrin was observed by Malla Reddy (1987). The oxygen consumption was reduced in all the tissues in an edible freshwater field crab, *Barytelphusa guerini* studied by Reddy *et al.* (1993) when exposed to cadmium chloride.

Studies on respiratory metabolism in the freshwater bivalve, *Corbicula striatllea* exposed to carbaryl and cypermethrin was reported by Jadhav and Sontakke (1997). They found a decrease in oxygen consumption with the increase in the exposure period. Asifa and Vasantha, (2001) studied effect of ammonia on respiratory activity of air breathing fish *C. batrachus* and showed that the acute and subacute treatment on animal lead to severe disruption of oxidation reduction process and suppressed tissue respiration. Tilak and Satyavardhan (2002), showed that the amount of oxygen consumption was initially increased and then gradually decreased when the fish, *Channa punctatus* were treated with fenvalerate.

The inhibition in oxygen consumption may be due to disintegration or rupture of respiratory epithelium and coagulation of mucus film over the gill surface (Jones 1947; Pones, 1947; Chinnayya, 1971). As a result, the absorption of oxygen by the gills from the external milieu might have adversely affected.

References

Chinnayya, B. and Kulkarni, G.K., 1971. Effect of heavy metals on the oxygen consumption by the shrimp, *Carolina rajadhari*. *Ind. J. Expt. Bio.*, 9(2): 277–278.

Dixon, G.D. and Leduc, G., 1981.Chronic cyanide poisoning of rainbow trout and its effects on growth, respiration and liver histopathology. *Arch. Enviro. Conta. and Toxicol.*, 10: 117–131.

Eller, L.L., 1971. Histopathological lessions in cut-throat trout *Salrno clarkii* exposed chronically to the insecticide endrin.

Jones, J.R.E., 1987.The oxygen consumption of *Gasterostes acculeatus* in toxic solutions. *J. Expt. Biol.*, 23: 291–311.

Khan, A.K., Patel, R.T. and Shaikh, F.I., 2000. Effect of mercuric chloride on the gills of the freshwater crab, *Barytelphusa guerini*. *International Conference on Probing in Biological Systems*, Mumbai Abstract, pp. 134.

Malla Reddy, P., 1987. Effect of fenvalerate and cypermethrin on the oxygen consumption of a fish, *Cyprinus carpio*. *Mendel*, 4: 209–211.

Nagbhushanam, R. and Kulkani, G.K., 1981. Freshwater palaemonid prawn, *Macrobrachium kistnensis*. Effect of heavy metal pollutants. *Proc. Ind. Sci. Acad.* B.47(3): 380–386.

Nagbhushanam, R. and Diwanm, A.D., 1972. Effect of toxic substance on oxygen consumption of the freshwater crab, *Barytelphusa cunicularis*. *Nat. Sci. J.*, 11: 127–129.

Newell, R.C., 1973. Factors affecting the respiration of intertidal invertebrates. *Am. Zool.*, 13: 513–528.

Nilkanth, G.V. and Savant, K.B., 1993. Studies on accumulation and histopathology of gills after exposure to sublethal concentration of hexavalent chromium and effect on oxygen consumption in *Scylla serrata*. *Poll. Res.*, 12(1): 11–18.

Rao, Ramanna and Ramamurthy, M.V., 1996. Histopathologieal effects of sublethal mercury on the gills of freshwater field crab, *Oziotelphusa senex* (Fabricus). *Ind. J. Comp. Ani. Physiol.*, 14(2): 33–38.

Reddy, S.L.N. and Venugopal, N.B.R.K., 1993. Effect of cadmium on oxygen consumption in freshwater field crab, *Barytelphusa guerini*. *J. Environ. Biol.*, 14(3): 203–210.

Sarojini, R. and Indira, B., 1990. Lethal and sublethal effects of antifouling organometallic compounds on the gills of *Caridina webberi*. *Himalay. J. Enviro. Zool.*, 4(1): 40–45.

Welsh, J. and Smith, R.T., 1960. *Laboratory Exercises in Invertebrate Physiology*. Burgess Publishing Co., Minneapalis.

Welsh, J.H. and Smith, R.I., 1959. *The Laboratory Exercise in Invertebrate Physiology*. Burgess Publishing Co., Minneapalis, USA, p. 10–53.

Chapter 28

Antibacterial Activity of Neem and Garlic Against *Penaeus monodon* Pathogens

☆ *Abhay Thakur, V.T. Mohite, G.J. Hande, R.B. Viadya*

Introduction

Prawn farming has encountered serious disease problems since past couple of years and as a consequence, the use of antimicrobial agents has increased significantly in aquaculture practices (Spanggaard *et al.*, 1993). However the use of antibiotics as a treatment has not only affected the aquatic ecosystem, but also the immune system of the aquatic animals (Richards *et al.*, 1991). Hence the current chemotherapy to control diseases need to be replaced with alternative disease control strategies. Available literature indicate that extract of plants *Allium sativum*, (garlic) *Eucalytpus globule(Niligiri)*, *Tamarindus indica* (*Tamrind*) exhibited inhibitory effects against human pathogenic bacteria (Srinivasan, 1995). Spices also have antibacterial properties in them and inhibit human pathogens (De *et al.*, 1999).

Similarly *Azadirachta indica* Neem (Division-Spermatophytes, Class-dicotyledonae, Family–meliacea) has been known for its medical properties since time immemorial. The earliest Sanskrit medical writings refer to the benefits of its fruits, seeds, oil, leaves, roots and barks. Each of these has been used in the Indian Ayurvedic and Unani systems of medicines. Neem leaves are now known to contain nimbin, nimbinene desacetylnimbinase, nimbandial, nimbolide and quercentin. In trials neem oil has suppressed several species of pathogenic bacteria. However, neem has many limitations as an antibiotic. In test, neem showed no antibacterial activity against *Citrobacter, Escherichia coli, Enterobacter, Klebsiella pneumoniae, Proteus mirabilis, Proteus morgasi, Pseudomonas aeruginosa, Pseudomonas EOI,* and *Streptococcus faecalis* (Neem Foundation, 2000).

Allium sativum (Garlic) (*Division–Spermatophyta, Class monocolyledonae, Family-Liliaceae*) is highly valued for its culinary effects it is also well known for its medicinal properties as is evidenced by its use in traditional medicines worldwide for over 5000 years. Garlic contains more than thirty compounds and elements. Among them are the amino acid, alliin (allylsulfinyl), adenosine, B-vitamins, C-vitamins, phosphorus, potassium, sulfur, s-allyl cysteine, iron, calcium, protein and hormones. Alliin is rapidly converted to allicin (allyl allylthiosulfinate) by the enzymatic action of alliinase when raw garlic is crushed or eaten (Stoll and Seebeck 1951). Allicin, a major sulfur-containing intermediate, which was isolated and identified as an antibacterial substance by Cavallito *et al.* (1944) and the resultant sulfide containing by-products are responsible for the well-known odour of garlic.

In view of antimicrobial properties of Neem and garlic the present communication deals with trials of Neem and garlic product against prawn pathogens–*Vibrio parahaemolyticus, V. angullarum, V. alginolyticus* and *V. vulnificus,* as alternative treatment.

Material and Methods

During the present investigation freshly collected leaves of *Azadirachta indica* were washed with distilled water and dried at 60°C. The dried leaves were extracted with methanol. For *Allium sativum* extract fresh cloves were washed with distilled water, homogenised and used for extraction with methanol. The extract obtained was completely dried to constant weight *in vacuo*. A stock solution containing 1,000 ìg/ml of plant extract was made in sterile distilled water. Further, dilutions were made to get the respective concentrations of 500 ìg/ml 250 ìg/ml 125 ìg/ml and 50 ìg/ml.

Petri dishes of Tryptic soya agar were prepared. The test organisms were inoculated on the petri dish. Using a sterile cork borer a well of approximately 1.5cm diameter was cut in the middle of the petri dish. The well was than filled with 0.2 ml of the plant extract. Same procedure was carried out on another petri dish, but filled with sterilised distilled water and maintained as control. All the dishes were incubated at 30° C for 24 hours. The effect of plant extract on the growth was observed after 24 hours, and the diameters of the inhibition zone recorded. Each experiment was done in duplicates, the zone of inhibition recorded is average of the two.

Results

The effects of extract of *A. indica* and *A. Sativam* have been depicted in Tables 28.1 and 28.2. The results indicate that *A sativum* is more effective against the isolated pathogens than *A. indica*. Only 250 µg/ml of *A indica* showed a higher zone of inhibition with *Vibrio parahaemolyticus*, when compared with *Allium sativum*.

Table 28.1: Antibacterial Activity of Neem (*Azadirachta indica*) Against Prawn Pathogen

Concentration	Zone of Inhibition (mm)			
	Vibrio alginolyticus	*Vibrio anguillarum*	*Vibrio parahaemolyticus*	*Vibrio vulnificus*
1000 µg/ml	28.5 ±0.7	31.5 ±2.12	31 ±1.4	33 ±1.4
500 µg/ml	26 ±1.4	26.5 ±0.7	27.5 ±0.7	27.5 ±0.7
250 µg/ml	23 ±1.4	23.5 ±2.12	24.5 ±0.7	25 ±1.4
125 µg/ml	0	0	0	0
50 µg/ml	0	0	0	0

The zone of inhibition with neem extract at a concentration of 250 µg/ml and the zone of inhibition with garlic extract at a concentration of 125 µg/ml were equal for three bacterial isolates *i.e.* (*V. alginolyticus, V. anguillarum* and *V. vulnificuxs*. The maximum zone of inhibition for neem was against

V. vulnificus, followed by *V. anguillarum* and *V. parahaemolyticus* with a concentration of 1000 µg/ml. A concentration of 125 ìg/ml neem extract did not show any zone of inhibition.

Table 28.2: Antibacterial Activity of Garlic (*Allium sativum*) Against Prawn Pathogen

Concentration	Zone of Inhibition (mm)			
	Vibrio alginolyticus	*Vibrio anguillarum*	*Vibrio parahaemolyticus*	*Vibrio vulnificus*
1000 µg/ml	53 ±2.82	56.5 ±0.7	34.5 ±2.12	54 ±1.4
500 µg/ml	28 ±1.4	46 ±1.4	30 ±2.82	47.5 ±2.12
250 µg/ml	25.5 ±0.7	28.5 ±0.7	22.5 ±0.7	32.5 ±2.12
125 µg/ml	23 ±1.4	23.5 ±0.7	0	25.5 ±0.7
50 µg/ml	0	0	0	0

Garlic at a concentration of 1000 µg/ml showed a maximum zone of inhibition against *V. anguillarum* while, *V. parahaemolyticus* showed minimum zone of inhibition at a concentration of 1000 µg/ml. Of the four isolates *V. parahaemolyticus* did not show any zone of inhibition at a concentration of 125 µg/ml of garlic extract. *V. vulnificus* showed a zone of inhibition of 25mm at that concentration. Activity of *Allium sativam* was minimum against *V. parahaemolyticus*. *Allium sativam* at lower concentrations (125 µg/ml) was effective against *V. anguillarum* and *V. vulnificus*.

A gradual increase in the antibacterial activity with higher concentration was recorded. With neem extract the difference between the zone of inhibition for concentrations of 250 µg/ml and 1000 µg/ml was only 4mm in *V. alginolyticus* and 11mm in *V. anguillarum*. Garlic extract showed relatively low sensitivity towards *V. parahaemolyticus*, and could not restrict its growth at a concentration of 125 µg/ml. *V. alginolyticus* showed maximum sensitivity towards garlic at 1000 µg/ml. At a concentration of 250 µg/ml and 500 µg/ml the inhibition zone was considerably lower than the other isolates, but at 1000 µg/ml it was comparable.

Discussion

The results obtained in this study indicate that clove extract of the plant *Allium sativam* and leaf extract of *Azardirachta indica* are effective against prawn pathogens of *Vibrio species*. The inhibition with garlic was more when compared with neem. No zone of inhibition was recorded with 50 µg/ml of garlic and 125 µg/ml of neem extract. This shows that garlic is more effective than neem against the *vibrios* tested. Many workers have reported similar type of antimicrobial activities from different plants (Chopra *et al.*, 1952; Das *et al.*, 1999; De *et al.*, 1999).

Earlier reports of use of *A indica* and *A sativam* as antibacterial agents are available, but their uses in controlling *vibrio* infection in fish or prawn are not many. Das *et al.* (1999) in their study with neem extract against fish pathogenic bacteria, reported inhibition of *Aeromonas hydrophila*, *Pseudomonas fluorescens*, *Escherichia coli* and *Myxobacteria sp* upto 70 per cent at 20 ppm. Comparative effect of various antibiotics incorporated with three plant extracts on the zone of inhibition of chitin degrading bacteria isolated from prawn *Metapenaeus dobsoni* indicated that most of the antibiotics in combination with extract of *Ocimum sanctum*, *A. indica* and *Phyllanthus niruri* are more effective than antibiotics alone (Rivonkar *et al.*, 1999).

In the present study a concentration of 250 µg/ml of neem extract inhibited the growth of *Vibrio sp*. Though not much literature is available on antibacterial activity of neem against marine *Vibrio sp*, inhibition of *Vibrio cholerae* due to *Melia azadirachta* at 200 µg/ml have been demonstrated by Patel and Trivedi (1962). Ahmed *et al.* (1995), studied antimicrobial activity of leaf and bark extract of *A. indica*. Aqueous as well as alcoholic extracts (200 µg/ml) showed good antibacterial activity. In this study concentration of 1000 µg/ml proved to be effective against the bacteria tested. The inhibition zone for all the tested organisms was nearly the same.

Garlic has often been called nature's penicillin because of its powerful antibiotic properties. "Healthline" reported that in a study at Boston University School of Medicine, researchers found garlic to be broadly effective against fourteen different strains of bacteria, even killing some that are resistant to commonly used antibiotics (Healthline, 1994).

In the present study garlic extracts of <50 µg/ml were non-inhibitory and those of 125 µg/ml were inhibitory to the four isolates tested. Ahsan *et al.* (1996) have also documented significant inhibitory activity of aqueous extract of *A sativam* on *Vibrio cholerae* and other bacterial strains. Dababneh and Al-Delaimy (1984) reported that 1 per cent garlic extract inhibited *Staphylococcus aureus*. Garlic extract >1 per cent in culture media was inhibitory to *Lactobacillus plantarum*, and >2 per cent was bactericidal (Karaioannoglou *et al.* ((1977). Similar results were obtained by Mantis *et al.* (1978) in which 2 per cent garlic extract in culture media had growth inhibitory effects against *Staphylococcus aureus*.

Salmonella typhimurium, Bacillus cereus, Clostridium botulinum, C. perfringens, Candida utilis and many other bacteria have been inhibited by garlic extract (Srivastava, *et al.,* 1982; Rees, 1993). *Staphylococcus aureus, Escherichia coli, Proteus mirabilis* and *Pseudomonas aeruginosa* which are multiple resistant to antibiotics including penicillin, streptomycin, doxycilline and cephalexin were also inhibited by garlic extract (Singh and Shukula, 1984).

In the present study *V. anguillarum* showed maximum zone of inhibition at a concentration of 1000 µg/ml and *V. parahaemolyticus* had the minimum zone of inhibition when tested with *A sativam*. Shin *et al.* (1999) evaluated the antimicrobial activity of garlic juice powder dehydrated by different methods. Freeze dried garlic juice powder showed high inhibitory effect on Gram positive bacteria (*Bacillus subtilis, Staphylococcus aureus* and *Streptococcus mutans*) having 0.3 to 2.0 per cent minimum inhibitory concentrations (MIC). But spray-dried garlic juice powder did not have inhibitory effect on Gram positive bacteria, except *Bacillus subtilis*.

The principal antimicrobial compounds of *Allium species* are those belonging to a group known as thiosulfinate. The antimicrobial activity of thiosulfinates has been explained as a general reaction between thiosulfinates and–SH groups of essential cellular proteins (Barone and Tansey, 1977). Small *et al.* (1947) mentioned that–S (O) S–was responsible for the antimicrobial activity and that reacted readily with cysteine to yield mixed disulfides. Fujiwara *et al.* (1958) showed essentially the same reaction between allicin and thiamine. The general reaction, as proposed by Small *et al.* (1947) can apply to where thiosulfinates are involved and the reaction is believed to be the common mechanism of antimicrobial activity of thiosulfinates.

Ajoene, a derivative of allicin originally reported for its potent antithrombotic activity (Block, 1984) exhibits a strong antibacterial activity toward Gram-negative bacteria and Gram positive bacteria (Yoshida *et al.* (1987; Naganawa, 1996). It does not possess thiosulfinyl group as allicin. Yoshida *et al.* (1987), found ajoene to be even more potent antifungal agent than allicin, assumed that ajoene may damage the cell walls of fungi, thus not expecting significant antibacterial activity, except for *Staphylococcus aureus*. Later, however, Naganawa *et al.* (1996) showed a different result concerning antimicrobial activity of ajoene. They reported that ajoene was strongly inhibitory against Gram-positive bacteria and yeasts and had various degrees of inhibition of Gram-negative bacteria like

Escherichia coli and *Pseudomonas aeruginosa*. Naganawa *et al.* (1996) postulated that the disulfide group in ajoene appears to be necessary for the antibacterial activity, since reduction by cysteine abolished its antimicrobial activity.

1. Study indicate that the antimicrobial activity of the two plants have different responce against the marine vibrios of prawn.
2. Both can be tried to control these pathogens however.
3. Garlic that gave a bigger inhibition zone at all the concentrations tested and could be a better option than neem for its use as antibacterial.
4. When given through feed these could help in controlling the growth of pathogenic bacteria, and hence reduce the occurrence of disease.

References

Adetumbi, M., Javor, G.T. and Lau, B.H.S., 1986.*Allium sativum* (garlic) inhibits lipid synthesis in *Candida albicans. Antimicrob. Agents Chemother.*, 30: 499–501.

Ahmed, I., Ahmad, F. and Hussain, S., 1995. *In vitro* antimicrobial activity of leaf and bark extracts of *Azardirachta indica* Juss. *Indian Veterinary Medical Journal*, 19(3): 204–206.

Ahsan, M., Chowdhury, A.K., Islam, S.N. and Ahmed, Z.U., 1996. Garlic extract and allicin: Broad spectrum antibacterial agents effective against multiple drug resistant strains of *Shigella dysenteriae* type 1 and *Shigella flexneri*, enterotoxigenic *Escherichia coli* and *Vibrio cholerae. Phytotherapy Research*, 10(4): 329–331.

Barone, F.E. and Tansey, M.R., 1977. Isolation, purification, identification, synthesis, and kinetics of activity of the anticandidal component of *Allium sativum*, and a hypothesis for its mode of action. *Mycologia*, 69: 793–825.

Cavallito, C.J., Buck, J.S. and Suter, C.M., 1944. Allicin, the antibacterial principles of *Allium sativim*. II. Determination of the chemical structure. *J. Am. Chem. Soc.*, 66: 1952–1954.

Chopra, J.C., Gupta, K.C. and Nazir, B.N., 1952. *Indian J. Med. Res.*, 40: 511.

Dababneh, F.A. and Al-Delaimy, K.S., 1984. Inhibition of *Staphylococcus aureus* by garlic extract, Lebensm. *Wissenschaft. Technol.*, 17(1): 29–31.

Das, B.K., Mukherjee, S.C., Sahu, B.B. and Murjani, G., 1999. Neem (*Azardicata indica*) extract as an antibacterial agent against fish pathogenic bacteria. *Indian Journal of Experimental Biology*, 37: 1097–1100.

DeWit, J.C., Notermans, S., Gorin, N. and Kampelmacher, E.H., 1979. Effects of garlic oil or onion oil on toxin production by *Clostridium botulinum* in meat slurry, *J. Food Prot.*, 42: 222–224.

Feldberg, R.S., Chang, S.C., Kotik, A.N., Nadler, M., Neuwirth, Z., Sundstrom, D.C. and Thompson, N.H., 1988. *In vitro* mechanism of inhibition of bacterial cell growth by allicin. *Antimicrob. Agents Chemother.*, 32: 1763–1768.

Focke, M., Feld, A. and Lichtenthaler, H.K., 1990. Allicin, a naturally occurring antibiotic from garlic, specifically inhibits acetyl-CoA synthetase. *FEBS*, 261(1): 106–108.

Ghannoum, M.A., 1988. Studies of the antimicrobial mode of action of *Allium sativum* (garlic). *J. Gen. Microbiol.*, 134: 2917–2924.

Healthline, 1994. *Scientists Say Garlic Killed Cold Germs*," March/April (1994)

Johnson, M.G. and Vaughn, R.H., 1969. Death of *Salmonella typhimurium* and *Escherichia coli* in the presence of freshly reconstituted dehydrated garlic and onion. *Appl. Microbiol.*, 17(6): 903–905.

Karaioannoglou, P.G., Mantis, A.J. and Panetos, A.G., 1977. The effects of garlic extract on lactic acid bacteria (*Lactobacillus plantarum*) in culture media, Lebensm. *Wissenschaft. Technol.*, 10: 148–150.

Mantis, A.J., Karaioannoglou, P.G., Spanos, G.P. and Panetos, A.G., 1978. The effect of garlic extract on food poisoning bacteria in culture media. I. *Staphylococcus aureus*, Lebensm. *Wissenschaft. Technol.*, 11: 26–28.

Mantis, A.J., Koidis, P.A., Karaioannoglou, P.G. and Panetos, A.G., 1979. Effect of garlic extract on food poisoning bacteria, Lebensm. *Wissenschaft Technol.*, 12: 230–232.

Naganawa, R., Iwata, N., Ishikawa, K., Fukyda, H., Fujino, T. and Suzuki, A., 1996. Inhibition of microbial growth by ajoene, a sulfur-containing compound derived from garlic, *Appl. Environ. Microbiol.*, 62(11): 4238–4242.

Neem Foundation, 2000. www.neemfoundation.com.

Patel, R.P. and Trivedi, B.M., 1962. *Indian Journal of Medical Research*, 50: 218.

Rees, L.P., Minney, S.F., Plummer, N.T., Slator, J.H. and Skyrme, D.A., 1993. A quantitative assessment of the antimicrobial activity of garlic (*Allium sativum*). *World J. Microbiol. Biotechnol.*, 9: 303–307.

Richards, R.H., Inglis, V., Frerichs, G.N. and Miller, S.D., 1991. Problems of chemotherapy from theory to reality. *Working Paper from the Conference*, Paris 12–15, March.

Rivonker, C.U., Abuvarajan, C.R. and Sangodkar, U.M.X., 1999. Chitin degrading bacteria from the prawn *Metapenaeus dobsoni* M and their control. *Indian Journal of Marine Sciences*, 28: 77–80.

Saleem, Z.M. and Al-Delaimy, K.S., 1982. Inhibition of *Bacillus cereus* by garlic extracts. *J. Food Prot.*, 45(11): 1007–1009.

Shin, D.B., Kim, Y.S. and Lee, Y.C., 1999. Effect of dehydration methods on the antimicrobial activity of garlic juice powder. *IFT Annual Meeting*, 98.

Singh, K.V. and Shukula, N.P., 1984. Activity of multiple resistant bacteria of garlic (*Allium sativum*) extract. *Fitoterapia*, 15(5): 313–315.

Small, L.D., Bailey, J.H. and Cavallito, C.J., 1947. Alkyl thiosulfinates. *J. Am. Chem. Soc.*, 69: 1710–ss1713.

Spanggaard, B., Jorgensen, L. Gram and Huss, H., 1993. Antibiotic resistance in bacteria isolated from three freshwater fish farms and an unpolluted stream in Denmark. *Aquaculture*, 115: 195–207.

Srinivasan, D., Krishnan, S. and Lakshmanaperumalsamy, P., 1995. Antimicrobial activity of some Indian medicinal plants used in folklaric medicine. International seminar on recent trends in pharmaceutical sciences. *Ootacamund*, Abstract No. A14.

Srivastava, K.C., Perera, A.D. and Saridakis, H.O., 1982. Bacteriostatic effects of garlic sap on Gram negative pathogenic bacteria: An *in vitro* study, Lebensm. *Wissenschaft. Technol.*, 15(2): 74–76.

Wills, E.D., 1956. Enzyme inhibition by allicin, the active principle of garlic. *Biochem. J.*, 63: 514–520.

Wunderlich, Ray C., 1995. *Natural Alternatives to Antibiotics*. Keats Publishing, Inc. New Canaan.

Yoshida, S., Kasuga, S., Hayashi, N., Ushiroguchi, T., Matsuura, H. and Nakagawa, S., 1987. Antifungal activity of ajoene derived from garlic. *Appl. Environ. Microbiol.*, 53(3): 615–617.

Chapter 29

Observation on Thermophilic Algal Communities in Unakeshwar Thermal Springs

☆ *A.M. Khole, J.M. Gaikwad, R.R. Rakh and R.S. Deshmukh*

Introduction

Phytoplankton are important component of aquatic flora, they serve as primary producer and forms base of the food chain. The estimation of phytoplankton provides good indices of water quality and capacity of water to sustain heterotrophic community. Algal floras play an important role in the primary productivity of any aquatic ecosystem and form the base of food chain. The importance of algae as a source of biofertilizers is already well established (Singh 1961, Venkataraman 1972).

Physico-chemical factors play an important role in qualitative distribution of phytoplankton during different seasons. Therefore, in the present investigation, an attempt has been made to survey the phytoplankton and physico-chemical parameters operating in thermal and cold spring of Unakeshwar

The Study Area

Unakeshwar is small town located in Nanded dist, Maharashtra. Unakeshwar thermal springs are situated on the bank of *Penganga,* lies between 19° 34' to 19° 40' N latitude and 78° 22' to 78° 34' E longitude. Unakeshwar is famous for its natural perennial two thermal springs and one cold spring (Figure 29.1). The thermal spring water possesses medicinal value, to cure skin disorders. Pilgrims when visit to *Mahur* temple (Holy spot, Godess Renukadevi temple) they also give visit to Unakeshwar to take bath in thermal spring water with the belief that their sins or diseases will be cured. Temperature of one of the thermal spring water stand 45°C and found contain sulphur more than average proportion.

(a) Cold Spring Tank

(b) Hot Spring Tank

Figure 29.1: Unakeshwar Springs

Material and Methods

The algal samples were collected from three sampling sites (TA, TB) for eight months on monthly basis from April to November 2007. The algal samples were preserved in 4 per cent formalin and taxanomic studies were conducted with the help of standard literature on the

Subject (Jana 1971, Forest 1959 and Randhawa 1959) and water samples were analyzed according to standard methods of Trivedy and Goel (1987), Kodarkar *et al.* (1992) and APHA (1975).

Result and Discussion

During the period eight months various physico-chemical parameters studied, the water from Unakeshwar springs seeps out a transparent, colorless and odorless. The water temperature observed between 41 to 45°C as compared to the air temperature of 23 to 39°C. Total Solids, pH, calcium, nitrate, phosphate and organic matters are important factors influencing the growth of algae (Puttaiah *et al.,* 1987). The sulphate were detected in high concentrations in the thermal springs throughout the period of investigation. Fluctuations in the values of sulphates found to be associated with the intensity of the algae (Moyle,1949).

The spring may classify as Euthermal (Vock, 1950) and provided stable environment to ecosystems which may have remained unchanged (Brock, 1967). The results of chemical analysis are summarized in Table 29.1. All the results presented are an average of data collected at three stations. The phytoplankton species was represented by chlorophyceae and cyanophyceae mentioned in Table 29.2. The other groups though represented, were scarce in number and poor in forms and, hence also considered in the present study.

The distribution of algae depends upon, besides many other factors, the nutrients and other organic and inorganic substances in the water, and relative adaptability of different species. The seasonal variations in water quality parameters of the tanks (TA,TB) has a marked influence on the numerical abundance of plankton. Phytoplankton community comprised algal groups described below:

Chlorophyceae

It was the most significant group form greenish scum on the surface of the stagnant water or grows firmly attached the submerged rocks (Vashishta *et al.,* 2004). It exhibited 47 per cent, maximum density during April, May, and December. The factors such high temperature and bright sunlight are favorable for the population of chlorophyceae (Rao, 1955). Jahangir *et al.,* 2001, observed the correlation between chlorophyceae and different physico-chemical factors. This group was represented by the genus, *Volvox,* and *Oedogonium, Spirogyra, Chara* as shown in Figure 29.2.

Cyanophyceae

Cyanophyceae provide a good example of the adaptability of life to extremes of environment, high temperature of hot spring and low temperature of Polar Regions (Vashsishta *et al.,* 2004). Cyanophyceae are found generally on rocks or soils, algae requirement for limiting nutrient concentration (Zahid, 1998), it was the second significant group form greenish scum and contributes 43 per cent to the annual production.

Tripathi and Pandey (1995) observed the maximum population of cyanophyceae during summer, while minimum in winter. Similar, results were recorded and the group was represented by genus, *Oscillatoria, Nostoc* and *Microcystis* (Figure 29.2).

Figure 29.2: Phytoplankton found at Unakeshwar

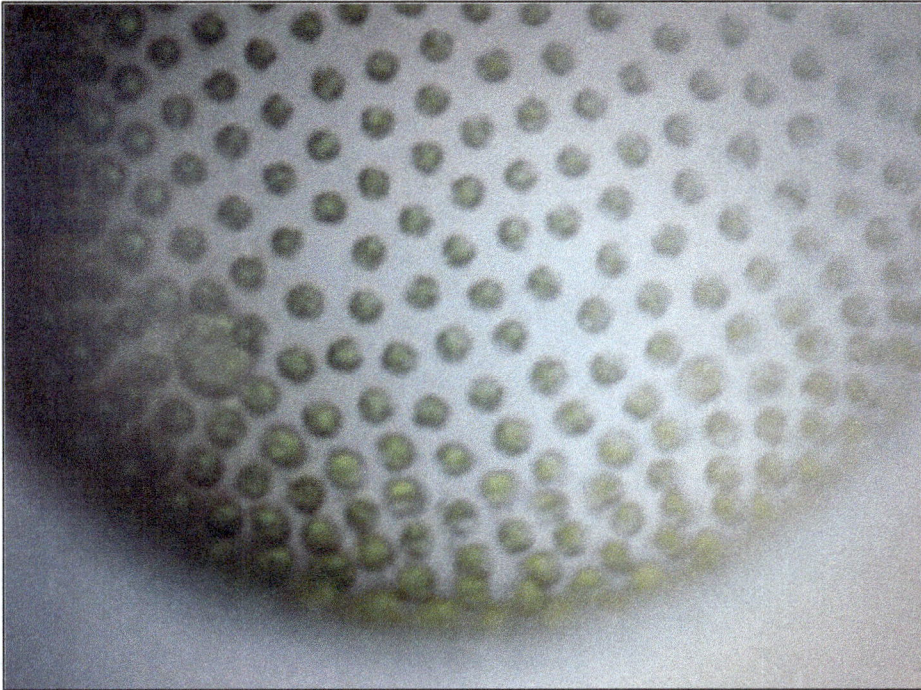

(a) *Volvox*

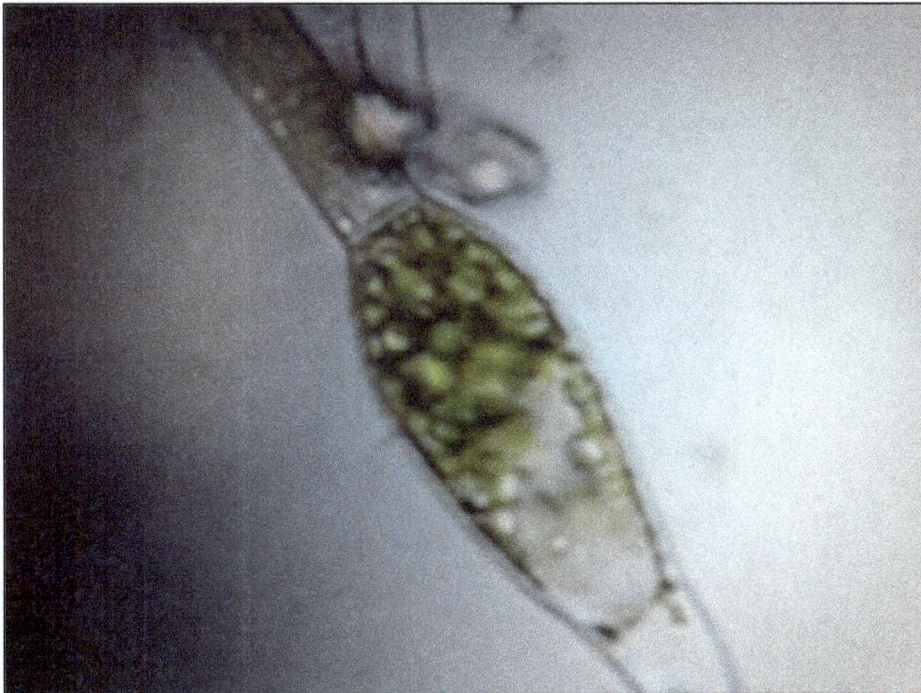

(b) *Oedogonium*

Contd...

Figure 29.2–Contd...

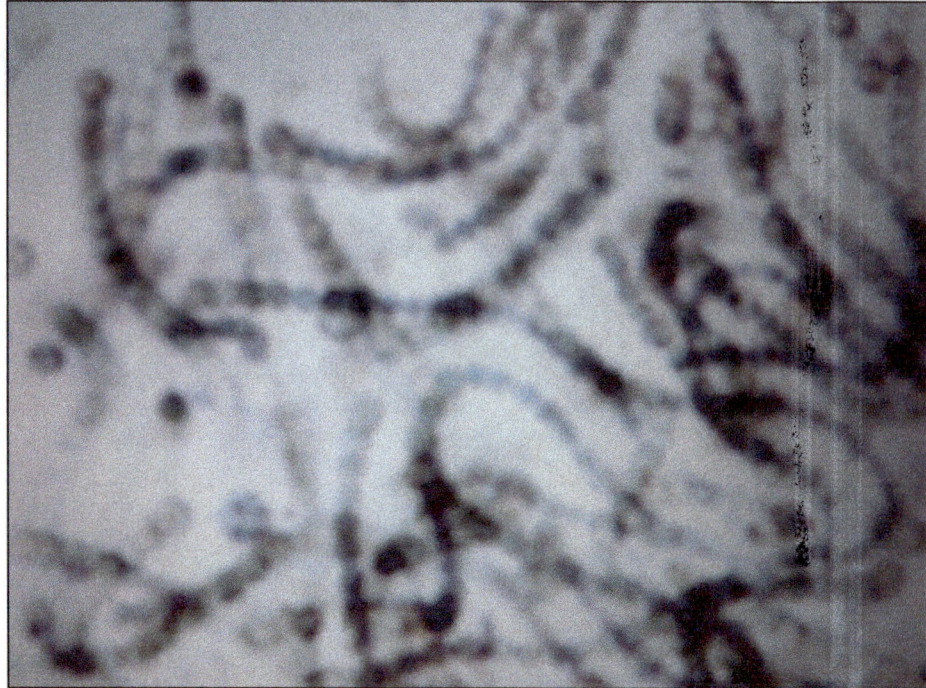

(c) *Nostoc*

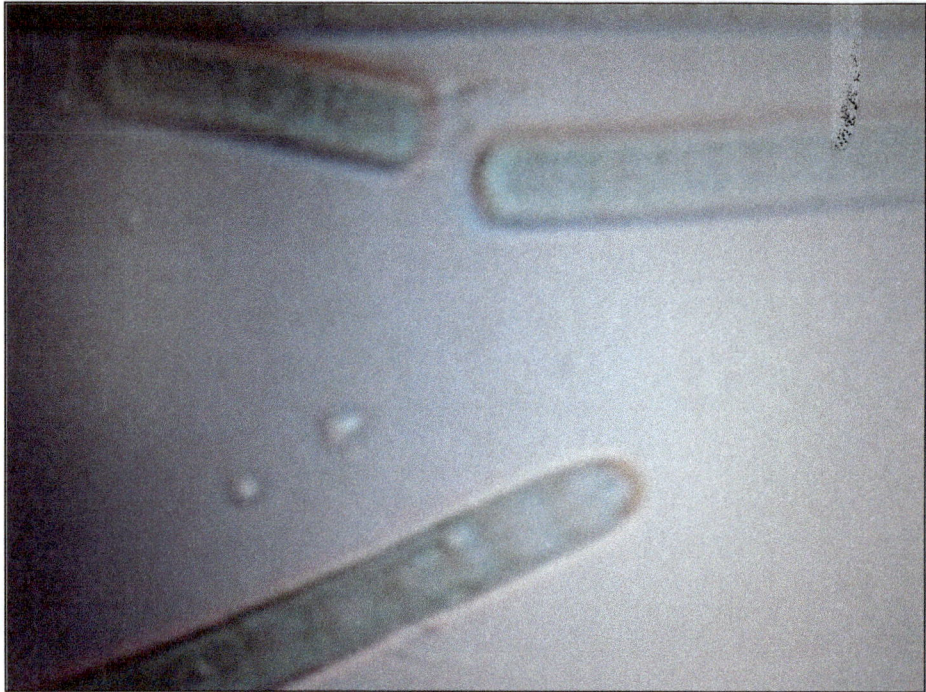

(d) *Oscillatoria*

Table 29.1: Physico-Chemical Characteristics of Unakeshwar Springs Water

Month	Atmos Temp (0°C)		Water Temp (0°C)		pH		TDS (mg/l)		Total Hardness TH		Ca		Mg		DO (mg/l)		DCO$_2$ (mg/l)		BOD (mg/l)		COD (mg/l)		Salinity (mg/l)		Sulphur (mg/l)	
	A	B	A	B	A	B	A	B	A	B	A	B	A	B	A	B	A	B	A	B	A	B	A	B	A	B
April	39	38	44	31	6.5	7.6	282	240	126	105	60	60	7.4	5.52	6.6	7.1	Nil	Nil	1.82	2.0	5.04	4.04	3.10	1.80	352	74
May	39	38	45	31	6.6	7.6	280	235	128	103	62	60	7.78	5.56	6.2	7.1	Nil	Nil	1.85	2.0	5.02	4.04	310	182	352	72
June	38	36	45	29	6.4	7.6	385	305	141	123	61	43	6.96	5.42	6.4	8.3	Nil	Nil	2.0	2.4	6.32	5.66	316	178	302	54
July	34	33	43	24	5.8	7.2	310	250	152	129	68	51	4.22	6.24	7.4	8.8	1.2	1.4	2.1	2.0	6.88	6.02	322	174	282	46
Aug.	29	27	44	21	5.8	7.2	330	220	163	146	71	59	7.22	6.78	7.6	8.6	Nil	1.8	2.0	2.6	4.02	4.01	302	172	280	40
Sept.	26	24	43	18	5.2	7.1	280	275	167	142	88	63	8.02	6.98	8.2	9.2	1.4	2.2	1.8	1.90	4.52	4.88	302	170	308	38
Oct.	25	23	41	16	4.8	7.0	208	248	178	145	88	71	8.78	7.12	8.4	9.4	1.6	1.8	1.4	1.32	5.24	4.98	300	71	300	58
Nov.	23	22	41	16	1.8	7.0	208	242	182	145	88	71	8.77	8.12	8.8	9.4	1.6	1.9	1.4	1.28	5.26	5.02	304	68	288	54

Site A: Warm Spring Water; Site B: Cold Spring Water.

Table 29.2: Occurrence of Phytoplankton Genera in Unakeshwar Springs

Phytoplankton Genera	Month of Max Occurrence	Period of Non-Occurrence	Percentage
Chlorophyta			47 per cent
Volvox	March, April	December	
Oedogonium	June, December	April	
Chara	May, December	November	
Spirogyra	April, December	April	
Cyanophyta			43 per cent
Oscillatoria	March,	June	
Nostoc	June	April	
Microcystis	May, December	April	
Others			10 per cent

The other species contributes 10 per cent to the annual population. Several authors have been emphasized the importance of water temperature in the periodicity of algae. (Jahangir 2001, Jha and Kumar 1990).

Conclusion

The eight months investigation shows that both the springs are dynamic and variable systems in terms of hydrochemistry. Marked seasonal, diurnal variations were observed. The seasonal and diurnal variations of spring tend to co-vary with temperature. Fluctuations occur during seasons depending on rainfall intensities and the features of geothermal activities.

The present day flora of the thermal spring is adapted high temperature, high salinity and few man-made polluted environments. Our investigation of thermal and cold spring supports the view of Brock (1978), Kar (2001) and Zahid (1998).

Unakeshwar sulphur springs are yet to be considered as a tourist spot since, these thermal springs have a special ecological niche, proper care and attention should be given to conserve them and protest the biodiversity.

Acknowledgements

We would like to thank to Principal, B. Raghunath College, Parbhani and Dr. S. D Ahirrao, Department of Fishery, Shri Shivaji College, Parbhani for their help and guidance during the course of the investigation.

References

APHA, 1985. *Standard Methods for Examination of Water and Wastewater*. Washington D.C.

Basahy, A.Y., 1993. Water chemistry of hot springs in Gizan area of Saudi Arabia. *Journal King Saudi Uni.*, 6(Science)(1): 23–29

Bendre, Ashok and Kumar, Ashok, 2003. *A Textbook of Practical Botany*. Rastogi Pub., Meerut.

Bora, Limpson, Kar, A. and Barauh, I., 2006. Hot springs of Twang and W. Kameng Dist. in A.P., India. *Current Science*, 91: 1011–1013

Brock, T.D., 1987. Life at high temperatures. *Science Journal*, 159: 1012–1019.

Chandrashekhar, K.R., Sridhar, K.R. and Kaveriappa, K.M., 1991. Aquatic hyphomycetes of a sulphur spring. *Hydrobiologia*, 218: 151–156.

Forest, H.S., 1954. *Zygnemaceae*. ICAR, New Delhi.

Jahangir, T.M., Khuhawar, M.Y., Leghari, S.M. and Laghari, Abdullah, 2001. Physico-chemical and biological study of Mangho Pir Euthermal springs of Karachi, Sindh Pakistan. *Journal Biological Sciences*, 1(7): 636–639.

Jana, B.B., 1971. The thermal springs of brackishwar, India: Physico-chemical conditions, flora and fauna. *Hydrobiologia*, 41(3): 291–307

Jana, B.B., 1978. The plankton ecology of some thermal springs in West Bengal, India. *Hydrobiologia*, 61(2): 35–43

Jha, M. and Kumar, H.D., 1990. Cyanobacteria flora and physico-chemical properties of Saptadhara and Brahma kund hot springs of Rajgir, Bihar, India. *Nova Hedwigia*, 50: 529–534.

Khuhawar, M.Y., Thebow, S.N. and Leghari, S.M., 1986. Physico-chemical investigation of the hot Laki springs of Sindh, Pakistan. *Physical Chemistry Journal*, 5: 43–49.

Leghari, S.M. and Thebow, S.N., 1983. Cyanophyceae of hot spring at Lakhi Shah Sadar, Sindh, Pakistan. *S.U. Res. Journal (Sci. Ser.)*, 15: 147–150.

Moyle, J.B., 1949. Some indices of lake productivity. *Trans. Amer. Fish. Soc.*, 76: 322–334.

Munawar, M., 1974. Limnological studies on freshwater ponds of Hyderabad, India. IV. The biocenose, periodicity and species compositions of unicellular and colonial phytoplankton in polluted and unpolluted environment. *Hydrobiologist*, 45(1): 1–32.

Puttaiah, E.T. and Somashekar, R.K., 1987. On the ecology of desmids in lakes of Mysore city. *Geobios News Report*, 6: 132–135.

Randhava, M.S., 1959. *Zygnemaceae*. ICAR, New Delhi.

Tripathi, A.K. and Pandey, S.N., 1995. *Water*. Ashish Pub. House, p. 92–286.

Trivedy, R.K. and Goel, P.K., 1986. *Chemical and Biological Methods for Water Pollution Studies*. Environment Publ., Karad.

Vashishta, B.R., Sinha, A.K. and Singh, V.P., 2005. *Botany for Degree Students: Algae*. S. Chand and Comp., p. 92–286.

Zahid, P.B., 1998. Studies on some thermophilic algae at Karachi, Pakistan. *Science Khyber Journal*, 2: 157–169.

Chapter 30

Plankton Diversity in Yeldari Reservoir, Maharashtra

☆ *P.K. Joshi, V.B. Sakhare and S.G. Jetithor*

Introduction

Among biotic communities phytoplankton constitute the first stage in trophic level by virtue of their capacity to transduce environmental radiant energy into the biological energy through photosynthesis. Also referred to as primary productivity, the magnitude of photosynthetic energy fixation depends primarily on diversity and biomass of phytoplankton. The planktonic photosynthesis plays a key role in conditioning the microclimate (zone around an ecosystem) as it helps in regulating the atmospheric level of oxygen and carbon dioxide. Apart from primary production, phytoplanktons also play an important role as food for herbivorous animals. They also are biological indicators of water quality in pollution studies. To summarize, due to their environment in cycling of energy and matter in an ecosystem, evaluation of phytoplankton population in terms of their diversity, density, biomass, spatial and temporal distribution, periodicity and productivity and population turnover, is vital in management of an ecosystem. Fishes consume the phytoplankton, which is found abundantly in ponds, lakes, streams and reservoirs. Phytoplankton also gives green colour to the water. It is due to the presence of chlorophyll. Growth and multiplication of phytoplankton is mainly dependent on temperature, solar illumination and the availability of certain essential nutrients such as nitrates, silicates and phosphates.

Zooplankton communities of freshwaters constitute an extremely diverse assemblage of organisms represented by different phyla of invertebrates. The most significant feature of zooplankton is its immense diversity over space and time. In an ecosystem 90 per cent of zooplankton species are herbivorous, remaining 10 per cent being carnivore. Since secondary production primarily depends on the biomass of herbivores, on-predatory zooplankter contributes significantly to the secondary productivity of an aquatic ecosystem.

The diversity of zooplankton is usually studied by enumeration of different taxa or species in a representative sample collected by towing standard plankton net over an adequate distance. Zooplankton diversity is one of the most important ecological parameters in water quality assessment. The zooplankton study has been a fascinating subject for a long time. Water bodies rich in phytoplankton are also rich in zooplankton diversity and biomass.

The present research paper is intended only to know the plankton diversity from Yeldari reservoir of Parbhani district in Maharashtra.The Yeldari reservoir a purely hydro-electric project was constructed in year 1962 in hilly area of Jintur tahsil.The reservoir is bounded by latitudes 19°43′N and longitudes 76°45′E and is included in the survey of India toposheet map no.56A/10.The area around reservoir comprise forest covered hills. The reservoir is having catchment area of 7,330 sq.km.The maximum level of reservoir is of 462.380m.

Materials and Methods

The planktonic samples were collected on monthly basis for the period of October 2000 to September 2001.Planktons were collected by filtering 100 liters of water through plankton net made up of bolting silk. The zooplanktonic samples were preserved in 5 per cent formalin and phytoplanktonic in lugol's solution. The preserved samples were brought to the National Foundation for Environmental Services, Karad (Maharashtra) for qualitative and quantitative analysis and the identification was done with the help of methods described by Pennak (1953), Arora (1963), Sehgal(1983), Battish (1992), Murugan *et al.* (1998), and Dhanapathi (2000).

Results and Discussion

Phytoplankton diversity: Altogether 14 species of chlorophyceae, 9 species of bacillariophycae, and 7 species of myxophyceae were recorded from the reservoir (Table 30.1). Among phytoplankton chlorophyceae dominated, followed by myxophyceae and bacillariophyceae (Figure 30.1).

Chlorophyceae was abundant in May with density of 2008/liter. Bacillariophycae was also more among the phytoplankton assemblage of the reservoir. The population was maximum (110/liter) in October. It was mainly represented by *Fragilaria sp*,*Navicula sp*,and *Nitzchia sp*.Myxophyceae dominate in many Indian reservoirs, the necessity of introducing fishes to utilize them as food was stressed by Natarajan (1975). The report of impressive growth of *Hypopthalmicthys molitrix* in Getalsud reservoir (Anon, 1977b) and the high growth rate of the same fish in Kulgarhi reservoir (Natarajan, 1975)

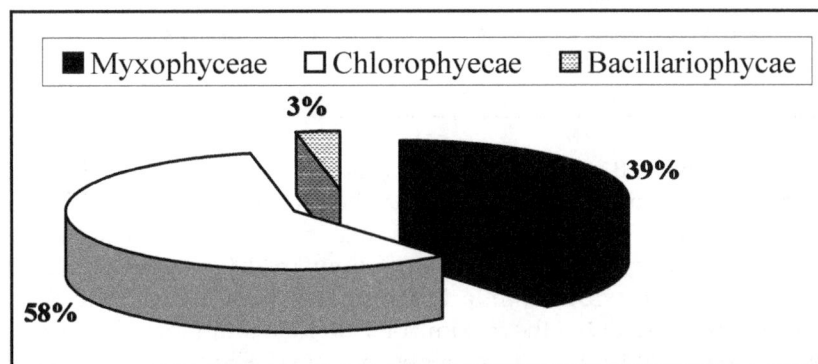

Figure 30.1: Phytoplankton Composition in Yeldari Reservoir, Maharashtra

indicate that the fish can be successfully introduced to other myxophyceae dominated lakes and reservoirs. During present investigation, the myxophyceae dominated the reservoir. Hence *Hypopthalmicthys molitrix* can be introduced in reservoir as discussed above. Myxophyceae was richly represented in months of September (1248/liter), followed by November (1183/liter). The group was dominated by *Anabaena spp*, and *Nostoc spp*. In the present investigation, the increased phytoplankton density in post monsoon period is due to rich nutrients received through rainwater, as reported by Sreenivasan (1974). A total of 30 phytoplankton species were recorded. The winter months showed higher phytoplankton density followed by summer and rainy months. The lower densities during rainy months may be due to high turbidity, low light intensity, cloudy weather, and more water coverage with rains.

Table 30.1: List of Phytoplankton in Yeldari Reservoir, Maharashtra

Myxophyceae:	*Anabaena* spp., *Arthrospira* spp., *Lygnbya majuscula, Microcystis areuginosa, Nostoc* spp., *Oscillatoria chlorina, Phormidium* sp.
Bacillariophyceae:	*Cyclotella operculata, Cymbella turgida, Fragilaria* sp., *Gomophonema gracile, Melosira* sp., *Navicula mutica, Nitzchia* sp., *Pinnularia viridis, Synedra ulna.*
Chlorophyceae:	*Eudorina* sp., *Pandorina morum, Volvox* sp., *Cosmarium microsporum, Ulothrix zonata, Microspora* sp., *Pediastrum duplex, Spirogyra margariata, Oedogonium* sp., *Chlorella vulgaris, Closterium* sp., *Cladophora* sp., *Stichococcus* sp., *Scenedesmus* sp.

The phytoplankton constituted 84.03 per cent of the total plankton, while the zooplankton was recorded at 15.97 per cent of the total plankton. Sugunan and Yadava (1991 b) reported 96.41 per cent of phytoplankton and 3.59 per cent of zooplankton in the Nongmahir reservoir of Meghalaya.

Zooplankton Diversity

Zooplanktons were represented by rotifera, cladocera, copepoda, and ostracoda. Among zooplankton copepoda dominated, followed by cladocera, rotifera and ostracoda (Figure 30.2). The seasonal variation in total zooplankton population is represented in Figure 30.3.

Table 30.2: List of Zooplankton in Yeldari Resevoir, Maharashtra

Rotifera:	*Branchionus calyflorus, B.diversicornis, B.flacatus, Euchianis dilata, Filina longiseta, Keratella tropica, Lecane bulla, Trichocera porellus*
Cladocera	*Ceriodaphnia cornuta, Moina micrura, Diaphanosoma sarsi, Diaphanosoma excisum, Alona rectangular, Biapertura karna, Indialona ganapati*
Copepoda:	*Cyclops viridis, Mesocyclops leukarti, M.hyalinus, Diaptomus marshianus, Phylladiaptomus annae, Neodiaptomus lindbergi, Nauplius larva*
Ostracoda:	*Cypris sp, Stenocypris sp, Cyclocypris globosa, Candocypria osborni*

Rotifers were represented by 8 species (Table 30.2) and accounted for about 7.09 per cent (Figure 30.1). The highest density of rotifers was observed in the month of January, Februray and April. Throughout the summer months, rotifer population was maximum. However during rainy and winter season, the rotifer population was comparatively less. Deshmukh (2001) reported 28 species of rotifera from Chhatri lake of Amravati with maxima in summer, which corroborate with the present investigation. Devi (1997) also observed the maxima of rotifera in Ibrahimbagh during pre-monsoon and in Shathamraj reservoir of Hyderabad during monsoon period. Sharma and Diwan (1993) reported the rotifers from Yeshwant sagar reservoir, which showed dominance during summer months. This

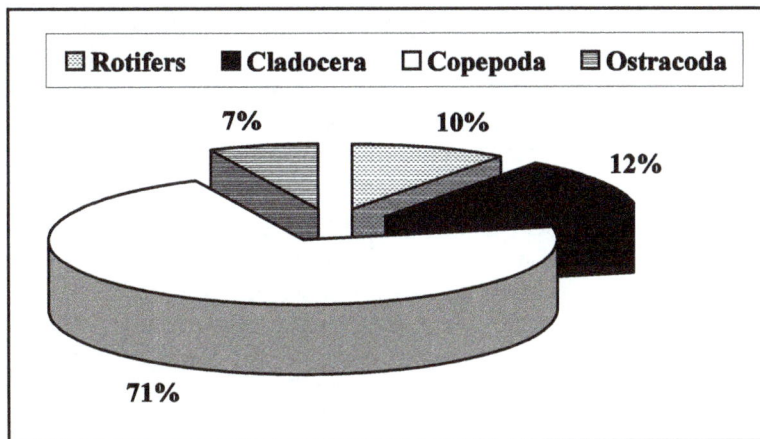

Figure 30.2: The Percentage of Composition of Zooplankton in Yeldari Reservoir, Maharashtra

observation also corroborate with present findings on rotifers in Yeldari reservoir. The peak period of rotifers was observed during month of January, February and April (80 organisms/liter) and it was minimum during month of October (10 organisms/liter).

Cladocera constitute an important group.The greater significance of cladocera in the aquatic food chain as food for both young and adult fish was emphasized much earlier (Pennak, 1978). The stomach content of young fish has 1 to 95 per cent cladocera by volume (Sharma, 1991). Out of 11 families of cladocera, 8 families have been reported from Indian waters, which represents about one fourth of the world cladoceran fauna (Rao and Choubey, 1993). During present investigation cladoceran were represented by 7 species and accounted for about 10.05 per cent. *Ceriodaphnia cornuta* was found for 11 months, followed by *Indiaalona ganapati* for 8 months. *Diaphanosoma excisum* present only during April. Diphanosoma sarsi during January and February, *Alona rectangular* and *Biapertura karna* during

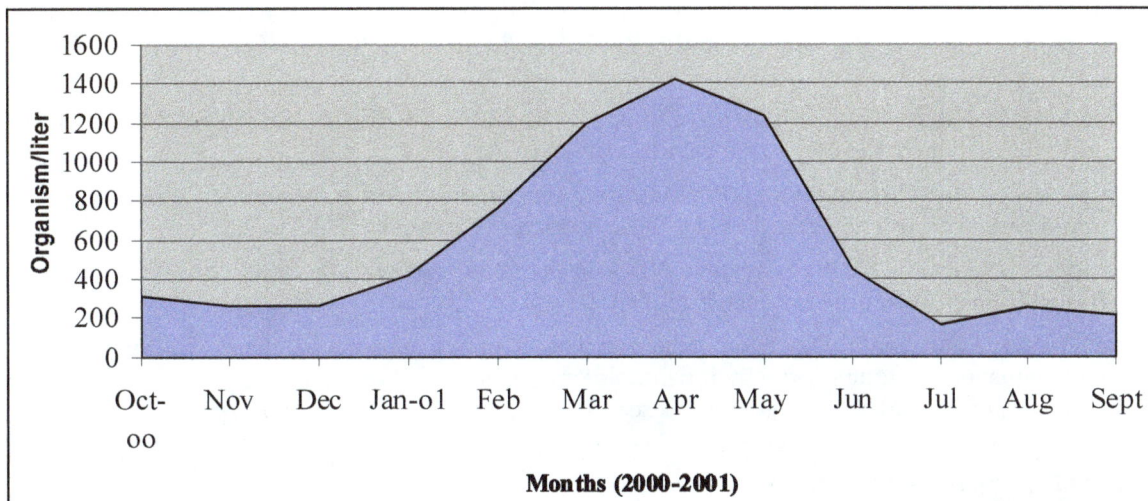

Figure 30.3: Monthly Fuctuations in Zooplankton in Yeldari Reservoir during Year 2000-01

January and Februrary, and *Moina micrura* during May and June. The maximum cladoceran population (210 organisms/liter) was recorded during April and minimum population (10 organisms/liter) during November. Sharma and Diwan (1993) accounted plankton population dynamics of Yeshwant sagar reservoir in which the cladocera showed maximum percentage density in June.Jhingran (1989b) mentioned the most abundant cladoceran population in Februrary, followed by July and October in Ramgarh reservoir of Rajasthan.

Free-living copepods are an essential link in food chain occupying the intermediate trophic level between bacteria, algae and protozoa on one hand and small and large plankton predators on the other. Though they are not as important as cladocerans in the diet of fish, they are well known as important intermediate host for helminth parasities. During present study, copepods were dominated by *Cyclops viridis, Mesocyclops leokarti, Mesocyclops hyalinus* and *Nauplius larvae*. The population was more during summer and least during the rainy season. This is in confirmation with the findings of Deshmukh (2001) and Bini *et al*. (1997). The summer peak may be due to the abundance of diatoms and blue green algae. During present investigation, all the seven species of copepods were present during Februrary, March and April.*Phyllodiaptomus annae* was not seen during the major period of monsoon and winter. *Neodiaptomous lindbergi* was absent in winter season and two months of monsoon period. *Cyclops viridis* and *Nauplia larvae* were found throughout the year, while *Mesocyclops leukarti and M.hyalinus* were recorded for eleven months.

During present study ostracoda were represented by four species with highest density in summer and least in winter. The findings show similarities with those of Deshmukh (2001) on occurrence of highest density in summer.

References

Anon, 1977b. *CIFRI Newsletter*, 1(3).

Arora, H.C., 1963. Studies on Indian rotifers (Part II). *J. Zool. Soc. India*, 15: 112–121.

Battish, S.K., 1992. *Freshwater Zooplankton of India*. Oxford and IBH Publishing Co. Pvt. Ltd., New Delhi.

Bini, L.M., Tundisi, J.G., Yundisi, T.M. and Matheus, C.E., 1997. Spatial variatiation of zooplankton groups in a tropical reservoir (Broa Reservoir: Sao Paulo State–Brazil). *Hydrobiologia*, 357(1–3): 89–98.

Deshmukh, U.S., 2001. Ecological studies of Chhatri lake, Amravati with special reference to planktons and productivity. *Ph.D. Thesis*, Amravati University, Amravati.

Devi, Sarla B., 1997. Present status, potentialities, management and economics of fisheries of two minor reservoirs of Hyderabad. *Ph.D. Thesis*, Osmania University, Hyderabad.

Dhanapathi, M.V.S.S.S., 2000. *Taxonomic Notes on the Rotifers from India (from 1889–2000)*. Indian Association of Aquatic Biologists, Publ. No. 10.

Jhingran, A.G., 1989b. Limnology and production biology of two man-made lakes in Rajasthan (India) with management strategies for their fish yield optimization. *Final Report, IDA Fisheries Management in Rajasthan*.Central Inland Fisheries Research Institute, Barrackpore, India, p. 1–63.

Murugan, M., Murugaval, P. and Kodarkar, M.S., 1998. *Cladocera: The Biology, Classification, Identification and Ecology*. IAAB Publ. No. 5.

Natarajan, A.V., 1975. Fish farming in man made lakes. *Indian Fmg.*, 25(6): 24–33.

Pennak, R.W., 1953. *Freshwater Invertebrates of United States*, 2nd Edn. John Willey and Sons Inc., New York, pp. 763.

Pennak, R.W., 1978. *Freshwater Invertebrates of United States*, 2nd Edn. John Willey and Sons Inc., New York, pp. 803.

Rao, K.S. and Choubey, Usha, 1993. Systematic and ecological studies on central Indian lentic cladocera. In: *Recent Advances in Freshwater Biology*, (Ed.) K.S. Rao. 1: 264–276.

Sehgal, K.L., 1983. Planktonic copepods of freshwater ecosystem. *Environ. Sci. Series.*, Interprint, New Delhi, pp. 1–69.

Sharma, Rekha and Diwan, A.P., 1993. Limnological studies of Yeshwantsagar reservoir. I. Plankton population dynamics. In: *Recent Advances in Freshwater Biology*, (Ed.) K.S. Rao. 1: 199–211.

Sreenivasan, A., 1974. Limnological features of a tropical impoundment, Bhavanisagar reservoir, (Tamil Nadu), India. *Int. Revueges. Hydrobiol.*, 59(3): 327–342.

Sugunan,V.V. and Yadava, Y.S., 1991b. Feasibility for fisheries development of Nongmahir reservoir. *CIFRI*, Barrackpore, pp. 30.

Chapter 31

Determination of Glucose Content in the Body Muscles of *Wallago attu* and *Mystus seenghala* from Yeldari Reservoir in Maharashtra

☆ *S.R. Shelke, R.S. Babare, M.P. Deshmukh and S.P. Chavan*

Indian reservoirs preserves a rich variety of fish species on the basis of Studies conducted so far, large reservoirs on an average harbour 60 species of fishes, out of which 40 species contribute to the commercial fisheries. Indian major carps occupy a prominent place among the commercially important fishes. More recently, number of exotic species have contributed substantially to commercial fishes.

Wallago attu and *Mystus seenghala* are commonly found silluroid fish species from the freshwater in parbhani district of Maharashtra. Yeldari is the largest reservoir in parbhani district.Selected two species of fishes found throughout year in the catch from yeldari reservoir, indicating that these two fish species is one of the bulk in food fish resource for the people in this region. The glucose is an important biochemical component in fish body along with the proteins and lipids.

Fish is an one of the most important sources of protein. It is consumed fresh and in dry condition. As it is well known that protein is very essential for the growth and development of human body, the present study is undertaken to analyse the protein and carbohydrate content of dry fish.

Fishes are available in the market are used for the study. The amount of protein and carbohydrate present in dry powder are analysed by various methods. India has rich water resource,especially marine water but the freshwater fish is comparatively more than sea water food fishes. In general,the biochemical composition of the whole body indicates the fish quality. Therefore, proximate biochemical composition of a species helps to assess its nutritional and edible value in terms of energy units compared to other species. Varition of biochemical composition of fish flesh may also occur within same species depending upon fishing ground, fishing season, age and sex of the individual and reproductive status. The spawning cycle and food supply are the main factors responsible for this variation (Love *et al.*, 1980). Therefore, the present study was undertaken to elucidate the dyanamics of biochemical composition of various tissues in *Wallago attu* and *Mystus seenghala*.

Material and Methods

Samples of both the fish species *i.e.* is *Wallago attu* and *Mystus seenghala* were collected from Yeldari reservoir (Parbhani district) during the period of July 2009–November 2009. The specimens were properly cleaned in laboratory the total length, total weight and sex were determined. Gonads and general viscera were removed and preserved in 5 per cent formaldehyde for further study.Body muscle samples (free from skin and scales) from four to six specimens were pooled together and used for the analysis of biochemical components every month.

The dried samples were finely powdered using mortar and pestle and stored in desiccator for further analysis.

Total carbohydrate was estimated by anthrone reagent method.

Weight 100mg of the sample and take it in test tube. Add 5ml of 2.5 HCl in test tube, keep the test tube in boiling water bath for 3 hours.

Remove the tube from water bath and cool it to room temperature.

Add solid sodium carbonate till the effervescence stop to come out.

Remove the content (all) from the test tube in a measuring cylinder and dilute the content to 100ml.

Take few ml (5-10ml) of the content diluted as above and centrifuge it.

Collect the supernant (upper layer from centrifuge tube) 0.5ml as a sample for further analysis.

Add 0.5ml distilled water +4ml Anthrone reagent.

Heat the tube (glass)for 8 minutes in boiling water bath.

Cool the tubes to room temperature.

Take the above cooled sample for calorimeter reading.

Absorbency was measured at 490nm. Draw a standard graph by plotting concentration of standard on the X-axis Vs absorbance on the Y-axis.

From it calculate the amount of carbohydrate present in the sample.

Calculation

$$\text{Amount of CHO present in 100mg of the sample} = \frac{\text{Mg of glucose}}{\text{Vol. of test sample}} \times 100$$

Results

Both seasonal and size wise variation in biochemical components were seen. Carbohydrate occurs in a very minute quantity in the fish tissues. In the present study,it varies between 0.31 per cent and

0.60 per cent with an average value of 0.50 per cent +-0.073 per cent. Monthly variation shows that carbohydrate fluctuated. In September it shows slightly increase that is 0.60 per cent and with slight decrease in july. Thereafter, there was a decline in carbohydrate values until December (0.42 per cent). The lowest value was recorded in july 2009 and the highest in September 2009. Percentage of carbohydrate with reference to size group was stable and did not show specific pattern.

Discussion

In fishes, the biochemical content is related to maturation of gonads and the food supply (Jacquot, 1961; Medford and Mackay, 1978). The biochemical constituents are also influenced by metabolism mobility of the fish and geographical area (Stansby, 1962). Marked changes were observed in the biochemical composition of the muscle of *Wallago attu* and *Mystus seenghala* and different tissues are also further studied now,which may be the result of the processes mentioned above. Carbohydrate formed a minor percentage or quantity of total composition in the tissues.

Conclusion

Carbohydrate are utilized for energy, thus the study of proximate composition of *Wallago attu* and *Mystus seenghala* revealed that it is rich in protein. Variation in biochemical composition in present study seems to be governed by spawning cycle and feeding activity.

References

Lowry, O.H., Rosebrough, N.J., Farr, A.L. and Randall, R.J., 1951. *J. Biol. Chem.*, 193: 265

Wilson, K. and Walker, J., 2000. *Practical Biochemistry: Principles and Techniques*. Cambridge University Press.

Chapter 32

Bioenergetic Study on the Post-Larvae of *Macrobrachium rosenbergii* Fed with Papain Supplemented Diet

☆ *D.W. Patil and H. Singh*

Introduction

Aquaculture is only mean to augment the aquatic food production in order to meet the growing demand for fin and shellfish all over the world. Among the cultured animals *M. rosenbergii* commonly known as scampi is commercially cultured in ponds, lakes, reservoirs, low laying areas and paddy fields in India. Thailand is a leading country in the world for *Macrobrachium* culture. Culture of prawn to marketable size within short period is achieved by using growth promoters and feed additives in the supplementary diets. Feed is the major constituent of variable cost in semi-intensive and intensive culture system comprising up to 60 per cent of the total production cost (Akiyama *et al.*, 1992; Sarac *et al.*, 1993). Therefore, it is necessary to develop nutritionally adequate and cost effective feed for prawn farming.

Dabrowska *et al.* (1979) stated that digestion efficiency of the cultured species can be increased by supplementing the enzymes to the feed However; there is very less scientific study on the use of enzymes in the diets of post-larvae of *M. rosenbergii*. Hence, the present study was designed to study the effect of papain supplemented diet on feed utilization of post-larvae of *M. rosenbergii*

Material and Methods

Test Animal

Post-larvae of *M. rosenbergii* were obtained from the freshwater prawn hatchery of Marine Biological Research Station, Ratnagiri and maintained in 500 L plastic pool. They were acclimatized for one week to the laboratory conditions and were fed three times per day with control diet (T_0) at the rate of 10 per cent body weight. Faeces and remaining feed were removed from the culture tank by siphoning out 25 per cent of water daily, which was replaced back with new freshwater. Aeration was provided throughout acclimatization period to avoid stress.

Diet Preparation

Control Diet

Control diet used for experimental purpose was prepared by using fishmeal and groundnut oil cake as dietary protein source while rice bran and wheat flour as basal sources. Diet contained 35 per cent protein on the basis of nutritional information available for *M. rosenbergii* (Balazs and Ross, 1976; Clarke *et al.*, 1990 and Shinde, 2001).

The required quantity of finely powdered and sieved ingredients were precisely weighed and mixed thoroughly in domestic mixer for one minute. Afterwards 350 ml water was added to it and further it was mixed in domestic mixer to form slurry. The slurry was steamed for 15 minutes and cooled to room temperature for 20 minutes. The cooled slurry was spread on polythene sheet in 2 to 3 mm thick layer with the help of soft brush (New Wilson, 100 mm size). Then it was dried for 3 hours to from flacks. After sun drying, the flaks were broken in to small pieces 2 to 3 mm size and stored in plastic container.

Experimental Diet

The enzyme papain (LOBA CHEMIE PVT LTD, Mumbai) was supplemented at the rate of 0.1, 0.2, 0.3 and 0.4 per cent in control diet and served as experimental diet *viz.* T_1, T_2, T_3 and T_4. The enzyme was dissolved in 50 ml water and mixed thoroughly in cooled slurry of diet in domestic mixer. The method of preparation of the experimental diets was similar to that of control diet.

Ingredients and Proximate Composition

Ingredients and proximate composition of control and test diets are give in Table 32.1. The diets were analyzed for moisture content (AOAC, 1990) and nitrogen content was determined by Micro-Kjeldhal method (AOAC, 1990). Protein was calculated by multiplying nitrogen content by a constant 6.25. Crude fat content was analyzed by Soxhlet extraction with petroleum ether (AOAC, 1990). Ash content was estimated after incarnation of sample at 550 ° C in muffle furnace for 5 hours (AOAC, 1990). The carbohydrate was computed by reminder method (Woods and Aurand, 1977). Gross energy in each of the diets was calculated using conversion factors carbohydrate, 4.1; fat, 9.5; and protein, 5.65 (El-Sayed, 1994).

Experimental Procedure

Post-larvae of *M. rosenbergii* (2.36 ± 0.07, 2.45 ± 0.02, 2.48 ± 0.03, 2.44 ± 0.01 and 2.36 ± 0.01 cm length and 0.119 ± 0.002, 0.123 ± 0.003, 0.122 ± 0.003, 0.122 ± 0.002 and 0.121 ± 0.002 g weight) were randomly stocked @ 30 PL/tub in circular plastic tubs with 15 L of water in it for T_0, T_1, T_2, T_3 and T_4 respectively. Experiment was conducted in duplicate. Animals were fed once a day (10.00 hrs.) at the rate of 10 per cent of their body weight. After 2 hrs. the feed was removed. Faeces were collected daily by filtering water through bolting silk. Unused feed and faeces were dried at 60° C in oven until constant weight was attained. Dried faeces and unconsumed feed were kept in airtight container at room temperature for subsequent analysis.

On termination of experiment the following observations were recorded and energy parameters were calculated as per Shinde, (2001).

1. Initial weight (W_1)
2. Final weight (W_2)
3. Mean weight (W) = $(W_1 + W_2)/2$
4. Production (P) = $W_2 - W_1$
5. Amount of feed given (a)
6. Remainder feed (b)
7. Consumption (C) = a–b
8. Fecal out put (F)
9. Assimilation (A) = C–F
10. Metabolism (R) = A–P
11. Assimilation efficiency = $A/C \times 100$
12. Gross growth efficiency (K_1) = $P/C \times 100$
13. Net growth efficiency (K_2) = $P/A \times 100$
14. Consumption/unit weight/number of experimental days = $(C/W)/$days
15. Conversion ratio = C/P
16. Conversion rate = $(P/C)/$days
17. Relative growth rate = $(P/W)/$days
18. Protein efficiency ratio = $\dfrac{\text{Production}}{\text{Dry protein consumed}}$
19. Protein digestibility coefficient = $\dfrac{\text{Feed protein} - \text{Fecal protein}}{\text{Feed protein}} \times 100$
20. Lipid digestibility coefficient = $\dfrac{\text{Feed lipid} - \text{Fecal lipid}}{\text{Feed lipid}} \times 100$

Statistical Analysis

Data obtained in feeding experiments were analyzed by one-way ANOVA. Significant difference was indicated at 0.05 level, the Newman Keul's multipal comparison test was used. (Snedecor and Cochran, 1967; Zar, 1974).

Reults and Discussion

In the present study, better FCR (1.63) and maximum PER (1.75) was observed in post-larvae fed with diet T_1 and it was found significantly different from diet T_0 but not significantly different from the other experimental diets.

Tagare (1992) reported better FCR for common carp fry fed with diet containing papain and crude papain in form of papaya leaf. Similarly better FCR also observed in present study in post-larvae of *M. rosenbergii* fed with papain supplemented diet. However, it is very much difficult to compare the results of present study with other study due to lack of information.

Table 32.1: Ingredients (per cent) and Proximate Composition (per cent dry weight basis) of the Diets Fed to Post-larvae of *M. rosenbergii* for 10 Days

Ingredients and Proximate Composition	Test Diet				
	T_0	T_1	T_2	T_3	T_4
Ingredients (%)					
Fish meal	29.21	29.21	29.21	29.21	29.21
Groundnut oil cake	29.21	29.21	29.21	29.21	29.21
Wheat flour	20.79	20.79	20.79	20.79	20.79
Rice bran	20.79	20.79	20.79	20.79	20.79
Papain	—	0.10	0.20	0.30	0.40
Proximate composition (%)					
Moisture*	10.89	9.93	10.57	10.23	9.65
Crude protein*	34.00	35.00	35.00	34.00	36.00
Crude lipid*	5.20	5.59	5.00	5.13	4.87
Ash*	10.50	9.44	10.10	9.50	10.25
Carbohydrate*	39.41	40.04	39.00	41.14	39.23
Gross energy (kcal/100 g)	407.02	419.02	409.05	413.62	414.43
Protein/energy ratio (mg protein/kcal)	83.53	83.53	85.56	82.20	86.87

*: Mean of two estimations.

Table 32.2: Feed Utilization and Nutrient Digestibility of Post-larvae of *M. rosenbergii* Fed with Diets during Experimental Period

Bioenergetics Parameters	Test Diet				
	T_0	T_1	T_2	T_3	T_4
FCR	20.60±0.03[a]	1.63±0.07[b]	1.77±0.09[b]	1.88±0.13[b]	1.89±0.04[b]
PER	1.13±0.01[a]	1.75±0.08[b]	1.62±0.08[b]	1.58±0.11[b]	1.47±0.03[b]
Assimilation	0.0172±0.0013[a]	0.0251±0.0019[b]	0.0248±0.0004[b]	0.0242±0.0019[b]	0.0239±0.0013[b]
Metabolism	0.0097±0.0008[a]	0.0086±0.0014[a]	0.0098±0.0006[a]	0.0102±0.0002[a]	0.0101±0.0008[a]
Assimilation efficiency (%)	88.22±0.12[a]	92.95±0.15[b]	93.59±0.26[b]	92.89±0.03[b]	92.57±0.35[b]
Gross growth efficiency (%)	38.49±0.40[a]	61.31±2.69[b]	56.55±2.71[b]	53.57±3.57[b]	53.01±1.16[b]
Net growth efficiency (%)	43.63±0.39[a]	65.96±3.00[b]	60.44±3.05[b]	57.68±3.86[b]	57.26±1.04[b]
Relative growth rate	0.0061±0.0003[a]	0.0126±0.0001[b]	0.0116±0.0005[b]	0.0108±0.0013[b]	0.0106±0.0002[b]
Consumption/unit weight/day	0.0159±0.0009[a]	0.0206±0.001[a]	0.0205±0.0002[a]	0.0201±0.0011[a]	0.0200±0.0008[a]
Conversion rate	0.039±0.00[a]	0.061±0.003[b]	0.057±0.003[b]	0.054±0.004[b]	0.053±0.001[b]
Protein digestibility coefficient	48.54±0.72[a]	65.49±0.36[b]	61.91±1.22[b]	64.29±0.21[b]	61.58±0.74[b]
Lipid digestibility coefficient	16.18±1.27[a]	14.63±1.63[a]	10.83±3.46[a]	12.27±0.27[a]	9.58±1.66[a]

*: Values are mean±S. E. of two replicates.

*n = 30 post-larvae per tub.

*Values within same row with same superscript are not significantly different at 0.05 probability level.

Maximum assimilation (0.0251), gross growth efficiency (61.31), net growth efficiency (65.96), relative growth rate (0.0126), conversion rate (0.061) and protein digestibility (65.49) observed for post-larvae fed with T_1 diet. While maximum assimilation efficiency (93.59) was observed for post-larvae fed with T_2 diet. However, diet T_1 was significantly different from T_0 but not significantly different from T_2, T_3 and T_4 for assimilation, assimilation efficiency, gross growth efficiency, net growth efficiency, relative growth rate, conversion rate and protein digestibility.

Maximum metabolism (0.0102) was observed for post-larvae fed with diet T_3. However, there was no significant difference among the diets for metabolism.

Present study showed the maximum consumption/unit weight/day (0.0206) and maximum lipid digestibility (16.18) for post-larvae fed with T_1 and T_0 diets respectively. One way ANOVA showed no significant difference among the control and experimental diets for consumption/unit weight/day and lipid digestibility.

This bioenergetic study concluded that the diet supplemented with 0.1 per cent of papain is found to be most suitable for better growth and maximum nutrient utilization from the feed for rearing of post-larvae of *M. rsenbergii*.

Acknowledgement

The authors wish to thank Dr. P. C. Raje, Associate Dean, College of Fisheries, Ratnagiri, Dr. S. G. Belsare, Senior Scientific Officer, Marine Biological Research Station, Ratnagiri for giving permission and providing the facilities to conduct the research work. The authors gratefully acknowledge the help given by Dr. A. M. Ranade, Professor and Head, Department of Aquaculture, College of Fisheries, Ratnagiri. The authors are thankful to Dr. S. T. Indulkar, Associate Professor, Department of Aquaculture, College of Fisheries, Ratnagiri for his guidance and encouragement during the course of present investigation.

References

Akiyama, D.M., Domny, W.G. and Lawrence, A.L., 1992. Penaeid shrimp nutrition. In: *Marine Shrimp Culture: Principles and Practices*, (Ed.) A.W. Fast and L.J. Lester. Elsevier, Amsterdam, pp. 535–568.

AOAC, 1990. *Official Methods of Analysis of the Association of Official Chemists*, 15th edn. Association of Official Analytical Chemists, Inc., Arlington. VA. pp. 1094.

Balazs, G.H. and Ross, E., 1976. Effect of protein source and level on growth and performance of the captive freshwater prawn, *Macrobrachium rosenbergii*. *Aquaculture*, 7: 200–213.

Clarke, A., Brown, G.E. and Holmes, L.S., 1990. The biochemical composition of eggs from *Macrobrachium rosenbergii* in relation to embryonic development. *Comparative Biochemistry Physiology*, 963: 503–511.

Dabrowska, H., Grundniewski, C. and Dabrowski, K., 1979. Artificial diets for common carps: Effect of the addition enzyme extracts. *Progressive Fish Culturist*, 41: 196–200.

El-Sayed, A.M., 1994. Evaluation of soybean meal, spirulina meal and chicken offal as a protein sources for silver sea bream (*Rhabdosargus sarba*) fingerlings. *Aquaculture*, 127: 169–176.

Sarac, H.Z., Gravel, M., Saunders, J. and Tabrett, S., 1993. Evaluation of Australian protein sources for diets of the black tiger prawn (*Penaeus monodon*) by proximate analysis and essential amino acid index. In: *Abstracts of the International Conference World Aquaculture'93*, May 26–28, (Eds.) M. Carrillo,

L. Dahle, J. Morales, P. Sorgeloos, N. Svennevig and J. Wyban. Torremolinos, Spain. European Aquaculture Society, Special Publication No. 19, Oostende, Belgium.

Shinde, M.M., 2001. Growth and survival of post-larvae of *Macrobrachium rosenbergii* (De man) fed on artificial diet supplemented with some hormones. *M.F.Sc. Thesis,* Dr. Balasaheb Sawant Konkan Krishi Vidyapeeth, Dapoli, 40.

Snedecor, G.W. and Cochran, W.G., 1967. *Statistical Methods,* 6th Edn. Oxford and IBH Publishing Co., New Delhi, pp. 593.

Tagare, M.N., 1992. Role of papain in growth, survival and improving feed efficiency of *Cyprinus carpio. M.F.Sc. Thesis,* Central Institute of Fisheries Education, Mumbai, p. 50.

Woods, A.E. and Aurand, L.W., 1977. *Laboratory Manual in Food Chemistry.* The AVI Publishing Company, Inc., pp. 1–72.

Zar, J.H., 1974. *Biostatistical Analysis.* Prentice Hall, Inc., Englewood Cliffs, N.J., U. S. A., pp. 620.

Chapter 33

Assessment of Physico-chemcial Parameters and Fish Diversity of Mehakari Reservoir in Beed District, Maharashtra

☆ *B.S. Khaire, A.D. Mohekar and R.J. Chavan*

Introduction

Water is essence of life and is also used for various activities such as drinking, irrigation, fish production, industrial purposes, power generation, transportation, recreation and many other purposes. Though the water covers about ¾ of our earth planet, it is said that about 70 per cent of the world's population survive without clean and pure water. In recent past years with unprecedented population growth and intensive agriculture surface and groundwaters are being exploited on increasing scales all over the world. Whatever freshwater collected in lakes, reservoirs, and other freshwater bodies are in endanger due to numerous anthropogenic activities. According to different surveys 70 to 80 per cent of Indian freshwater sources are polluted there fore now a days water quality and safety have become major issue in public health.

Reservoirs fishery in India is important from social-economic point of view as it is excellent food for man. It has the potential of providing the employment to many peoples. The considerable studies on water quality and fish diversity of some freshwater bodies have been carried out during the recent years by many workers. (Trivedy and Goel (1988), Talwar and Jhingran (1991), Salaskar (1997), Kodarkar (1998), Agarwal (1999), Wagh (1999),Pejawar and Somani (2002), Sakhare (2003),Megha Rai (2004), Chavan and Mohekar (2005), Yeole and Patil (2005) and Kadam and Gaikwad (2006).

Materials and Methods

Water samples were collected for one year from October 2005 to September 2006 in the first week of each month between 7.30 to 9.00 am for analysis of physico-chemical parameters. Physicochemical parameters like water temperature (with thermometer), pH (with Hanna's pH probe), transparency (with Secchi disc), were recorded at the sampling site. DO, Free CO_2, total hardness, magnesium hardness, calcium hardness, alkalinity, chlorides, phosphates, nitrates etc. were analysed in the laboratory as per standard methods for examination of water (APHA 1989, Trivedy and Goel 1984 and Kodarkar *et al.*, 1998).

Fishes were collected with the help of local fishermen and fish bazaar and identified by using standard methods of taxonomy suggested Day (1878), Jayram (1981), Menon (1964) and Talwar and Jhingran (1991).

Results and Discussion

Table 33.1 shows monthly mean values of different physico-chemical parameters during October 2005 to September 2006.

Table 33.1: Monthly Values of Physico-chemical Parameters of Water from Mehakari Project Reservoir during October 2005–September 2006

Parameters	Oct.	Nov.	Dec.	Jan.	Feb.	Mar.	Apr.	May.	Jun.	Jul.	Aug.	Sept.
Atm. Temp	19.2	17.5	16.6	19.2	22.5	25.8	28	30	27.2	27	25.6	25.4
Water Temp	20.6	19.6	19.2	21.3	21.2	23	26	27	26.2	25.2	23.4	22.6
pH	8.3	8.2	8.1	8.3	8.5	8.6	8.7	8.6	8.4	8.3	8.3	8.2
Transparency	104	118.5	131	133	136	140	130	122	106	66	72	96
Conductivity	260	268	268	282	292	298	305	321	290	280	256	252
TDS	168	140	122	110	172	168	195	196	148	186	165	178
DO	6.41	6.8	6	7	6.4	5.8	4.6	3.8	4	6.4	6	5.8
Free CO_2	3.74	5.7	6.8	6.1	3.74	4.8	3.4	3.52	3.3	4.8	4.4	4.2
T. Alkalinity	135	120	115	145	185	180	155	145	145	120	135	125
Bicarbonates	135	120	115	145	185	180	155	145	145	120	135	125
T. Hardness	78	82	94	102	108	122	130	126	118	98	80	76
Calcium	18.43	23.24	33.66	33.66	38.47	39.27	38.47	39.27	34.46	33.66	23.24	24.06
Magnesium	7.79	5.8	7.3	7.3	10.23	10.72	9.74	9.25	9.25	7.79	7.3	7.3
Chlorides	17.04	19.88	18.46	24.14	28.4	29.82	26.98	25.56	24.14	26.98	25.56	18.46
Phosphate	0.46	0.62	0.34	0.09	0.12	0.09	0.24	0.42	0.32	0.34	0.22	0.44
Nitrate	0.32	0.52	0.14	0.12	0.22	0.34	0.16	0.08	0.22	0.3	0.34	0.34

During study period atmospheric temperature ranged between 16.6 °C to 30 °C and water temperature ranged between 19.2 °C to 27 °C. Water temperature was observed more than atmospheric temperature during winter season. It might be due to the early morning sampling. The difference between atmospheric temperature and water temperature was 1 °C to 3 °C throughout study period. pH of water ranged between 8.1 to 8.7. It was found to be minimum in the winter and maximum in summer months. Temperature brings out changes in pH. Singhai (1986) has obtained a direct relationship between water temperature and pH. The present studies showed pH range favorable for aquatic life. Transparency ranged between 66 cm to 140cm during study period. Over all transparency was observed maximum during late winter and early summer months while it was observed minimum during the monsoon months. Transparency decrease in monsoon is due to sewage discharge with rain water from the surrounding area.Prakasam and Joseph (2000).The specific conductivity of present water bodies varied between 252 µ Mhos/cm to 321 µ Mhos/cm. Its maximum values were observed during summer season and least in winter. The conductivity might be higher during summer due to increased chlorides and dissolved solids due to evaporation of water resulting in increased concentration of salts. Rise in conductivity is due to increased TDS. Megha Rai and Shrivastava (2006), Nalina and Puttaiah (2006). The range of TDS reported was 110 to 196 mg/lit. TDS values were observed maximum during the summer season followed by monsoon and winter. These were observed maximum (196 mg/lit) in the month of May and minimum (110 mg/lit) during the month of Dec. High values of TDS in summer might be due to evaporation of water resulting in increased concentration of salts and in monsoon it was due to addition of sewage, domestic wastes with the influx of monsoon run off. Similar results were recorded by Deshmukh and Ambore (2006).

Dissolved oxygen in water is one of the most important abiotic parameters to indicate water quality. The presence of O_2 in water is mainly due to diffusion from air and photosynthetic activity of algae and submerged plants. In the present investigation the DO ranged between 3.8 to 7.0 mg/lit. throughout study period. Peak values of dissolved oxygen were observed in winter and least in summer. High values of DO in winter would be due to low temperature of water (Mathew Varghese, 1992; Dutta *et al.*, 2001; Ahirrao *et al.*, 2001 and Chavan and Mohekar, 2005). The CO_2 in water is mainly due to diffusion from air, from inflow groundwaters, due to decomposition of organic matters and respiration of aquatic organisms. Welch (1952). It was ranged between 3.3 to 6.8 mg/lit. throughout study period. The CO_2 values were recorded maximum in winter and minimum in summer (Salaskar, 1997). Total alkalinity of the water was due to bicarbonates. It was ranged between 115 mg/lit. to 185 mg/lit. The alkalinity values of present water bodies generally remains higher than 100 mg/lit. Thus these water bodies seems to have moderately polluted due to domestic sewage and agricultural runoff. Total alkalinity was observed maximum during summer and minimum in winter. Calcium and magnesium are the principal cations imparting hardness. The anions responsible for hardness are mainly carbonates, bicarbonates, sulphates, chlorides, nitrates silicates etc. Total hardness was ranged 76 to 130 mg/lit. The total hardness was observed maximum in summer followed by monsoon and least in winter throughout study period. The maximum values of hardness in summer may be due to presence of higher concentration of carbonates and bicarbonates. During the present investigation calcium hardness ranged between 18.43 to 39.27 mg/lit. Total hardness and calcium hardness shows very positive correlation, indicates that total hardness is mainly due to presence of calcium salts. The concentration of magnesium is generally lower than the concentration of calcium in natural waters. The main sources of magnesium are rocks, sewage and industrial wastes in natural waters. It was recorded between 5.8 to 10.72 mg/lit. Similar results were observed by Mathew Verghese (1992), Chandrashekhar (1996) Lendhe and Yeragi (2004) etc. Chloride as chloride ions (Cl-) are major anions present in natural water. Leaching of natural rocks and domestic sewage are the main sources of chloride in water. In the present investigations the chlorides ranged between 17.04 to 29.82 mg/lit. Maximum values of chlorides recorded in summer may be due to increased organic decomposition of animal origin or it may be due to loss of water by evaporation (Wagh, 1999 and Yeole and Patil, 2005). In natural water phosphates are present in small quantities. Excess of phosphate mixed in to natural water through untreated domestic sewage and agricultural runoff. These are important nutrients

(Trivedy, 1998). It was ranged between 0.09 to 0.62 mg/lit. The nitrates present in the water bodies were within permissible limit. Nitrates were ranged between 0.08 to 0.52 mg/lit.

During the present study fish fauna was studied for their diversity. The result of present investigation confirmed the occurrence of 17 fish species belonging to five orders. Order Cypriniformes was observed dominant with nine species (Table 33.2).

Table 33.2: Fish Diversity of Mehakari Reservoir in Beed District, Maharashtra

Order: Cypriniformes		
Family: Cyprinidae	1. *Labeo rohita* (Ham)	2. *L. calbasu* (Ham)
	3. *Catla catla* (Ham)	4. *Cirrhinus mrigala* (Ham)
	5. *Cyprinus carpio* (Linn)	6. *Puntius sophore* (Ham)
	7. *Puntius sarana sarana* (Ham)	8. *Puntius ticto ticto* (Ham)
	9. *Puntius kolus* (Ham)	
Order: Siluriformes		
Family: Siluridae:	10. *Wallago attu* (Sch)	11. *Ompak bimaculatus* (Bloch)
	12. *Clarius batrachus* (Linn)	
Family: Bagridae	13. *Mystus seenghala* (Skyes)	
Order: Channiformes		
Family: Channidae	14. *Channa marulius* (Ham)	15. *Channa punctatus* (Bloch)
Order: Osteoglossiformes		
Family: Notopteridae	16. *Notopterous notopterous* (Ham)	
Order: Mugiliformes		
Family: Mugilidae	17. *Rhinomugil corsula.* (Ham-Sri.)	

The fishes which get favorable and optimum pH, temperature, DO, and availability of food in water results in better growth and attains proper weight. Natural water supports the fish life. The temperature of water bodies under investigation ranges between 16.6 °C to 30.0 °C indicates favorable conditions for growth of fishes. The aquarium species like *P. ticto*, *P. sophore* can grow better at 24 °C to 29 °C range of temperature. Common carp, *Cyprinus carpio* grows best at temperature range between 20 °C to 25 °C. Pailwan *et al*. (2006) Many species of fishes lives better comfortable in the range of pH between 7 to 8.6. The result of present investigation showed that the pH range was observed between 8.1 to 8.7 which was favorable for growth of fish. The range of pH between 6 to 8 is preferable to aquarium species like *P. ticto, P. sarana and P. sophore* (Pandey and Shukla 2007).

The fishes require adequate concentration of DO for survival and growth. The need of DO is vary according to fish species. For cyprinid its range is favorable between 6 to 7 mg/lit. The present study showed the wide range of DO from 3.8 mg/lit. to 7.0 mg/lit.(Banerjee *et al*., 1990). Study reveals that the DO was always with a favorable range with few exceptions. Presence of CO_2 is considered as an indicator of biogenic condition of water, especially its suitability to the fish. During present study no significant effect of CO_2 was observed because all the time it was ranged below the toxic limit.

Dissolved nutrients play an important role in fish production. The food species like *Catla catla, Labeo rohita, Wallogo attu* etc. were observed abundantly due to favorable conditions.(Banerjee *et al*., 1990).

The species diversity reported in the present study shows marked similarity with earlier studies on reservoirs from same geo-climatic region of the Maharashtra State. Sakhare (2001) observed 23 fish species belong to seven orders from Jawalgaon reservoir in Solapur district of Maharashtra. Kadam and Gaikwad (2006) observed 23 fish species from Masooli reservoir in district Parbhani (M.S.). Sakhare and Joshi (2003) reported the ichthyofauna of Bori reservoir in Maharashtra. Overall less diversity of fish fauna was recorded in water body under investigation might be due to continuos draught for two to three years before study period.

The study reveals that with few exceptions all the parameters were within the permissible limits and indicates that the water body is more productive.

References

Agarwal, C.S., 1999. *Limnology.* Prabha Prakashan, Kanpur.

Ahirrao, S.D. and Chaudhari, P., 2001. Assessment of physico-chemical parameters of water under influence of some synthetic organic insecticides at different pH level. *J. Aqua. Biol.,* 16(1): 57–60.

APHA, 1989. *Standard Methods for the Examination of Water and Wastewater.* American Public Health Association and Water Pollution Control Federation, Washington D.C.

Banerjee, R.K. and Babulal, 1990. Role of soil and water in fish farming with special reference to primary production. In: *Technologies for Inland Fisheries Development,* (Eds.) V.V. Saguna and Bhaumik Utpal. Central Inland Capture Fishing Research Institute, Barrackpore, West Bengal, pp. 123–129.

Chandrashekhar, S.V.A., 1996. Ecological studies on Sarrornagar Lake, Hyderabad. *Ph.D. Thesis* Submitted to Osmania University, Hyderabad. (Dr. Kodarkar M.S.).

Chavan, R.J., Mohekar, A.D., Savant, R.J. and Tat, M.B., 2005. Seasonal variations of abiotic factors of Manjra project water reservoir in Dist. Beed, Maharashtra, India. *Poll. Res.,* 24(3): 705–708.

Day, F.S., 1878. *The Fishes of India.* William Dowson and Sons Ltd., London.

Deshmukh, J.U. and Ambore, N.E., 2006. Seasonal variations in physical aspects of pollution in Godavari river at Nanded. (M.S., India). *J. Aqua. Biol.,* 21(2): 93–96.

Dutta, S.P.S., Kaur, H. and Bali, J.P.S., 2001. Hydrobiological studies on River Basantar, Samba, Jammu (J&K). *J. Aqua. Biol.,* 16(1): 41–44.

Hynes, H.B.N., 1970. *The Ecology of Running Waters.* Uni. Toranto Press, Toranto, pp. 555.

Jayram, K.C., 1981. *The Freshwater Fishes of India, Pakisthan, Bangladesh, Burma and Sri Lanka: A Handbook.* Zoological Survey of India, Calcutta, p. 1–475.

Kodarkar, M.S., 1998. *Methodology for Water Analysis: Physico-chemical, Biological and Microbial.* IAAB Publication, Hyderabad.

Lendhe, R.S. and Yeragi, S.G., 2004. Seasonal variations in primary productivity of Phirange Kharbhav lake, Bhiwandi, district Thane, M.S. *J. Aqua. Biol.,* 19(2): 49–51.

Mathew, Varghese and Naik, L.P., 1992. Hydrobiological studies of a domestically polluted tropical pond. II. Biological characteristics. *Poll. Research,* 11(2): 101–105.

Megha, Rai and Shrivastva, R.M., 2004. Effect of fertilizer industry on surface and groundwater quality, Raghogarh, Madhya Pradesh. *J. Aqua. Biol.,* 21(2): 101–104.

Menon, A.G.K., 1964. Monograph of the Cyprinid fishes of the Genus *Gorra Hammilton*. *Memoirs India Mus.*, 14: 173–260.

Nalina, E. and Puttaiah, E.T., 2006. Studies on the groundwater quality of Kadur and its surrounding areas, Karnataka: A statistical analysis. *J. Aqua. Biol.*, 21(2): 105–110.

Pailwan, I.F. and Muley, D.V., 2006. Limnology and fishery status of freshwater perennial tank at Kaneriwadi near Kolhapur, (M.S.) India. *J. Aqua. Biol.*, 2(1 and 2): 72–76.

Pandey, Kamleshwar and Shukla, J.P., 2007. *Fish and Fisheries*, Revised 2nd Edn. Rastogi Publication, Meerut.

Patel, N.G. and Bhadane, V.V., 2004. Comparative account of planktonic diversity of Tadya Nallah pond at Amalner district Jalgaon (Maharashtra), India. *J. Aqua. Biol.*, 19(1): 53–56.

Pejawar, Madhuri, Somani, Vaishali and Borkar, M., 2002. Physico-chemical studies of lake Ambegosale, Thane, India. *J. Ecobio.*, 14 (4): 277–281.

Prakasm, V.R. and Joseph, M.L., 2000. Water quality of Sasthamcotta lake, Kerala (India) in relation to primary productivity and pollution from anthropogenic sources. *J. Envir. Biol.*, 21(4): 305–307.

Raghunathan, M.G.S. and Mahalingam, Vanithadevi, 2000. A study of physico-chemical characteristics of Otteri lake and Polar river water in Vellore Town (T.N.). *J. Aqua. Biol.*, 15(1 and 2): 56–58.

Sakhare, V.B. and Joshi, P.K., 2002. Ecology and ichthyo fauna of Bori reservoir in Maharashtra. *Fishing Chimes*, 22(4): 40–41.

Sakhare, V.B. and Joshi, P.K., 2003. Physico-chemical limnology of Papnas: A minor wetland inTuljapur town, Maharashtra. *J. Aqua. Biol.*, 18(2): 93–95.

Salaskar, Promod and Yeragi, S.G., 1997.Studies on water quality characteristics of Shenala lake, Kalyan, M.S. India. *J. Aqua. Biol.*, 12(1 and 2): 28–31.

Singhai, S.G., 1986. Hydrobiological and ecological studies of newly made Tawa reservoir at Ranipur. *Ph. D. Thesis*, H.S. Sagar Univ., Sagar .

Talwar, P.K. and Jhingran, A.G., 1991. *Inland Fishes India and Adjacent Countries*, Vols. I and II. Oxford and IBH Publishing Co. Pvt. Ltd., New Delhi, 2(xix): 1158.

Trivedy, R.K. and Goel, P.K., 1984. *Handbook of Chemical and Biological Methods for Water Pollution Studies*. Enviromedia Publications, Karad, 1–247.

Wagh, Nitin, 1999. Hydrobiological parameters of Harsul Dam in relation to pollution. *Ph.D. Thesis*, Dr. B.A.M. University, Aurangabad.

Welch, Paul S.S., 1952. *Limnology*, 2nd Edn. McGraw Hill Book Company, New York, p. 1–538.

Yeole, S.M. and Patil, G.P., 2005. Physico-chemical status of Yedshi lake in relation to water pollution. *J. Aqua. Biol.*, 20(1): 41–44.

Chapter 34

Analytical Studies of Certain Parameters of Two Reservoirs of Katihar (Bihar)

☆ *Birendra Kumar*

Introduction

Katihar is a North district of Bihar. The district Katihar come into existence in the year 1974. Earlier it was the part of the district Purnea.It is surrounded by West Bengal in east.Purnea district in North,Purnea and Bhagalpur district in the west, and the River Ganga in the South.Katihar is the main town and administrative head quarter of the district. It is situated on 8704 Eastern Longitude and 2503 Northern Latitude. The climate is subtropical. The rainfall of this district is about 30-40″ from May to October of the year. Katihar is one of the most important industrial district of Bihar state.

In the north Bihar and especially in the districts of Katihar and numerous annual and perennial ponds lakes and tanks of different dimensions. Such water bodies emerged or have been formed by derelict channel of rivers, which have been blocked by the accumulation of silt or by the deep natural depression in which the surface water gets collected without finding an adequate outlet.

Materials and Methods

The analysis of water was done at the spot and few parameters in the laboratory by using standard methods of APHA (1989).The effect of polluted water on the loan habitants bears estimated by personnel survey of the neighboring area.

1. Biological Oxygen Demand (B.O.D.)
2. Chemical Oxygen Demand (C.O.D.)
3. Dissolved Oxygen (D.O.)

Biochemical Oxygen Demand (B.O.D.)

Biochemical oxygen demand (B.O.D.) is defined as the amount of O_2 required by microorganisms while stabilizing biologically decomposable organic matter in a waste under aerobic conditions. The B.O.D. test is widely used to determine 1) the pollution load of wastewater 2) the degree of pollution in lakes and streams at any time and their self purification capacity and 3) efficiency of wastewater treatment methods.

Principle

The dissolved oxygen content of the sample is determined before and after five days at $2°C$. Sample devoid of oxygen, or containing less amount of oxygen and diluted several time with special type of dilution water saturated with oxygen in order to provide sufficient amount of oxygen for oxidation.

Procedure

Preparation of Dilution Water

100 ml of distilled was aerated in a container bubbling compressed air for 1-2 days to attain DO saturation maintaining the temperature $2OC$. 1 ml each of phosphate buffer magnesium sulphate, calcium chloride and ferric chloride was added in the dilution water and mixed well. Now 2 ml settled sewage was added as seed to the dilution water.

The BOD of the sample was calculated as follows:

Let, DO= DO in the sample bottle on 0th day.

D1= Do in the sample bottle on 5th day

CO= Do in the blank bottle on 0th day

C1= Do in the blank bottle on 5th day

CO–C1= DO depletion in the dilution water alone

D.O.–D1= DO depletion in the sample dilution water

(D.O.–D1)= (Co–C1) = Do depletion due to microbes

BOD (Mg/l)= (Co–D1) – (Co–C1) mg x decimal fraction of sample used.

As the sample was seeded the B.O.D. of the seed was determined in the above manner.

Chemical Oxygen Demand (C.O.D.)

The oxygen required for chemical oxidation of organic matter with the help of strong chemical oxidant and the test is known as C.O.D. The test can be employed for the same purpose as the B.O.D. test after taking into account its limitations.

The intrinsic limitation of the test lies in its inability to differentiate between the biologically oxidizable and biologically inert material.

C.O.D. determination has one advantage over B.O.D. determination in that the result can be obtained in about 5 hours as compared to 5 days required for B.O.D. test. Further, the test relatively easy, gives reproducible results and is not affected by interferences as the B.O.D. test.

Principle

The organic matter gets oxidized completely by $K_2Cr_2O_7$ in the presence of H_2SO_4 to produce CO_2 + H_2O. The excess $K_2Cr_2O_7$ remaining after the reaction is titrated with $Fe(NH_4)(SO_4)_2$. The dichromate consumed gives the O_2 required for oxidation of the organic matter.

Proceudre

100 ml of the sample was placed in a refluxing flask. A suitable quantity of mercuric sulphate was added in it. 70 ml H_2SO_4 containing Ag_2SO_4 was added slowly and mixed thoroughly. A few granules of pumice stone, followed by 10 ml standard. $K_2Cr_2O_7$ solution was also added. Now the refluxing flask was connected to condenser and refluxed for a minimum of 2 hours. The flask was allowed to cool and washed the condenser with distilled water into the flask. The contents of flask was transferred to a 500 ml conical flask and fluted to about 350 ml with distilled water. After the solution in the conical flask was cooled down, 2 to 3 drops of Ferro in indicator was added and titrated with 0.1N ferrous ammounium sulphate sharp colour change form blue green to wine red indicated and point of the titration. A blank was also conducted using 50 ml distilled water, instead of sample, in the above manner.

$$C.O.D. (mg/l) = \frac{(a-B)N \times 8000}{ml \text{ of sample taken for estimation}}$$

where,

a: ml Ferrous ammonium sulphate solution for blank

b: ml Ferrous ammonium sulphate for sample

N: Normality of Ferrous ammonium sulphate.

Dissolved Oxygen

Living organisms need oxygen to maintain their metabolic processes. Dissolved oxygen (D.O.) is also important in precipitation and dissolution of inorganic substances in water. The solubility of oxygen in water depends upon its temperature, pressure and salinity. Among the methods available for the determination of dissolved oxygen, Iodometric method is most useful.

Iodometric Method

Principle

When maganous sulphate is added to the sample containing alkaline potassium iodide, manganous hydroxide is formed, which is oxidized by the dissolved oxygen of the sample to basic maganic oxide. On addition of H_2SO_4 the basic magnesium oxide liberates iodine equivalent to that of dissolved oxygen originally present in the sample. The liberated iodine is titrated with a standard solution of sodium thiosulphate using starch as the indicator.

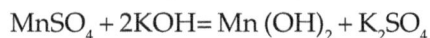

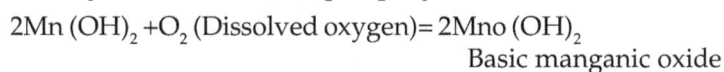

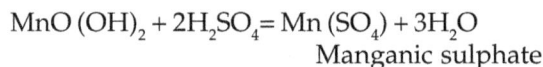

$$MnSO_4 + 2KOH = Mn(OH)_2 + K_2SO_4$$

$$2Mn(OH)_2 + O_2 \text{ (Dissolved oxygen)} = 2MnO(OH)_2$$
$$\text{Basic manganic oxide}$$

$$MnO(OH)_2 + 2H_2SO_4 = Mn(SO_4) + 3H_2O$$
$$\text{Manganic sulphate}$$

$$2Na_2S_2O_3 + I_2 = Na_2S_4O_6 + 2NaI$$

Procedure

A sample was collected in a BOD bottle using DO sampler. 2 ml $MnSO_4$ followed by 2 ml $NaOH + KI + NaN_3$ was added in the sample. The above solution was mixed well by inverting the bottle 2-3 times and allowed the precipitate to settle leaving 150 ml clear supernatant. At this stage, 2ml concentrated H_2SO_4 was added and mixed well till the precipitate passed in to solution. Now 203 ml of the above solution was taken in a conical flask and titrated against $Na_2S_2O_2$ using starch as an indicator. When 2ml $MnSO_4$ followed by 2 ml NaOH +KI+NaN_3 was added to sample, the above 4 ml of the original solution was lost. Thus 203 ml taken for titration would correspond to 200 ml of the original sample.

$$200 \times 300/(300\text{-}4) = 203 \text{ ml}$$

Calculations

1ml of 0.025 (N)$Na_2S_2O_2$= 0.2 mg of O_2

Since the sample was 200 ml $= \dfrac{0.02 \times 10000}{2000}$

1ml of thisulphate=D.O. mg/l

Each mole of gas at NTP occupies 22.4 liters.Since the mole of O_2

Weight 16, the value (D.O. mg/l) was multiplied by 0.698 to give ml of O_2

Present in the sample at °C and 760 mm. (16/22.4=0.698)

Table 34.1: Physico-chemical Parameters of Water Samples of 'Bijay Pokhar'

Parameters	Rainy		Winter		Summer	
	Min	Max	Min	Max	Min	Max
pH	7.65	8.05	7.86	8.20	7.65	8.25
T.A.	86.05	135.05	122.05	150.05	122.06	196.00
B.O.D.	22.40	64.35	24.40	44.25	41.35	158.25
C.O.D.	500.14	710.45	640.30	820.50	700.20	1562.30
Ca	22.45	34.05	44.85	56.10	16.00	26.08
D.O.	5.85	7.84	8.06	9.10	5.44	10.05
Na	10.25	19.25	16.85	24.15	16.45	28.35
Mg	5.05	12.35	8.82	14.85	10.15	17.32
T.O.	0.275	0.995	0.280	0.596	0.485	0.995
PO4	0.035	0.285	0.025	0.075	0.055	0.136
SO4	20.30	140.24	72.65	175.25	162.20	220.40
C1	32.8	48.5	28.4	45.6	36.1	45.6
W.T. (°C)	26.6	34.6	20.4	23.0	25.0	36.0

Parameters except pH are in mg/l otherwise stated.

BOD: Biological Oxygen Demand; COD: Chemical Oxygen Demand; DO: Dissolved Oxygen; TA: Total Alkalinity; T.N.: Total Nitrogen; W.T.: Water Temperature.

Results and Discussion

Biological Oxygen Demand (B.O.D.)

B.O.D. is an appropriate measure of the amount of biochemically degradable matter present in the water samples. High B.O.D. may cause oxygen depletion to the level, detrimental to the aquatic life.

The water samples under investigation were analyzed for B.O.D. values in rainy, winter and summer season. The values ranged (22.50–65.50) mg/l during rainy season (24.5-44.30) mg/l in winter (41.35-158.25) mg/l in summer season.

The highest value of B.O.D. in summer may be attributed to the reduced water flow, high organic pollution, direct discharge of waste products and maximum water temperature which in turn increases the activity of micro-organisms.

Chemical Oxygen Demand (COD)

The measurement of C.O.D. is one of the most important factors in the water pollution studies. The portion of the organic matter is oxidisable by the activity of the micro-organisms is measured by B.O.D. test and the organic content inclusive of the portion that is not oxidized by the micro-organisms is measured by C.O.D. test. High C.O.D. may cause oxygen depletion to level detrimental to the aquatic life.

The water samples under investigation were analyzed for C.O.D. values in rainy, winter and summer seasons. The values ranged (500.05-710.40) mg/l in rainy season) 640.35-820.54)mg/l winter and (700.2-1562.35)mg/l in summer seasons.

The C.O.D. values found maximum in summer may be due to minimum flow of water and high organic pollution. The high organic pollution may be attributed to direct discharge of waste products *i.e.,* animal excreta.

Dissolved Oxygen (D.O.)

Dissolved oxygen (D.O.) concentration is an important gauge of existing water quality and the ability of water body to support aquatic life. It is due to the fact that living organisms need oxygen to maintain their metabolic processes. Several works have been carried out by different workers on different water reservoirs.Banerjee (1967) observed that oxygen below 5ppm may be considered unfavorable for the productivity of fish ponds. According to him under normal condition, concentration of dissolved oxygen above 7 ppm is suitable for productivity of water bodies. Ellis *et al.* (1946) has further observed that a varied fish fauna could be expected only in waters with dissolved oxygen content above 5 mg/l.

The water samples under investigation were analysed for D.O. concentration during rainy, winter and summer seasons. The values of D.O. were found in the range of 5.5 to 7.5 mg/l during rainy season,8.5 to 9.5 mg/l during winter and 5.5 to 10.02 mg/l in summer season. It is evident that lowest values were recorded during rainy season, higher values in winter and highest values in summer season.

These observation suggest that the oxygen level was above the standard (5.0 mg/l) set by W.H.O. for supporting and maintaining the healthy aquatic life including the presence of self purification trend in water.However, the other factors,in spite of the availability of oxygen may affect the water of 'Bijay Pokhar' and turn it polluted.

The minimum value (5.5 to 7.5) mg/l in rainy season may be due to several factors such as increase in water temperature, decrease in photosynthesis activity and increase in biochemical degradation.

The maximum value (5.5 to 10.00) mg/l in winter may be due to water temperature and release amount of oxygen by algal flora present in water. Similar observation were made by Sengu (1985).

References

Banerjee, S.S., 1967. Water quality and Soil conditions of fish ponds in some states of India in relation to fish productions. *Indian J. Fish.*, 14(1812): 114-115.

Ellis *et al.*, 1996. Determination of water quality. *Rs. Rep. U.S. Fish Serv.*, p. 1–9.

Mandal, T.N., 2000. Limnological studies of Gogabeel Lake Katihar (Bihar). *Ph.D.Thesis*, B.N.M.U., Madheapura (Bihar).

Sengu, R.P.S., 1985. *Indian J. Environ. Hlth.*, 27(3): 257–261.

Chapter 35

Effect of Temperature on the Level of Oxygen and pH in Dharmapuri Pond Water of Beed District, Maharashtra

☆ *D.B. Sirsath, J.S. Pulle, N.G. Kashid and N.E. Ambore*

Introduction

The Dharmapuri pond is minor project made up before 19th century. Its depth is about 15 to 200 feet. This pond is constructed on small canals. These canals arise from hill region. The pond is situated near Dharmapuri village Beed district in Maharashtra State. It is located on East longitude 76° 37' 00'' and North latitude 18° 44' 00'' with catchments area about 38 hectares. The pond water is used for domestic, agricultural, fish culture and drinking purposes after giving proper treatment. Fort and pond both were made as same time. The pond is completely surrounded by different types of herbs, shrubs, weeds and large tree. The present investigation have been undertaken to study the influence of temperature on dissolved oxygen and pH.

Many researchers have done studies on physico-chemical and biological characteristics of river waters (Shaikh and Yeragi, 2003; Bandela *et al.*, 1998; Basulu *et al.*, 1967; Chakraborty *et al.*, 1977; Joshi and Bisht, 1993; Gill *et al.*, 1998).

Materials and Methods

Water samples were collected over a period of 12 months from November 2001 to October 2002 from five stations namely D_1, D_2, D_3, D_4, D_5. Water temperature was recorded with the help of centigrade temperature, D.O. was estimated by Winkler's Iodometrically (Trivedi and Goel, 1995). pH of water sample was determined by pH meter in the laboratory. The temperature and DO were determined at the time to sampling.

Results and Discussion

Temperature

In liminological studies temperature is an important parameter required to get an idea of self purification of rivers, reservoirs and control of treatment plant.Dissolved oxygen in water depends on the temperature. Higher temperature, lower, the oxygen contents of water. Water temperature plays an important role in influencing the periodicity, occurrence and abundance of phytoplankton. The temperature of drinking water has an influence on its taste.

Naik *et al.* (2002) observed the water temperature value ranged between 26°C to 27°C from Puina lake near Kalamnuri. Bhalerao *et al.* (2002) observed the water temperature values ranged between 18°C to 28°C from Penganga river of tribal area of Kinwat. Kedar and Patil (2002) observed the water temperature value ranged between 20.5°C to 28.3°C from Rishi lake of Karanja. Temperature is one important physical parameter which directly influence some chemical reaction in aquatic ecosystem and in his investigation temperature values at both station varied from 15°C to 38°C and 15.9°C to 38.7°C respectively (Jakher and Rawat, 2003). In the present study the water temperature were recorded between 18.5°C to 31.0°C at station D_1, 19.5°C to 29°C at station D_2, 17°C to 32.5°C at station D_3, 18°C to 31°C at station D_4 and 18°C to 29.5°C at station D_5.

Table 35.1: Monthly Mean Value of Temperature, DO and pH

Months	Temperature in °C					Dissolved Oxygen in mg/l					pH				
	D_1	D_2	D_3	D_4	D_5	D_1	D_2	D_3	D_4	D_5	D_1	D_2	D_3	D_4	D_5
Nov.	21.7	22.0	21.5	21.0	21.7	2.4	2.4	2.6	4.0	2.6	7.3	7.4	7.3	7.4	7.0
Dec.	18.9	20.0	18.9	18.9	18.2	5.6	8.0	4.8	4.8	4.0	7.5	8.2	7.6	8.1	8.1
Jan.	18.5	20.0	17.0	18.0	18.0	4.8	6.4	4.0	4.0	2.4	8.1	8.6	7.7	8.7	8.6
Feb.	20.0	19.5	19.9	19.9	20.0	4.8	4.0	5.6	3.2	3.2	8.3	8.7	9.2	8.3	8.5
Mar.	22.0	20.9	20.0	21.5	21.5	3.2	4.8	4.0	4.0	4.8	8.8	8.8	8.1	8.7	8.6
Apr.	26.0	24.5	27.5	26.0	24.5	2.6	2.6	2.4	2.4	3.0	7.7	7.8	7.6	7.7	7.9
May	31.0	29.5	32.5	31.0	29.5	1.6	1.6	2.0	2.1	2.0	8.1	8.2	8.0	8.2	8.1
Jun.	25.0	24.0	25.0	25.0	24.5	4.8	5.6	5.6	5.6	4.0	7.2	7.7	7.8	7.8	7.1
Jul.	23.0	23.0	23.0	23.0	23.0	4.8	4.0	4.0	4.0	6.4	8.0	8.1	7.9	8.1	8.0
Aug.	26.0	25.5	25.5	26.5	26.0	7.2	7.2	6.4	8.0	6.4	9.1	8.9	9.0	9.1	9.0
Sep.	28.5	28.0	28.0	28.0	28.5	3.2	3.2	4.0	4.0	4.0	9.2	9.4	9.3	8.9	9.1
Oct.	27.0	26.0	26.0	26.0	26.0	4.0	5.6	6.4	4.8	4.0	8.7	9.2	9.0	8.6	8.8

Dissolved Oxygen

Dissolved oxygen in water is essential to aquatic life. Deficiency of DO in receiving water gives rise to odoriferous product of anaerobic decomposition. Oxygen saturated water have a pleasant taste. While the water lacking oxygen have an inspire taste. Sreenivasan (1972) and Bahura (1998) reported an inverse relation of dissolved oxygen with temperature.

Bhalerao *et al.* (2002) observed the DO value ranged between 1.2 mg/l to 5.5 mg/l from Penganga river of tribal area in Kinwat. Kedar and Patil (2002) noted the DO value ranged between 2.8 to 8.8 mg/

l from Rishi lake of Karanja. Thakare *et al.* (2002) observed the DO value ranged between 1.4 mg/l to 7.2 mg/l from Dhaswadi lake. Shaikh and Yeragi (2003) noted the DO value ranged between 4.9 to 11.7 mg/l and observed that in summer season increase with atmospheric temperature. The water temperature rises and dissolved oxygen decreases while in winter due to low temperature, the dissolved oxygen increase which ultimately decrease the pH of Tansa river of Thane. Khedkar and Dixit (2003) observed the DO value ranged 0.00 to 7.80 mg/l at domestic wastewater of Amravati.

In present study of Dharmapuri pond, dissolved oxygen shows inverse relation with temperature. DO value ranged between 1.6 to 7.2 mg/l at station D_1, 1.6 to 8.0 mg/l at station D_2, 2 to 6.4 mg/l at station D_3, 2.1 to 8 mg/l at station D_4 and 2 to 6.4 mg/l at station D_5. The minimum DO was observed during summer season in the month of May while maximum in the post monsoon and winter season.

pH

pH is defined as the negative $\log_{10}$ of the hydrogen ion concentration. According to WHO the standard pH of drinking water is in between 6.5 to 9.2. pH when the value pH of water is less or more than that of the standard pH, then such water is not good for public use or domestic purpose. The fish and other aquatic organism prefer pH value between 6.7 to 8.4 and pH value below 5.0 and above 8.8 may be detrimental or even lethal to aquatic life (Ellis, 1937).

Barul dam being an impoundment having its pH range which complies with the above statement. The pH ranging between 7.04 to 8.36 was high recorded in summer season. The lower range of 7.1 to 7.2 was observed during winter season (Bandela *et al.*, 1998). Dhaswadi lake, pH of water was range between 6.9 to 7.5 (Thakare, *et al.*, 2002). Naik *et al.* (2002) noted the pH value 7.5 from Puina lake near Kalamnuri. Kedar and Patil (2002) noted the pH value ranged between 7 to 8.2 from Rishi lake of Karanja. Shaikh and Yeragi (2003) observed the pH value ranged 7.04 to 8.43 in Tansa river of Thane. Khedkar and Dixit (2003) observed the pH value ranged 6.5 to 8.4 in domestic wastewater of Amravati.

In present study, pH range was recorded between 7.2 to 9.2 at station D_1, 7.4 to 9.4 at station D_2, 7.3 to 9.3 at station D_3, 7.4 to 9.1 at station D_4 and 7.0 to 9.1 at station D_5. High pH value recorded in September month because due to the increase in temperature and decrease in dissolved oxygen level. Low pH value was observed in November month due to the lower temperature.

References

Bahura, C.K., 1998. A study of physico-chemical characteristics of highly eutrophic temple tank, Bikaner. *J. Aqua. Biol.*, 13(1 and 2): 47–51

Bandela, N.N., Vaidya, D.P. and Lomate, V.S., 1998. Seasonal temperature changes and their influence on the level of carbon dioxide and pH in Barul dam water. *J. Aqua. Biol.*, 3(1): 43–46.

Basul, K.R., Arora, H.C. and Aboo, K.M., 1967. Certain observation on self purification of Kher river and its effects on Kridhna river. *Ind. J. Env. Hlth.*, 9(4): 275–296.

Bhalerao, A.P., Khan, A.M. and Bemrekkar, S.K., 2002. Studies on physico-chemical and biological parameter causing pollution of river Penganga of tribal area of Kinwat (M.S.). *Nat. Conf. Umarkhed Abst.*, p. 73.

Chakraborty, R.D., Roy, P. and Singh, S.S., 1977. A quantitative study of plankton and physico-chemical conditions of river Jamuna at Allahabad in 1954–55. *Ind. J. Fish*, 6(1): 186–203.

Ellis, M.M., 1937. Detection and measurement of stream pollution. *Bull. US Fish.*, Washington, 48(22): 385–437.

Jakher, C.R. and Rawat, M., 2003. Studies on physico-chemical parameters of a tropical lake Jodhpur (Rajasthan, India). *J. Aqua. Biol.*, 18(2): 79–83.

Joshi, B.D. and Bisht, R.E., 1993. Some aspects of physico-chemical characteristics of Western Ganga Canal near Jwalapur, Hardwar. *Himalayan J. Env. Zool.*, 7(1): 76–82.

Kedar, G.T. and Patil, G.P., 2002. Water quality of Eutrophic Rishi lake of Karanja (Lad). *Nat. Con. Umarkhed Abst.* p. 90.

Khedkar, D.D. and Dixit, A.J., 2003. Physico-chemical analysis of domestic wastewater of Amravati. *J. Aqua. Biol.*, 18(1): 69–72.

Naik, N.B., Patil, P.M., Sirsikar, A.N., Deshpande, V.D., Nomular, P., Singh, S. Premchandra and Shubhachandra, M., 2002. Analysis of some physico-chemical parameters of water sample and ecological study of Puina lake near Kalamnuri (M.S.). *Nat. Conf. Umarkhed Abst.*, p. 59.

Shaikh, Nisar and Yeragi, S.G., 2003. Seasonal temperature changes and their influence on free carbon dioxide, dissolved oxygen (DO) and pH in Tansa river of Thane district, Maharashtra. *J. Aqua. Biol.*, 18(1): 73–75.

Takare, B.G., Pampatwar, N.G. and Gaikwad, V.B., 2002. Studies on physico-chemical and biological parameter relation to blue revolution with reference to Dhaswadi lake, Latur District. *Nat. Conf. Umarkhed Abst.*, p. 56.

Chapter 36

Polychaetes: An Overview

☆ *P. Murugesan*

The marine and estuarine environs are rich in benthic faunal communities which not only support the fishery potentiality by producing millions of larvae in the form of meroplankters, but also maintain the state of equilibrium of the ecosystem by producing burrows, pumping air and waters within the soil. A majority of the intertidal benthic fauna prefer such brackish water situations where environmental conditions are markedly different from those of terrestrial or freshwater habitats. Animals in such habitats acquire a certain degree of steno-euryhalinity as a security against every oscillating environmental variables. One among the dominant groups of the macro-benthic communities is polychaetes. In this regard, key characters for identification, method of collection and importance of polychaetes are detailed here.

Systematic Position of Polychaetes

Phylum	:	Annelida
Class	:	Polychaeta
Order	:	Chaetopoda
Group	:	Errantia
	:	Sedentaria

Polychaetes, segmented bristle-bearing worms of class Polychaeta in phylum Annelida, are usually the most abundant animals living within sands and mud on the seashore. These worms cannot often be seen on the surface but sometimes they may create subtle signs and traces of their presence. Only a few may be found exposed on bare rock surfaces, but they are common cryptic animals within rock crevices on the seashore and under boulders. Those that can survive the harsh environmental conditions on top of rock usually live in colonies, protected inside tubes. The class polychaeta contains most of the living marine species. Why the name polychaete? Poly–means many; Chaeta–having many bristles/ setae. The division of the body into similar parts which are arranged in a linear fashion along the antero-posterior axis called Metamerism. This is the most distinguishing characteristic feature of the phylum Annelida. The segmented part is always limited to the trunk. In most polychaetes the additional segments are added throughout the life (Figures 36.1a&b). The youngest segment occurs at the posterior end of the series. The head, represented by the prostomium and containing the brain is not a segment, nor is the pygidium, the terminal part of the body which carries the anus. Polychaetes are very common marine animals. Majority of the species are 5–10 cm long with the diameter ranging from 2 to 10 mm. Deep water forms are no longer than 1 mm whereas one species attains a length of 3 meters. Polychaetes can be divided into two groups as errant (free-moving) forms and sedentary forms, although the distinction between the two groups is not always sharp. The errant polychaetes or errantia, include some species that are strictly pelagic, some that crawl about beneath rocks and shells, some that are active burrowers in sand and mud and many species that construct and live in tubes. The sedentary polychaetes or sedentaria, are largely tube-dwellers, or inhabit permanent burrows. Usually only the head of the worm ever emerges from the opening of the tube or burrow.

The most distinguishing feature of polychaetes is the presence of parapodia, the paired lateral appendages extending from the segments. A typical parapodium is a fleshy projection extending from the body wall and is more or less laterally compressed. The parapodium is basically biramous, consisting of an upper division, the notopodium, and a ventral division, the neuropodium. Each division is supported internally by one or more chitinous rods, or acicula. A cirri form process projects from the dorsal base of the notopodium and from the ventral base of the neuropodium. The notopodia and neuropodia assume various shapes in different families and may be subdivided into several lobes or one may be somewhat reduced. In fact, reduction in some polychaetes has resulted in uniramous parapodia. This is true, for example, in most phyllodocids, in which all the notopodia have disappeared except the broad flattened dorsal cirrus. The distal ends of the parapodial rami are invaginated to form pockets or setal sacs in which are located many projecting chitinous bristles or setae.

Many polychaetes are strikingly beautiful and are coloured red, pink, or green or possess a combination of colors. Some are iridescent, owing to the presence of crossed layers of collagen fibers in the cuticle.

Regarding feeding, the polychaetes are mostly detrital feeders. They include members of many families of surface dwelling, pelagic groups and tubicolous groups. The prey consists of various small invertebrates, including other polychaetes, which are usually captured by means of an eversible pharynx (proboscis). A scavenger or omnivorous habit has also evolved in many polychaetes. Apart from this, few members are categorized under non–selective deposit feeders and selective feeders. The non–selective feeders consume sand or mud directly when the mouth is applied against the substratum. In the selective feeders lack a proboscis. Special head structures extend out over the substratum. Deposit materials adhere to mucous secretions on the surface of the feeding structure which is then conveyed to the mouth. Gills are common among the polychaetes, but they vary greatly in both structure and location, indicating that they have arisen independently within the class a number of times. Most commonly the gills are associated with the parapodia and most cases are modified parts of the parapodium. The process of excretion in polychaetes is either by protonephridia or metanephridia. There may be one pair of excretory organ per segment or one pair for the entire animal.

Polychaetes have relatively great powers of regeneration. Tentacles, palps and even heads ripped off by predators are soon replaced. Most polychaetes reproduce only sexually, and the majority of species are dioeciously. Polychaete gonads are not distinct organs but are masses masses of developing gametes, which develop as projections or swelling of the peritoneum in different parts of some segments.

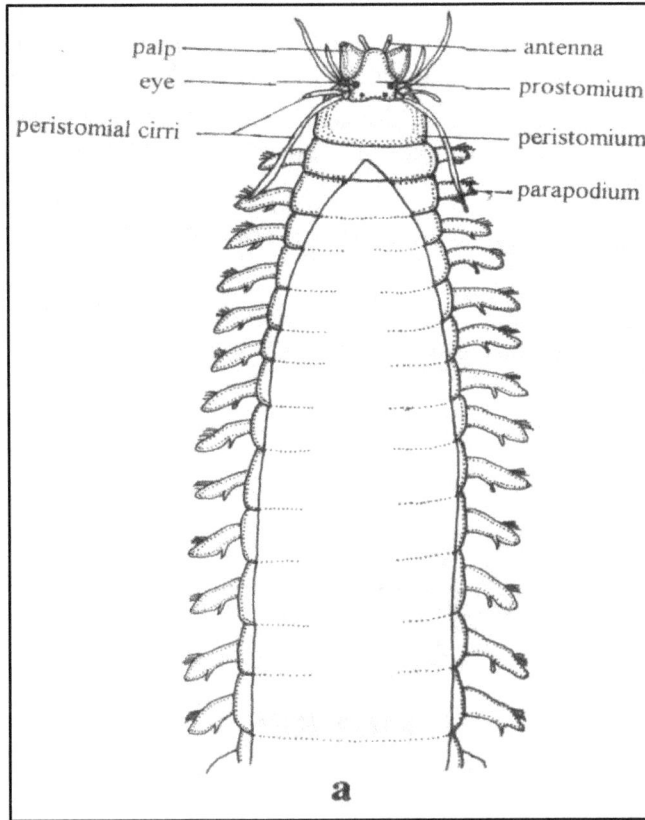

Figure 36.1a: Diagnostic Characters of a Polychaete

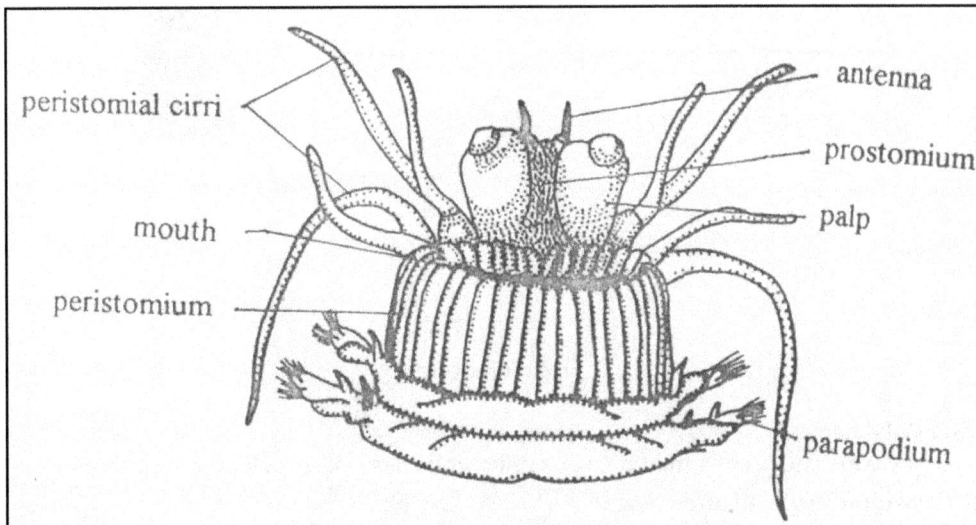

Figure 36.1b: Mouth of a Polychaete

Epitoky and swarming means of reproduction can also be seen in polychaetes. There are some hermaphroditic polychaetes. The larval stage involved in the life history is the trochophore. Asexual reproduction is also known in some polychaetes; it takes place by budding or division of the body into two parts or number of fragments.

Based on its adaptation, the polychaetes can be divided into the following:

1. Pelagic polychaetes
2. Burrowing polychaetes
3. Tubicolous polychaetes

Pelagic Polychaetes
☆ They are transparent
☆ Enormous eyes
☆ Exclusively pelagic
☆ Setae rudimentary, instead, membranous pockets

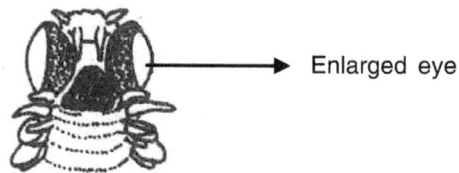

Enlarged eye

e.g. Alciopid

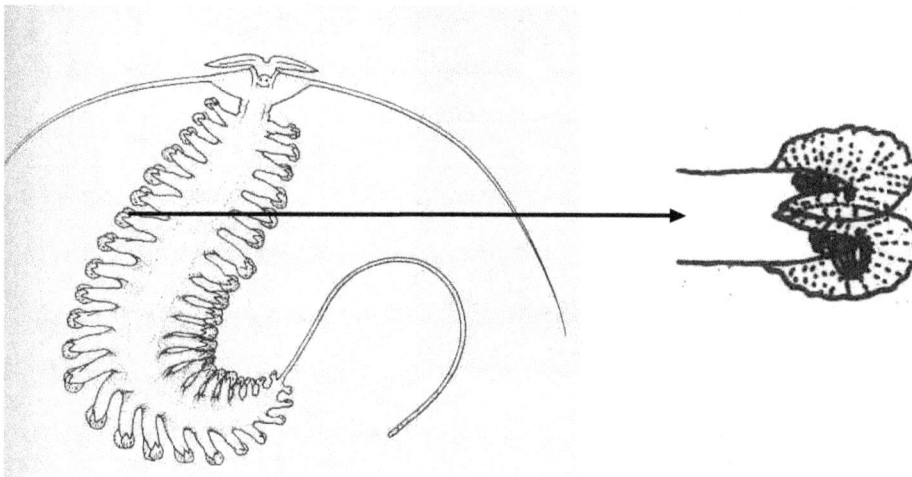

e.g. Tomopterid

Burrowing Polychaetes
☆ They construct a system of mucous lined galleries within which the organisms move
☆ Prostomium small and conical

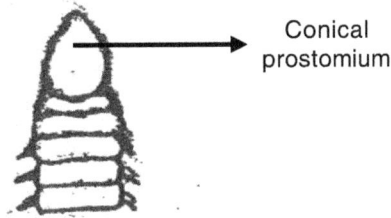

***e.g.* Lumbriconereids**

☆ Parapodia are greatly reduced, the setae modified into hooks called 'uncini'

***e.g.* Glycerids, Opheliids**

Tubicolous polychaetes

    ☆ This habit is evolved in many polychaete families
       e.g. Eunicids, Maldanids, Sabellids and Serpulids

    ☆ This tube serves as a lair for catching passing prey

    ☆ The tube composed of sand grains cemented together

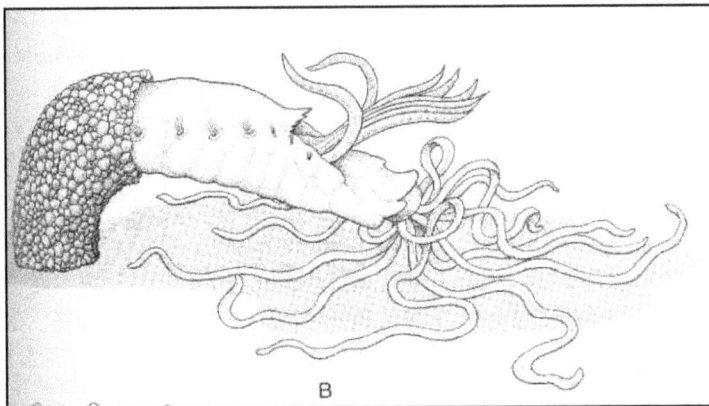

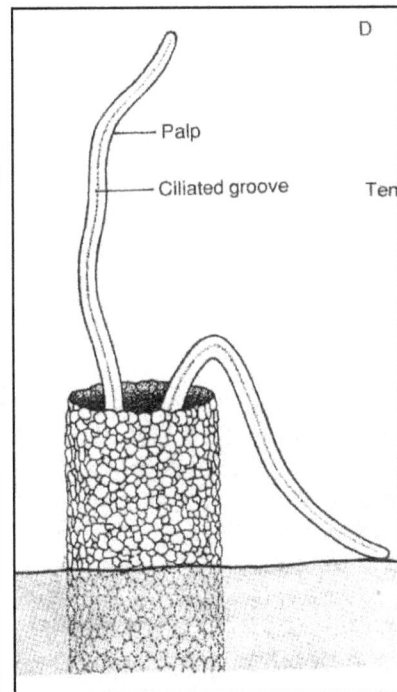

Collection Techniques

There have been a number of extensive comparative studies concerning sampling devices and their efficiencies. Key works are Ankar (1977), Eleftheriou and Holme (1984), Gage (1975), Rosenberg

(1978) and Rumohr (1990). To efficiently sample the benthic macro fauna, a sampler must be able to penetrate to a depth sufficient to capture the organisms present. Most report say that the majority of species and individuals to be present in the upper 5 to 10 cm, although large burrowing mollusks and crustaceans may be found deeper.

Pelagic polychaetes can be collected from the plankton sample. Some of the forms can be collected from floating logs, intertidal region, bottom mud, gastropod and bivalve shells and other hard substrates like stones, cement boulders, wooden piers and boat hulls in the coastal zone. Pelagic forms will be estimated by filtering 200 liters of surface water and the sample will be counted under a binocular microscope using Sedgwick rafter and the number of animals per cubic meter of water will be calculated. In the intertidal region, 625 cm^2 wooden quadrate is placed and the substrate is dug out for 10 cm. Sub tidal forms can be quantitatively estimated using Peterson grab covering an area of about 625 cm^2 5–10 cm depth and calculated to a square meter. The Peterson grab is found to be effective in shallow water environments while a long armed Van–veen grab and box corer are effective in collecting sediments in deep water bodies. Besides sediment grabs, dredges, trawls and traps are also used for collecting benthos for qualitative purpose.

Size and Number of Samples

The size and number of samples taken depend largely upon the aim of the study. For instance, in a general sub littoral grab sampling survey designed to map the invertebrate communities notwithstanding polychaetes of an area of sea bed, a single 0.1m^2 sample per station may be adequate (Cuff and Coleman, 1979), though many workers take two or three replicates (Mackie *et al.*, 1985). Alternatively, an investigation concerning temporal change or quantitative comparison will require perhaps 5 to 10 such samples per station in order to meet an acceptable level of statistical precision (Lie, 1968; McIntyre *et al.*, 1984). But still the optimization of sample size and number has been subject to much theoretical debate and some practical investigations. Similarly the depth of sediment collected by samplers varies according to sampler and the nature of the substrate. So, benthic workers have to decide the minimum 'acceptable' for their locality and the aims of the investigation.

Sieving, Treatment and Staining Samples

The size of sieve used is effectively a compromise between cost and accuracy. Once the mesh size has been set, the sediment samples collected will be sieved through 0.5 mm aperture size and the number of polychaetes is counted. In the case of damaged animals, the heads alone are taken into account. The polychaetes can be preserved in 5 per cent formalin and stained with Rose Bengal solution (0.1 gm in 100 ml distilled water) for the enhanced visibility at the time of identification. Although the use of rose Bengal is depreciated by many taxonomists, it is certainly beneficial in aiding sorting and identification. After adding it, the container is to be turned upside down gently for thorough mixing. As it is acidic in nature, leaving animals in it for too long may deteriorate the animals. To do away with this, the animals are washed thoroughly with freshwater and later preserved it in 75 per cent alcohol. The concentration of alcohol should be increased step by step *i.e.* 30 per cent, 50 per cent, 75 per cent. MgCl$_2$ or menthol or propylene phenoxyetol is also added as narcotics before adding alcohol to prevent the distortion of the animals. The use of these relaxants can greatly improve the condition of the specimens obtained. After staining, the sample is allowed to be kept for a day or two.

Sorting and Identification

After at least 24 hr. fixation, the sieved and preserved samples are gently but thoroughly washed in freshwater. This removes the formalin and salt, preventing the former from dissolving the shells of

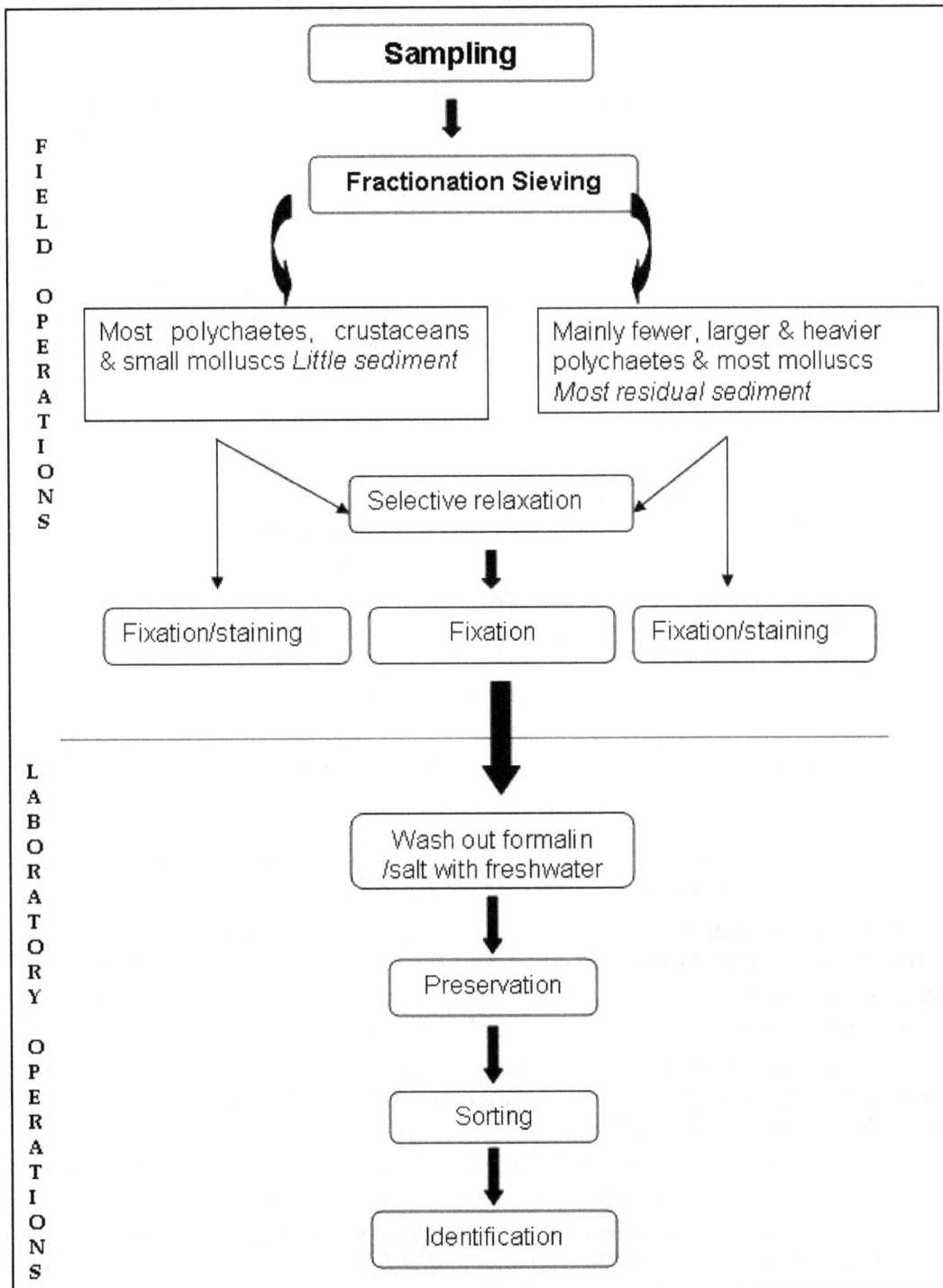

Figure 36.2: Protocol for Sampling Benthic Fauna

delicate mollusks. For initial sorting, white background trays with bench lights, needles and pen brushes will be sufficient. Dissection microscope can also be used for sorting purpose. The key factor for success of sorting is not to place too much sample in the sorting tray at any one time. More the materials, greater the chances of overlooking the specimens, therefore care should be taken in all the stages since there is every possibility of missing of many species. To cape it all, workers must always be aware of the purpose of the animals. Since important things are the animals. Unfortunately this is often forgotten and the processing of samples focuses on getting the material through the sieve as fast as possible. The protocol for sampling shallow and sub littoral benthos in general and polychaetes is given in Figure 36.2.

After sorting, all containers should be identifiable by marking the outside with indelible pen and robust chemically stable lebels must be placed inside. After this lengthy process, animals are enumerated and identified to the most advanced level possible using standard references. Electronic keys can also be used besides the monographs and field guides for identification.

Importance of Polychaetes

☆ Polychaetes serve as food for bottom feeding fin and shell fishes.

☆ Used as 'bait organisms' in fish angling industry.

☆ Important resource in aquaculture for Crustacea–since it provides correct balance of PUFA which are very much essential to maturation and egg production of shrimps.

☆ Can consume as food–*e.g. Eunice viridis*.

☆ Used as indicator species of pollution since it attach so much importance in EIA (Environment Impact Assessment) studies.

☆ Play a key role in recycling of nutrients between pelagic and benthic realms

☆ Used as toxicological test organisms

☆ Used to remove organic wastes from aquaculture systems.

References

Ankar, S., 1977. Digging profile and penetration of the Van Veen grab in different sediment types. *Contribution from the Asok Laboratory*, 16: 1–22.

Balasubrahmanyan, K., 1960. Studies on the ecology of the Vellar estuary. I. A preliminary survey of the estuarine bottom and its fauna collected on 11.9.1959. *J. Zool. Soc., India*, 12: 209–215.

Cuff, W. and Coleman, N., 1979. Optimal survey design: Lessons from a stratified random sample of macrobenthos. *Journal of the Fisheries Research Board of Canada*, 36: 351–361.

Eleftheriou, A. and Holme, N.A., 1984. Macrofauna techniques. In: *Methods for the Study of Marine Benthos*, 2nd edn, (Eds.) N.A. Holme and A.D. McIntyre. IBP Handbook, Blackwell Scientific Publication, Oxford, 16: 140–216.

Gage, J.D., 1975. A comparison of the deep-sea Epibenthic sledge and Anchorbox Dredge Samplers with the van veen grab and hand coring by diver. *Deep Sea Research*, 22: 693–702.

Lie, U., 1968. A quantitative study of benthic infauna in Puget sound, Washington, USA in 1963–1964. *Fiskeridirektoratets Skrifter (Havunders Okelser)*, 14: 223–556.

McIntyre, A.D., Elliott, J.M. and Ellis, D.V., 1984. Introduction: Design of sampling programmes. In: *Methods for the Study of Marine Benthos*, 2nd edn, (Eds.) N.A. Holme and A.D. McIntyre. IBP Handbook, Blackwell Scientific Publication, Oxford, 16: –26.

Rosenberg, D.M., 1978. Practical sampling of freshwater macrozoobenthos: a bibliography of useful texts, reviews and recent papers. *Technical Report*, Fisheries and Marine Service, Canada, 790: 1–15.

Rumohr, H., 1990. Soft bottom macrofauna: collection and treatment of samples. *ICES, Copenhagen, Techniques in Marine Environmental Sciences*, 8.

Srikrishnadhas, B., 1977. Studies in Polychaete larvae of Portonovo water, South India. *Ph.D. Thesis*, Annamalai University, 205 pp.

Srikrishnadhas, B., Jayabalan, N. and Ramamoorthi, K., 1981. Ecology of the population of polychaetes in the intertidal regons of the Vellar estuary. *Proc. Symp. Ecol. Anim. Popul. Zool. Suv., India*, p. 73–81.

Srikrishnadhas, B., Murugesan, P. and Khan, S. Ajmal, 1998. *Polychaetes of Parangipettai Coast*. CAS in Marine Biology, Annamalai University, Parangipettai, 110 pp.

Chapter 37

Present Status and Future Chances of Pond Fishery Development in Parbhani District of Maharashtra

☆ *S.D. Niture and S.P. Chavan*

Introduction

Global production from freshwater aquaculture has touched 25.8 million tons in 2004. The cyprinids alone contribute 18.2 million tons valued 16.3 billion USD. India is in second position in global freshwater fish production; the low quantum of total production is deplorable considering the large size of the country, vast water resources and large human capital. The enterprising farmers in Andhra Pradesh have done outstanding work in carp production in the country and have shown a way for utilizing the huge untapped potential of India for freshwater fish culture (Sakthivel M, 2007). Fish culture pond in the agriculture land is constructed by excavating the land either manually or by using machines like excavator. The construction of pond in non-percolating land with good water holding capacity is essential. The land must be in low line or in depression area where water usually remain stagnant.

Depending on availability of the land, artificial ponds of various sizes like 100x50x3 m, 70x40x3m, 50x30x3 m etc. of the square shape, rectangular shape or irregular shape can be constructed. It is not necessary that the depth of pond should be uniformly equal the depth may be variable. The uniform depth is useful for easy harvesting of fishes. To prevent the percolation of water from sidewalls of ponds pitching work by masonry, stone or bricks in concentrate is done. However, the pond bottom must be with soil for the formation of fish food like algae and planktons. Using cow dung manure, an organic fertilizers the pond bottom is manured for the development of fish food. Using calcium carbonate, the pond liming is carried out to kill the pathogens and maintain the pH of the soil.

Catla catla, Labeo rohita, Cirrhina mrigal, Cyprinus carpio and silver carp species are stocked in the fish culture ponds In Maharashtra.

Material and Methods

To study the present status of artificial pond culture in various Taluca of Parbhani district, survey was carried out throughout the district to visit the artificial pond sites to determine the status of pond culture fishery. The status of fish culture ponds were discussed with the pond owners. To determine the number of farmers involved in artificial pond fishery, the reference of F.F.D.A. (Fish Farmers Development Agency) for district Parbhani was considered as standard reference data.

Method of Survey, study and observation of every artificial fish culture farm was adapted. The status of pond culture was discussed individually in the form of interviews of every individual pond owner by using video recordings and photography during July 2005 to June 2008.

Present Status of Fish Culture in Artificial Pond in Parbhani District

It was found that, 28 artificial fish culture farms were present in Parbhani district of Maharashtra. From the observed 28 cases, 04 fish farms were in the actively working stage, 23 fish farms were sick and 01 fish farm was facing some problems of weed infestation hence, it was in the semi-working state. The details of the pond fish culturist are given in the Table 37.1.

Table 37.1: Present Status of Artificial Pond Fish Culture in Parbhani District (2005–2008)

Sl.No.	Name and Address of Fish Pond Culture Farmers	Year of Construction	Size of Pond	Present Status
1.	Mr. Gautam Kishanrao More. Selu	1995	61.69 × 25.75 m	Sick
2.	Mr. Pradip Dinkar Mane. Warpud. Tq. Parbhani	1996	27 × 19 m, 37 × 33 m 37 × 33 m, 78 × 36 m 78 × 35 m, 78 × 19 m 54 × 17 m	Sick
3.	Mr. Anil Ambadas Warpudkar. Tq. Parbhani	1996	105 × 40 m,105 × 40 m 78 × 34 m,78 × 34 m 105 × 34 m, 116 × 36 m 116 × 24m, 58 × 51m	Partially Working
4.	Mr. Pradip Dinkar Mane. Warpud. Tq., District Parbhani	1996	15 × 12m,98 × 21m 100 × 34m, 100 × 28m 100 × 27m, 60 × 30m 60 × 30m, 60 × 30m	Sick
5.	Mr. Bhagvan Supadu Mahajan. Warpud. Tq. Parbhani	1996	134 × 29m, 134 × 36m 134 × 41m, 134 × 55m 88 × 47m, 48 × 38 m 66 × 55 m.	Sick
6.	Mr. Karbhari Vitthal Solunke. Rumana Tq. Gangakhed	1996	59 × 51m, 45 × 31m 45 × 40m	Sick
7.	Mr. Sahebrao Kondiba Solunke. Rumana Tq. Gangakhed	1996	59 × 29m, 65 × 26 m	Sick
8.	Mr. Aashish Wamanrao More. ChinchTakali Tq. Gangakhed	1996	33 × 26m, 63 × 25 m 36 × 45m 86 × 50 m	Sick
9.	Mr. Abdul Kadar Mohammad (Adanwala) Ansari. Babhulgaon Tq Pathri	1996	25 × 16m, 23 × 17 m 35 × 15 m, 32 × 16m	Sick

Contd...

Table 37.1–Contd...

Sl.No.	Name and Address of Fish Pond Culture Farmers	Year of Construction	Size of Pond	Present Status
10.	Mr. Prafulchand Devman Gajire. Niwali(45) Tq Pathri	1996	52 × 19m	Sick
11.	Mr. Kundlik Devrao Shelke. Kawalgaon wadi Tq Purna	1996	33 × 47m, 37 × 44m 37 × 53m, 37 × 47m	Sick
12.	Mr. Kisan Sampatrao Magar. Gogalgaon. Tq Selu	1997	40 × 43m,40 × 40m	Sick
13.	Mr. Keshav Balabhau Magar. Gogalgaon. Tq Selu	1997	50 × 20m, 50 × 26m	Sick
14.	Mr. Sharad Haribhau Kedar. Gogalgaon. Tq Selu	1997	60 × 30m, 11 × 8m	Sick
15.	Mr. Ratan Narayan Kedar. Gogalgaon. Tq Selu	1997	70 × 26m	Sick
16.	Mr. Bhaskar Punjaji Wamkede. Yetoli. Tq Jintur	1997	30 × 23m, 16 × 28m 15 × 46m, 29 × 56m	Sick
17.	Mr. Madukar Raghunath Mohite. Sonkhed. Tq Sonpeth	1997	40 × 22m, 75 × 22m	Sick
18.	Mr. Sambhaji Dagudi Dake. Kawalgaonwadi Tq Purna	1998	43 × 39 m	Sick
19.	MRs Kalawatibai Laksmanrao Shelke. Kawalgaonwadi Tq Purna	1998	57 × 47 m	Sick
20.	Mr.Keshav Vitthalrao Jodhale. Kamkhed. Tq Purna	1998	30 × 45 m,	Sick
21.	Mr. Arun Vishwanath Jukate. Lohgaon.tq Parbhani	2000	50 × 20 m	Sick
22.	Mr. Kerba Gyanoji Soudagar. Sarangi Tq Purna	2003	72C40 m	Sick
23.	Mr. Manik Bapurao Soudagar. Sarangi Tq Purna	2003	49 × 51 m	Sick
24.	Mr. Zotig Sadashiv Kalaskar, Dharkhed. Tq Gangakhed	2003	60 × 43 m	Sick
25.	Mrs Kavita Shivhar Jogwadkar (Deshpande), Kanha, Charthana tq Jintur	2003	189 × 55 m=10395 24 × 62m=1488	Working
26.	Mr Udhav Narayan Dombe. Pokharni. Tq Parbhani	2005	47 × 26m=1363 47 × 26m=1363 47 × 26m=1363	Working
27.	Mrs Shilabai Mijahari Bhalerao. Rampuri. Tq Pathri	2005	110 × 38m=4180	Working
28.	Mr. Dattrao Tukarm Renge At Jamb Tq. Parbhani	2001	4 Ponds of Size 100 × 50 × 1.5 each and 16 Ponds of size 12 × 18 × 1.5 M	Working

Source: F.F.D.A, Parbhani and Survey and Interviews of Fish culturist during 2005-2008.

Present Status of Successful Artificial Ponds

It was observed that, four farmers were actively involved in pond fish culture. The remarkable feature of culture ponds which are in working is that, all are medium in size as given in the table1., therefore the management of these ponds for Maintenance of the water level, manuring, artificial feeding, fish harvesting, control on blooms and weeds was easy in these ponds during fish culture. Another important factor behind the success of these cases was the updated and complete knowledge of fish culture towards the pond owners. For all the successful cases, the availability of water to store in the culture pond was either from bore-well, dug-well or from the nearest reservoir. There were no algal blooms and weeds in the culture ponds due to medium and manageable size. The examples of the successful cases of Pond culturists in the district are elaborately explained as below.

Case No. 1: Sheelabai Munjahari Bhalerao Fish Farm, at Rampuri, Taluka Pathri, District Parbhani

In this case, the pond owner regularly stock the fish seed of Indian major carps, individually the owner is involved in post stocking management of pond like netting, pond ploughing, manuring, eradication of algal blooms, weeds, aquatic insects and marketing of fishes from the pond. From the all observed cases, this was most successful fish farmer to run the farm ideally and get the benefit from it.

Case No. 2: Kavita Shivhar Jogwadkar, At Kanha, Post Charthana, Taluka Jintur, District Parbhani

This farm is located 5 Km. east of village Charthana near small locality 'Kanha' in Jintur Taluka of Parbhani district. The source of water for this fish culture pond was from bore well. The farmer also has a modified type of Chinese hatchery for the fish seed production of Indian Major Carps, Cyprinus and silver carp. The main intension of this farmer is maintenance of fish brooders in the culture pond (Rearing type of pond). This farmer is involved actively in marketing the fish seed to stock in reservoirs from this region. The reservoir fishery lease holders from Jintur Taluka purchase the fish seed from the hatchery established by this fish culturist in this region, which provide pure culture batch of individual fish species *i.e.* one can get the fish seed of only of Catla, Rohu, Mrigal, Cyprinus or silver carp.

The farmer also stock and maintain the *Channa straitus* and *Channa punctatus* species along with brooders of Indian major carps in the same pond successfully and get the additional source of income from the marketing of *Channa* species *i.e.* Murrels, which cost for Rs.80-110/Kg (Cost in year 2006).

Case No. 3: Udhav Narayan Dombe, Fish Farm At Pokharni Tq. Dist: Parbhani

This farm is located near village Pokharni (Narshiha) in Parbhani district. The farmer has three ponds. The source of water for these ponds is bore well and farm small branch of irrigation canal and stagnant water from a stone mine located near the pond.

Indian Major Carps and *Cyprinus carpio* were stocked into the pond. This farmer stocks the fish seed annually and market the catch by wholesale marketing to fish Merchant. 750 gm to 1.25 Kg fish weight was marketing size of the catch. The pond was not well managed, no artificial feeding and no any special efforts of post stocking management.

Case No. 4: Sudarshan Fisheries Development Corporation Fish Culture Pond, At Jamb Tq. Dist Parbhani

This fish culture farm is not specially constructed for fish culture but the stocking pond of the hatchery unit is used to culture the fishes and prawns along with Maintenance of fish brooders. This farm is located within the area of Sudarshan Fisheries Development Corporation, near village Jamb at Pedgaon reservoir, 07 Km. West of Parbhani city, exactly the hatchery and pond is located 01 km west of village Jamb near Pedgaon reservoir. The harvested catch was packed in crushed ice using thermocoel containers and marketed to Kolkata, Mumbai, Amritsar and Delhi fish markets. The prawn from this farm was also exported to Dubai, Malaysia, and Singapore fish markets.

1.B. Fish Farm in Semi Working Condition of Anil Ambadasrao Warpudkar-Friend's Fish Farm At Warpud Tq. Dist. Parbhani

This farm is located close to village Warpud, near Singnapur in Tq. and Dist. Parbhani. This was a famous and actively working fish farm in the initial period from its construction up to 7 years *i.e.* (1996-2003).

Unfortunately, the Poisoning by Endosulfan pesticide happened from the unknown source and mass mortality of the fish stock happened in year 2003. The farmer suffered from huge loss. Since the time of this poisoning accident, the fish farm undergone into semi-active state due to disinterest of the pond owner. Another mistake done by this farmer was mixing of small ponds to convert into single large pond, resulted in the conversion of manageable controlled structure to non-manageable uncontrolled structure. In the mean time in this vast pond, there was uncontrolled growth of weed species *Typha* sp., which acted as hideouts for fishes resulted in less catch. The pond owner tried to their level best to eradicate the typha by pond drying and removal of rhizomes and root system of the weed typha from the pond by excavation and burning but they could not succeed in getting control on the flourishment of *Typha* weed.

At present nearly ¾ area of the pond is infested by Typha. The major problem behind the failure of this pond is uncontrolled growth of Typha. It is Semi-active because there is availability of fishes in the pond for fish harvesting but due to fear of poisonous snakes, harmful insects and silt in the pond it is difficult to harvest the fishes. Therefore it is reaching total sickness stage in coming future.

In the initial stage of it's construction, it was an ideal visiting site for the fishery students, fish farmers and businessmen in this region but the failure of this pond for culture of fish has spread negative signals in the society as' Fish culture is a failure business', but actually the lack of knowledge on fish culture and careless attitude of the farmer and poisoning accident made the system failed.

The interviews of culture pond owners to determine the success and failure of fish culture in the artificially constructed or excavated ponds for fish culture were taken by questionnaire method. The explanation of interviews of pond owners which were involved in the fish culture and their business was still continued up to June 2008, the details are explained in successful case studies in this regard. The interviews of the pond owners, which were involved in fish culture and their fish culture ponds are not working still today were also taken into consideration, but for these cases only the reasons of failure of their project are given in the Table 37.2.

Six Reasons of Failure of Fish Culture in Excavated Ponds

1. Water percolation from the pond 03
2. Scarcity of water 04
3. Water scarcity and percolation problem 08
4. Faulty Size of the constructed ponds
5. Lack of technical guidance and knowledge 08
6. Economically not good as compare to the agriculture crop 04

7. Negligence by the pond owner 12

8. Non availability of fish seed 03

9. Problems of weeds, predatory animals, poisoning and poaching.

10 Plentiful availability of water but farmers prefer to use the water for agriculture instead of fish culture.

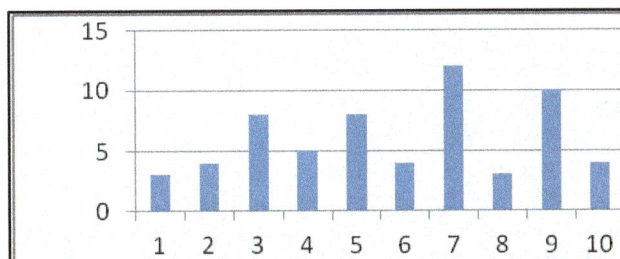

X-axis: Reasons of Failure (Mentioned as above)

Y-axis: Number of fish culture ponds

Figure 37.1: Reasons of Failure of the Fish Culture in Constructed Ponds with Respective Number

Table 37.2: Artificial Pond Owners and Reasons of Failure of their Project of Fish (2005–2008)

Sl.No.	Name and Address of the Pond Owner which are Not Working Presently	Working Duration in Years After Excavation	Reasons of Failure									
			1	2	3	4	5	6	7	8	9	10
1.	Mr. Gautam Kishanrao More. Tq. Selu	1995			√		√		√	√		
2.	Mr. Pradip Dinkar Mane. Warpud. Tq. Parbhani	1996									√	
3.	Mr. Anil Ambadas Warpudkar. Tq. Parbhani	1996									√	
4.	Mr. Pradip Dinkar Mane. Warpud. Tq. Parbhani	1996									√	
5.	Mr. Bhagvan Supadu Mahajan. Warpud. Tq.Parbhani	1996									√	
6.	Mr. Karbhari Vitthal Solunke. Rumana Tq. Gangakhed	1996		√			√	√	√			
7.	Mr. Sahebrao Kondiba Solunke. Rumana Tq. Gangakhed	1996		√			√	√	√			
8.	Mr. Aashish Wamanrao More.Chinch Takali Tq. Gangakhed	1996		√			√	√	√			
9.	Mr. Abdul Kadar Mohammad (Adanwala) Ansari. Babhulgaon Tq Pathri	1996			√		√	√	√			
10.	Mr. Prafulchand Devman Ghahire. Niwali(45) Tq Pathri	1996							√			
11.	Mr. Kundlik Devrao Shelke. Kawalgaon wadi Tq Purna	1996		√		√		√		√		

Contd...

Table 37.2–Contd...

Sl.No.	Name and Address of the Pond Owner which are Not Working Presently	Working Duration in Years After Excavation	Reasons of Failure									
			1	2	3	4	5	6	7	8	9	10
12	Mr. Kisan Sampatrao Magar. Gogalgaon. Tq Selu	1997	√									
13.	Mr. Keshav Balabhau Magar. Gogalgaon. Tq Selu	1997	√									
14.	Mr. Sharad Haribhau Kadar. Gogalgaon. Tq Selu	1997			√							
15.	Mr. Ratan Narayan Kadar. Gogalgaon. Tq Selu	1997			√							
16.	Mr. Bhaskar Punjaji Wankede. Yetoli. Tq Jintur	1997	√								√	√
17.	Mr. Madukar Raghunath Mohite. Sonkhed. Tq Sonpeth	1997		√								
18.	Mr. Sambhaji Dagudi Dake. Kawalgaonwadi Tq Purna	1998			√		√		√		√	
19.	Mrs Kalawatibai Laksmanrao Shelke. Kawalgaonwadi Tq Purna	1998			√		√		√		√	
20.	Mr. Keshav Vitthalrao Jodhale. Kamkhed. Tq Purna	1998	NA									
21.	Mr. Arun Vishwanath Jukate. Lohgaon.tq Parbhani	2000			√							
22.	Mr. Kerba Gyanoji Soudagar. Sarangi Tq Purna	2003							√	√	√	√
23.	Mr. Manik Bapurao Soudagar. Sarangi Tq Purna	2003							√	√	√	√
24.	Mr. Zotig Sadashiv Kalaskar, Dharkhed. Tq Gangakhed	2003							√			√

Right (√) indicates the reasons of failure of fish culture shown as above.

NA: Not Available.

Source: The interviews of the above mentioned pond owners during 2005-2008.

1.C. Future Chances of Pond Culture Fishery Development

The future chances of development of artificial fishery in Parbhani district is explained as below.

The region has good potential for this type of fish culture practice but there should be strict follow up of the plans of development suggested as below:

1. *Creation of Ideal Model Pond or Pilot Project in Every Taluka of Parbhani District by State Fisheries Department*: Inland fishery development activity in Maharashtra State is governed by state fisheries Department under Ministry of Agriculture, Dairy, Animal Husbandry and Fisheries Development Officer recently called as Assistant Commissioner of fisheries and their assistant staff to promote the fish culture in artificially constructed pond culture activity there should be at least one ideal model of constructed pond in every taluka of the district. Such Ponds

Figure 37.2: View of Stocking Ponds Used for Carp Culture in Parbhani District

should be constructed and maintained by State Fisheries Department. The ideal ponds will act as visiting sites to understand the fish culture activity in the ponds for those who are interested in this business. Due to visit to such ponds, the interested person will get an idea about possibility of success and failure of the plan of construction of fish culture pond in his own agriculture land. The interested farmers who are willing to construct the culture pond should report and discuss with District Fisheries Development Officer/Assistant Fisheries Commissioner Officer. Then the office should manage the visiting day for a group of farmers interested in pond culture to the ideal or model pond site, which will be present in every taluka, therefore the access of repeated visit will be possible for the farmers to the model site. The model ponds construction and maintenance, monitoring, extension, education, training etc. should be from the Government.

2. *Extension Activities by State Fisheries Department to Promote the Farmers to Involved in Fish Culture*: The State Fisheries department should have to involve in the extension activities to promote the inland fishery development by artificial pond construction. The extension activities may include preparation and distribution of information bulletin about fish culture in pond. There must be special budget for the distribution of such information in every

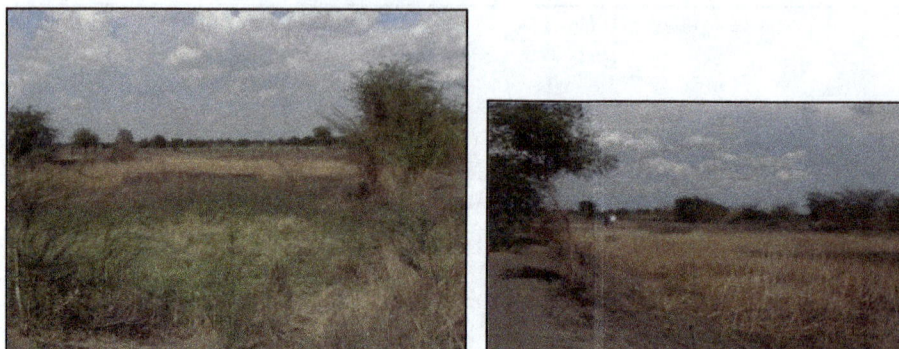

Figure 37.3: Non-Working Pond at Rumna in Gangakhed Taluka

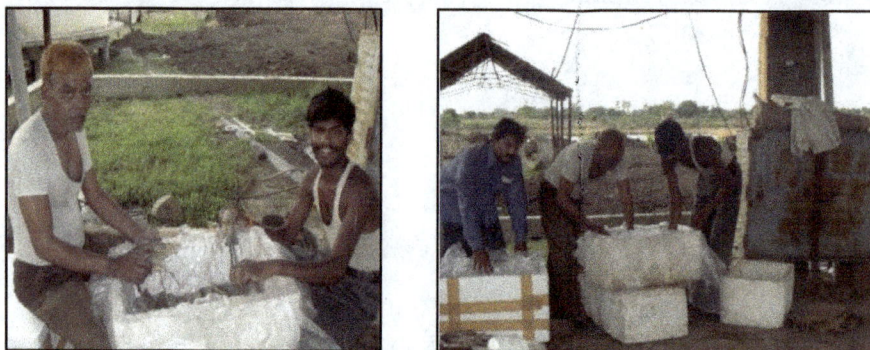

Figure 37.4: Fish and Prawn Packing at Sudarshan Fish Hatcheries at Pedgaon in Parbhani District

village; such activity is regularly done by District Fisheries Development office but for the distribution of this literature, it was found that the office does not take special efforts.

Radio talk is one of the best, for providing information towards the farmers. The interested farmers in artificial pond fish culture should be trained theoretically and practically on the ideal or Model Fish Culture ponds under Government control and then only the farmers will be allowed to construct the one. Due to training to the farmers, the farmers themselves will decide whether to involve in this business. At present, the fish farmers receive the fish seed from the Government hatcheries, usually it is a mixed fish seed of random proportion of the species, some time the fish seed contain only one dominant species while other is less is number. When fish seed is stocked into the constructed pond then the farmer, do not have an idea of how much actual seed is stocked, of what kind, and of what species. Therefore, during seed supply the farmers must get the correct information about type, species and number of fish seed to be stocked according to the size of pond.

The interested fish farmers should get correct information about pre-stocking and post-stocking management of the constructed pond.

3. *The farmers having 5 to 10 ha irrigated land area could be diverted to fish culture in pond in Small Scale.*

4. *Purna taluka in Parbhani district has good potential in the form of shallow groundwater level, water lodged soil, and canal system for getting water source, therefore in this area the fish culture practice by artificial pond construction could be promoted.*

5. *Need of survey for identification of potential sites for the fish culture by pond construction:* The state fisheries department in co-ordination with state geological survey department by taking the reference of toposheet map should identify the suitable area for artificial pond construction; such possible sites should be evaluated for the suitability of the land to construct the pond.

   The information about the availability low line area, the land where water pools develops during Monsoon, such information could be collected from the farmers in the villages and such sites are necessary to visit by the state fisheries officers to check out the possibility of formation of fish culture pond. This will be easiest but effective method to determine the possible sites of pond construction, in this business there will be co-operation and co-ordination form Agricultural assistants, Gramsevak, and Talathi to collect the information.

6. *Use of durable polythene as pond bottom cover to prevent water percolation from the pond:* If there is plenty of water available in a region but the bottom of pond and sidewalls, have fast and greater water percolation rate. Then such ponds could be covered with high quality durable thick polyethylene sheets. It is a recently used technique for fish culture.

7. *Stocking of fingerling stages of fast growing species after peak monsoon in artificial pond and total harvesting during summer:* In major area of the districts, the groundwater level during summer is very deep upto 150 feet similarly the ponds may get evaporated, therefore instead of stocking spawn stages or fry stages the fingerling stages of fast growing species like catla could be stocked during monsoon and harvested during summer.

8. *Few hector land in watershed area, near every reservoir as a good potential site for pond culture:* It was observed that, nearly every reservoir develop water shed close to the embankment but in this region, there was no fish pond. In such area, a pond for fish culture could be constructed.

Result and Discussion

In District Parbhani, working artificial fish culture pond maintained by private parties under promotion of F.F.D.A were only 03 out of enrolled 28 in total and one private pond named Sudarshan fisheries Development Corporation Fish Culture Pond, At Jamb Tq. Dist Parbhani. Rest of the enrolled ponds to the records of the F.F.D.A were found closed due to various reasons like water percolation, less available of water has main reasons. For culture, pond development there is need of C.I.F.A. assisted Projects of Composite carp culture in Maharashtra, due to which transfer of fish culture technology generated by C.I.F.A. will definitely helpful to the state in fish production (Sangama *et.al*). The other reasons behind the fewer establishments of the culture ponds in the districts are, no proper survey of suitable sites for pond construction, lack of knowledge of the farmers about pond culture fishery (Veerapa G.H.S.,2009), and no feedback from state fisheries department about the working/ non working of established but closed culture ponds, if all these things cleared then there is high possibility of new establishment of success of fish culture ponds in the districts. Sarangi N and P. C. Das (2007) mentioned the need of diversified freshwater fish farming systems to boost up present 6.4 million tones fish production up to 8.2 million tones of fish per annum to cater to the domestic demand as per the WHO prescription of per capita requirement at 11.2 kg fish in India; They further mentioned the results of species diversification in culture ponds at CIFA that the technology of intensive carp culture involving stocking of advanced fingerlings of 25-50mm at higher densities of 15,000-25,000/ha and provision of balanced supplementary feed, fertilization, aeration, water exchange and fish health management has demonstrated higher production levels of 10-15 tones/ha/yr. Monoculture of giant freshwater prawn (*Macrobranchium* rosembergii) at stocking densities of 30,000-50,000/ha has production levels of 1.0-1.5 tones/ha during a farming period of 7-8 months. The prawn form excellent component species for poly culture with Indian Major Carps. Prithviraj Jha and Sudip Bharat (2007) reported Ornamental fish (*Pterophyllum scalare* and *Carrassius auratus*) Culture by the involvement of Women Self-Help Groups in Darjeeling District of West Bengal. They further explained that if women are trained to breed and rear ornamental fish, an individual could earn a profit of Rs 500 per head. There is a need of such type of projects, training extension in culture pond fishery in this region. The details explained earlier about pond culture fishery is the first record during this investigation in the Parbhani district. Earlier to this work, no one has carried out such type of work in these districts.

References

Ahirao, S.D. and Mane, A.S., 2000. The diversity of ichthyofauna, taxonomy and fisheries from freshwater of Parbhani district, Maharashtra State. *J. Aqua. Biol.*, 15(1 and 2): 40–43.

Anand, P.E Vijay, 2009. Indian carp farms: Application of improved feeds–Status and scope. *Fishing Chimes*, 29(1): 169–172.

Biswas, K.P., 1990. *A Text Book of Fish, Fisheries and Technology*, Narendra Publishing House, Delhi, p. 143–164.

Chakraborty, N.M. *et al.*, 2004. Aquaculture development in north-east India: CIFE's role. *Fishing Chimes*, 24(1): 114–116.

Dash, P.C., 2004. Role of NABARD in aquaculture development with special reference to Andhra Pradesh. *Aqua. International*, March, p. 39–42.

Jha, Prithviraj and Bharat, Sudip, 2007. Ornamental fish (*Pterophyllum scalare* and *Carassius auratus*) culture: Involvement of women self-help groups in Darjeeling district of West Bengal. *Fishing Chimes*, 29(1): 72–75.

Martyshew, F.G., 1983. *Pond Fisheries*. Amerind Publishing Co. Pvt. Ltd., New Delhi.

Sangma, C.T., Chakraborti, N.M., Chakraborti, P.P. and Mondal, S.C., 2007. Composite carp culture: CIFA-Assisted projects in Meghalaya, Boon to the State. *Fishing Chimes*, 27(1): 92–96.

Santhanam, R., Sukumaran, N. and Natrajan, P., 1987. *A Manual of Freshwater Aquaculture*. Oxford and IBH Publishing Co. Pvt. Ltd.

Sakthivel, M., 2007. Enhancement of aquaculture production in India: A few key issues. *Fishing Chimes*, 27(1): 152–157.

Sarangi, N. and Das, P.C., 2007. Diversified freshwater fish farming systems: Present status in relation to research inputs. *Fishing Chimes*, 27(1): 58–63.

Srivastava, U.K., Dholkia, B.H., Sreenivasrao, S. and Vatsala, S., 1993. *Freshwater Aquaculture in India*. Oxford and IBH Publishing Co. Pvt. Ltd.

Veerapa Gowda, H.S., 2009. New initiatives for the development of culture fisheries of Karnataka. *Fishing Chimes*, 29(1): 69.

Chapter 38

Role of Bat Guano in the Bioremediation of Aquatic Ecosystem

☆ *C.M. Bharambe, N.B. Patil, N.R. Rahate and B.V. Patil*

Introduction

Lonar crater is situated in village Lonar in the Buldhana District of Maharashtra, India. It has an almost perfectly circular shape and accumulated with water in the deeper parts of basin. Rocks in the crater reveal many characteristic features of the moon rocks. There are many old temples on the peripheral boundary of the crater which have now become roosting places for bats. Ramgaya Temple has become the source of sweet drinking water, as this is the only sweet water stream available in the crater; rest of the crater water is highly saline. Kamalja Devi temple is situated at the southern base of the crater. Morache temple (Peafowl's temple) is now famous for existence of thousands of bats and peacocks. Waghache temple (Leopards temple) is also famous for bats and people have seen leopard found in it many times.

Bat Guano

The word guano originated from the Quichua language of the Inca civilization and means "the droppings of bat". The bats forage at night for insects over a particular area, and they return to the old temples during the day to sleep and care for their young. They attach themselves to ceiling, and their excrement accumulates on the floor below. In some situations the guano can reach a depth of feet in many years and appeared as guano-hip, and it has a valuable importance.

Bioremediation and Bat Guano

One of the most serious universal, international problems facing us today is the removal of harmful compounds from industrial and municipal waste. If it is discharged into lakes and rivers, a process called eutrophication occurs (Prince, 2003).

Environmental contamination whether it is from industrial or municipal toxic waste that degrades the various environments is a vital concern to the public. Thus it is crucial to develop and implement accurate means to clean and preserve our precious and deteriorating environment. Although there are many techniques in cleaning environmental contaminations, one process has the most potential, namely bioremediation. Bioremediation, or commonly referred to as biodegradation, is a process in which microbes such as bacteria, fungi, yeast, or micro algae are involved in degrading toxic wastes (Pace, 1997 and Knezevich, 2006).

A marvelous symbiosis exits between the microorganisms and bat guano. Bacteria in the mammalian intestinal tract aid in the breakdown of food during digestion. These organisms synthesize enzymes capable of degrading a vast array of substances. Innumerable microbes are regularly excreted along with waste products and together with other organisms; they constitute the microbial population of a bat guano deposit (Steele, 1989).

Large populations of bat deposit thousands of kilograms of dropping annually. An ounce of bat guano contains billions of bacteria, and a single guano deposit may contain thousands of bacterial species. Guano being rich in bioremediation microbes cleans up toxic substances, (Barry *et al.*, 1997). At present we do not know these species.

Materials and Methods

To study the role of bat guano in the degradation of water pollutants, 10 mg bat guano was dissolved in 100 ml of experimental water (10: 100 proportions) for both times. After addition of bat guano in water, then the water was analyzed for the change in its pH, chloride, nitrate (NO_2), phosphate (PO_4) and sulphate (SO_4) contents. The change in water parameters were noted after every two hour upto 24 hours. Thereafter, the samples were kept undisturbed and analyses were carried out for 30 days at an interval of 5 days. The water was analyzed by using standard methods for water analysis suggested by APHA (1998).

Table 38.1: Impact of Bat Guano on Water Content at an Interval of 2 hrs

Ps	Sg	Time (Hrs)												
		0	2	4	6	8	10	12	14	16	18	20	22	24
pH	W1	5.00	6.15	6.49	6.55	6.65	6.82	6.55	6.84	6.85	7.91	7.95	7.02	7.25
	W2	7.19	7.25	7.30	7.34	7.41	7.46	7.50	7.56	7.60	7.63	7.65	7.66	7.66
Cl⁻	W1	201	187	173	160	155	143	132	125	115	107	98	93	91
	W2	211	205	202	196	189	180	173	160	143	136	122	107	106
NO_2	W1	56.5	52.4	48.8	47.0	44.8	41.5	39.5	29.5	27.8	26.1	25.4	24.8	24.4
	W2	34.0	33.0	32.4	31.6	30.7	29.7	29.0	28.2	28.1	27.6	27.1	26.6	26.0
PO_4	W1	57.5	56.5	55.0	45.0	44.0	43.5	39.5	35.5	32.0	31.5	30.5	29.5	29.5
	W2	62.0	56.0	51.0	48.0	45.5	41.5	38.0	32.6	31.4	27.6	25.5	24.5	24.5
SO_4	W1	46.8	46.0	45.6	44.5	43.8	42.3	39.5	37.2	34.8	33.2	30.5	29.6	29.6
	W2	54.2	51.2	48.2	47.0	45.5	43.6	40.4	38.2	37.2	34.6	34.4	34.2	34.2

All values are the mean of five replicates.

Ps: Parameters; Sg: Sampling; W1: Water from industrial waste; W2: Water from lake, Wadali.

Observations and Results

When bat guano was dissolved in industrial and lake water with pH 5.00 and 7.19. After 2 hours the pH was found to be changed to 6.15 and 7.25 respectively and after 4 hours increased gradually and it reached to 7.25 and 7.66 after 24 hours (Table 38.1). The industrial and lake water was kept undisturbed till 30 days and the pH was noted after every 5 days upto 30 days. After 5 days the pH was seen to be increased upto 20 days and then it remained constant during 25 to 30 days of observations (Table 38.2).

Table 38.2: Impact of Bat Guano on Water Content at an Interval of 5 Days

Ps	Sg	Time (Days)							
		0	1	5	10	15	20	25	30
pH	W1	5.00±0.37	7.25±0.39 (+45.00)	7.42±0.24 (+48.40)	7.38±0.30 (+47.60)	7.40±0.32 (+48.00)	7.43±0.40 (+48.60)	7.55±0.45 (+51.00)	7.55±0.40 (+51.00)
	W2	7.19±0.52	7.66±0.34 (+4.73)	7.75±0.42 (+6.54)	7.81±0.45 (+9.18)	7.85±0.57 (+8.48)	7.80±0.50 (+8.21)	7.78±0.56 (+7.93)	7.78±0.58 (+7.93)
Cl⁻	W1	201±7.60	91±8.83 (−54.73)	86±10.95 (−57.21)	84±10.09 (−47.26)	83±9.73 (−58.21)	82±9.41 (−58.71)	81±11.06 (−59.70)	81±9.68 (−59.70)
	W2	211±11.40	106±10.38 (−49.76)	101±11.96 (−52.13)	99±10.09 (−48.34)	97±10.49 (−54.03)	96±9.98 (−54.50)	95±11.29 (−54.98)	95±10.11 (−54.98)
NO$_2$	W1	56.5±2.71	24.4±1.12 (−56.81)	24.0±1.25 (−57.52)	23.6±1.49 (−58.23)	23.1±1.28 (−58.94)	22.8±1.13 (−59.29)	22.2±1.29 (−59.29)	22.2±1.40 (−59.29)
	W2	34.0±1.63	26.0±1.20 (−23.53)	25.5±1.33 (−25.00)	24.8±1.56 (−27.06)	24.2±1.33 (−28.82)	23.8±1.17 (−30.00)	23.6±1.32 (−30.59)	23.6±1.44 (−30.59)
PO$_4$	W1	57.5±3.05	29.5±1.68 (−48.70)	27.5±1.96 (−47.83)	26.4±1.45 (−54.09)	25.3±1.29 (−56.00)	23.0±1.15 (−60.00)	21.4±1.28 (−62.78)	21.4±1.20 (−62.78)
	W2	62.0±3.29	24.5±1.40 (−60.48)	25.5±1.66 (−58.87)	23.6±1.30 (−61.94)	21.2±1.08 (−65.81)	20.5±1.03 (−66.94)	19.2±1.15 (−69.03)	19.2±1.08 (−69.03)
SO$_4$	W1	46.8±2.48	29.6±1.68 (−37.18)	28.9±1.59 (−38.25)	28.1±1.43 (−39.96)	27.8±1.39 (−40.60)	26.9±1.61 (−42.52)	26.1±1.46 (−44.23)	26.1±1.70 (−44.23)
	W2	54.2±2.87	34.2±1.95 (−36.90)	32.5±1.79 (−40.04)	30.6±1.56 (−43.54)	27.2±1.36 (−49.82)	26.5±1.59 (−51.11)	25.2±1.41 (−53.51)	25.2±1.64 (−53.51)

All values are the mean±SE of five replicates.

Figures in parenthesis indicate percent change over the result on 0 day.

Ps: Parameters; Sg: Sampling; W1: Water from industrial waste; W2: Water from lake, Wadali.

When bat guano was dissolved in industrial water with chloride (201); nitrate (56.5); phosphate (57.5) and sulphate (62.0), after 2 hours the parameters was found to be changed to chloride (187), nitrate (52.4), phosphate (56.5) and sulphate (46.0) and after 4 hours decreased gradually to chloride (91), nitrate (24.4), phosphate (29.5) and sulphate (29.6) upto 24 hours (Table 1). The industrial water was kept undisturbed till 30 days and the chloride, nitrate, phosphate and sulphate was noted after every 5 days upto 30 days. After 5 days the parameters was seen to be decreased upto 20 days and then it remained constant during 25 to 30 days of observations (Table 38.2).

When bat guano was dissolved in lake water with chloride (211); nitrate (34.0); phosphate (62.0) and sulphate (54.2), after 2 hours the parameters was found to be changed to chloride (205), nitrate

(33.0), phosphate (56.0) and sulphate (51.2) and after 4 hours decreased gradually to chloride (106), nitrate (26.0), phosphate (24.5) and sulphate (34.2) upto 24 hours (Table 38.1). The lake water was kept undisturbed till 30 days and the chloride, nitrate, phosphate and sulphate was noted after every 5 days upto 30 days. After 5 days the parameters was seen to be decreased upto 20 days and then it remained constant during 25 to 30 days of observations (Table 38.2).

Discussion

Tilak *et al.* (2005) reported a number of bacterial species associated with the bat guano belonging to genera, *Azospirillum, Alcaligens, Arthrobacter, Acinetobacter, Bacillus, Burkholderia, Enterobacter, Erwinia, Flavobacterium, Pseudomonas, Rhizobium* and *Serratia*. He also suggested that this bacterium has high bioremediation capacity. Hutchens *et al.* (2004) had demonstrated aerobic methane oxidizing bacteria, Methylomonas and Methylococcus in bat guano.

The bacterial enzymes capable of degrading a number of substances (Martin, 1991; Dvorak *et al.*, 1992; Edenborn *et al.*, 1992; Bechard *et al.*, 1994; White and Chang, 1996; Frank, 2000; Kaksonen, *et al.*, 2003; Vallero *et al.*, 2003; Boshoff *et al.*, 2004; Miranda, 2005; Seena, 2005; Tilak *et al.*, 2005). Murphy (1989) demonstrated a nutritious broth formation when the bat guano was added in water and further he proved that this broth supported the growth of numerous microbes.

Alley and Mary (1996) stated that an ounce of bat guano contains billions of bacteria and thousands of bacterial species and these bacteria are important to bioremediation. Sridhar *et al.* (2006) and Pawar *et al.* (2004) examined the fungal fauna of bat guano and used for bioremediation of lack soil.

Conclusions

Other than municipalities, various industries disposing off the industrial effluents are the worst polluters of the aquatic resources. It is of utmost importance, hence, to prevent the pollution of aquatic resources by all possible means to control its quality from further deterioration. Applying microorganisms for industrial pollution control is an area of interest all over the world.

In the present investigation is an attempt to study the impact of bat guano with its rich microbial flora on bioremediation of industrial effluents and water from lake. The results revealed that within a period of 30 days, there was a remarkable reduction in the physico-chemical parameters of industrial effluents, thus stabilizing the industrial effluents, suggesting that industrial effluents can be properly treated by bat guano.

No much work has been carried out on the bat guano in India and hence it was thought to study the impact of bat guano from and to assess the feasibility of the bat guano as supplementary bioremediatant.

References

Aaranson, S., 1970. *Experimental Microbial Ecology*. Academic Press, New York, pp. 236.

APHA, 1998. *Standard Methods for the Examination of Water and Wastewater*, 20th Edn. APHA, AWWA and WEF New York, Washington DC.

Boyd, S.A. and Patricia, E.G., 2005. An Approach to evaluation of the effect of bioremediation on biological activity of environmental contaminants: Dechlorination of polychlorinated biphenyls. *Environmental Health Perspectives*, 113(2): 180–185.

Chapelle, F.H. *Bioremediation: Nature's Way to a Cleaner Environment*. U. S. Geological Survey. URL: http://water.usgs.gov/wid/html/bioremed.html.

Conde-Costas, C., 1991. The effect of bat guano on the water quality of the Cueva EL Convento stream in Gauayanilla. *Puerto Rico. Nss. Bull.*, 53(1): 15.

Dash, M.C., Mishra, P.C., Kar, G.K. and Das, R.C., 1986. Hydrobiology of Hirakund Dam Reservoir. In: *Ecology and Pollution of Indian Lakes and Reservoirs*, Mishra Publishing House, New Delhi, p. 317–337.

Dilip, K.M. and Markandey (Eds.), 2002. *Microorganisms in Bioremediation*. Capital Pub., New Delhi.

Dvorak, D.H., Hedin, R.S. and McIntire, P.E., 1992. Treatment of metal contaminated water using bacterial sulphate reduction: results from a pilot-scale reactor. *Biotechnol. Bioeng.*, 40: 609–616.

Edenborn, D.H. and Hedin, R.S., 1992. Treatment of water by using sulphate reducing bacteria. *Biotech. Bioeng.*, 30: 512–516.

Everett, J.W., Gonzales, J. and Kennedy, L., 2004. Aqueous and mineral intrinsic bioremediation assessment: Natural attenuation. *Journal of Environmental Engineering*, 130(9): 942–950.

Faison, B.D. and Knapp, R.B., 1997. A bioengineering system for *in situ* bioremediation of contaminated groundwater. *Journal of Industrial Microbiology and Biotechnology*, 18(2–3): 189–197.

Keleher, S., 1996. *Guano: Bats' Gifts to Gardeners*, 14(1): 15–17.

Knezevich, V., Koren, O., Ron, E.Z. and Rosenberg, E., 2006. Petroleum bioremediation in seawater using Guano. *Bioremediation Journal*, 10(3): 83–91.

Pace, N.R., 1997. A molecular view of microbial diversity and the biosphere. *Science*, 276: 734–740.

Pawar, K.V. and Deshmukh, S.S., 2004. Bioremediation of Lack soil using bat guano. *Indian J. Environ and Ecoplan.*, 8(3): 699–704.

Pierce, W., 1999. Speech on 'Bat guano' Sept., 1999. Cassette from National Vanguard Books, P.O. Box 330, Hillsboro, WV 24946.

Prince, R.C., 2003. *Bioremediation in Marine Environments*. Prince RC. Exxon Research and Engineering, Annandale, NJ 08801. Bioremediation.

Steele, D.B., 1989. Bats. *Bacteria and Biotechnology*, 7(1): 3–4.

Tuttle, M.D., 1986. Endangered gray bats benefits from protection. *Bat.*, 4(4).

Vidali, M., 2001. Bioremediation. An overview. Dipartimento di Chimica Inorganica, Metallorganica, e Analitica, Università di Padova Via Loredan, 435128 Padova, Italy. *Pure Appl. Chem.*, 73(7): 1163–1172.

Walecha, V., Vyas, V. and Walecha, R., 1993. Rehabilitation of the twin lakes of Bhopal. In: *Ecology and Pollution on Indian Lakes and Reservoir*. Ashish Publishing House New Delhi, p. 317–337.

Chapter 39

Effect of Pollutants from Car Washing Centre on Oxygen Consumption in Freshwater Fish *Channa punctatus*

☆ *A.R. Jagtap, Shaikh Afsar, S.D. Kothole and R.P. Mali*

Introduction

The primary source of pollution is wastewaters containing toxic substances in the form of pesticide residue, heavy metal salts, oils etc. as reported by Akberali *et al*. (1981). Modern civilization with its rapidly growing industrial units and an increase in the population, has lead to an accelerate degradation of the freshwater resources. The water bodies are subjected to a wide variety of human activities such as washing, swimming, bathing and waste disposal, disposal of industrial effluents etc. These pollutants are likely to affect the biological systems in different ways according to their chemical properties. The sum of physiological changes created particular pollutants is likely to be characteristics of these pollutants. Thus by observing the effects of polluted water and a set of physiological parameters. It might be possible to establish specific responses of that pollutant. From this it is easy to identify a pollutant on the basis of its physiological effect pattern (Sastry *et al*., 1979).

Water resources are said to be polluted, when because of man's activity in adding or causing the addition of matter to the water or altering the temperature, the physical, chemical or biological characteristics of the water are changed such an extent that its utility for any reasonable purpose or its environmental value is demonstrably depreciated. The aquatic animals are susceptible to such various pollutants, but they have to adjust to these new circumstances by changing their metabolic activities. The higher concentration of toxicants brings about the adverse effects of an freshwater fishes which causes gills damage, skin of fishes, lack of availability of natural food to fishes, depletion of dissolved oxygen, reduction in maturation of oocytes, necrosis of seminiferous tubules and hypertrophy of cell etc.

The objectives of the present study are to evaluate the effect of pollutant on oxygen consumption on freshwater fish *i.e. Channa punctatus*. Respiration is an endless oxidative process in a living animal resulting in consumption of O_2 and production of CO_2. Therefore the calculation of oxygen consumed especially with reference to energy utilization. by fishes can by expected o throw lighten the physiological mechanism in animals,

Material and Methods

Medium sized freshwater fishes *i.e. Channa punctatus* were collected from Godavari River, Nanded. The fishes were acclimatized in the laboratory for 2-3 days prior to experiment. The fishes were feed with pieces of bivalve and earthworm. Feeding was stopped one day before the commencement of experiment. The fishes were divided into five sets I, II, III, IV, and V. Set V fishes were maintained as a control. The set I to IV were exposed to different concentration of polluted sample.

Figure 39.1: Graphs Showing Rate of Oxygen Consumption Vs Exposure Time in hrs for I, II, III, IV, and V Sets

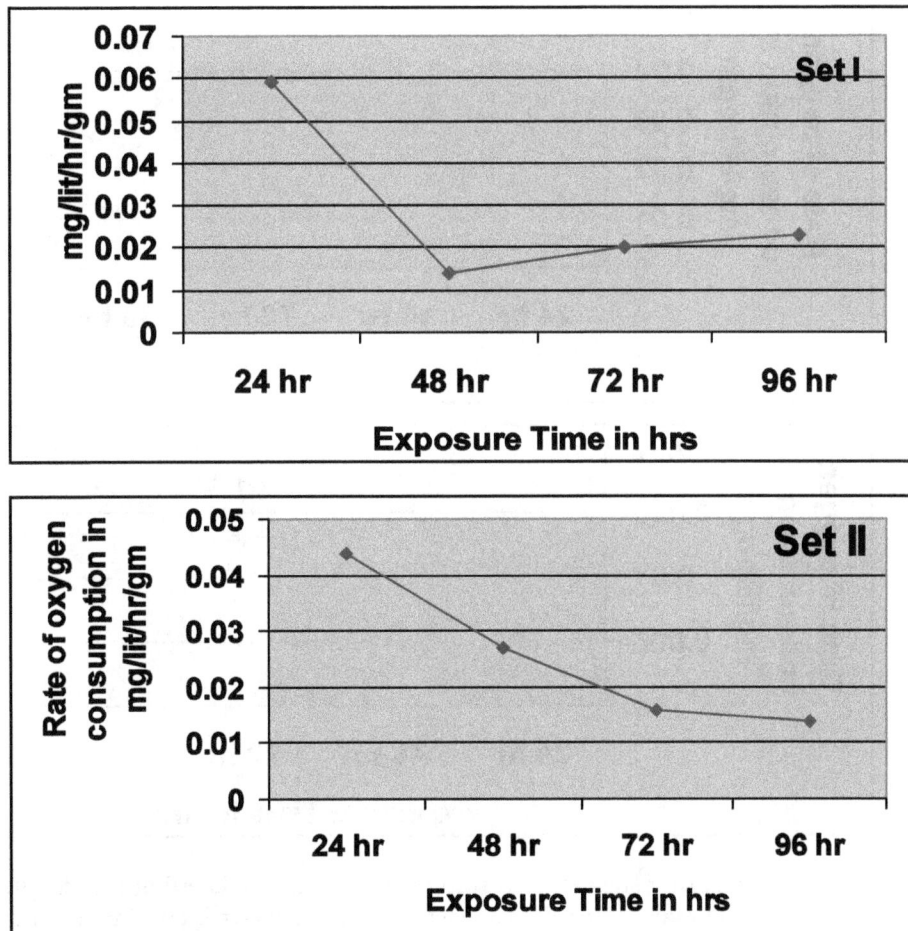

Contd...

Figure 39.1–Contd...

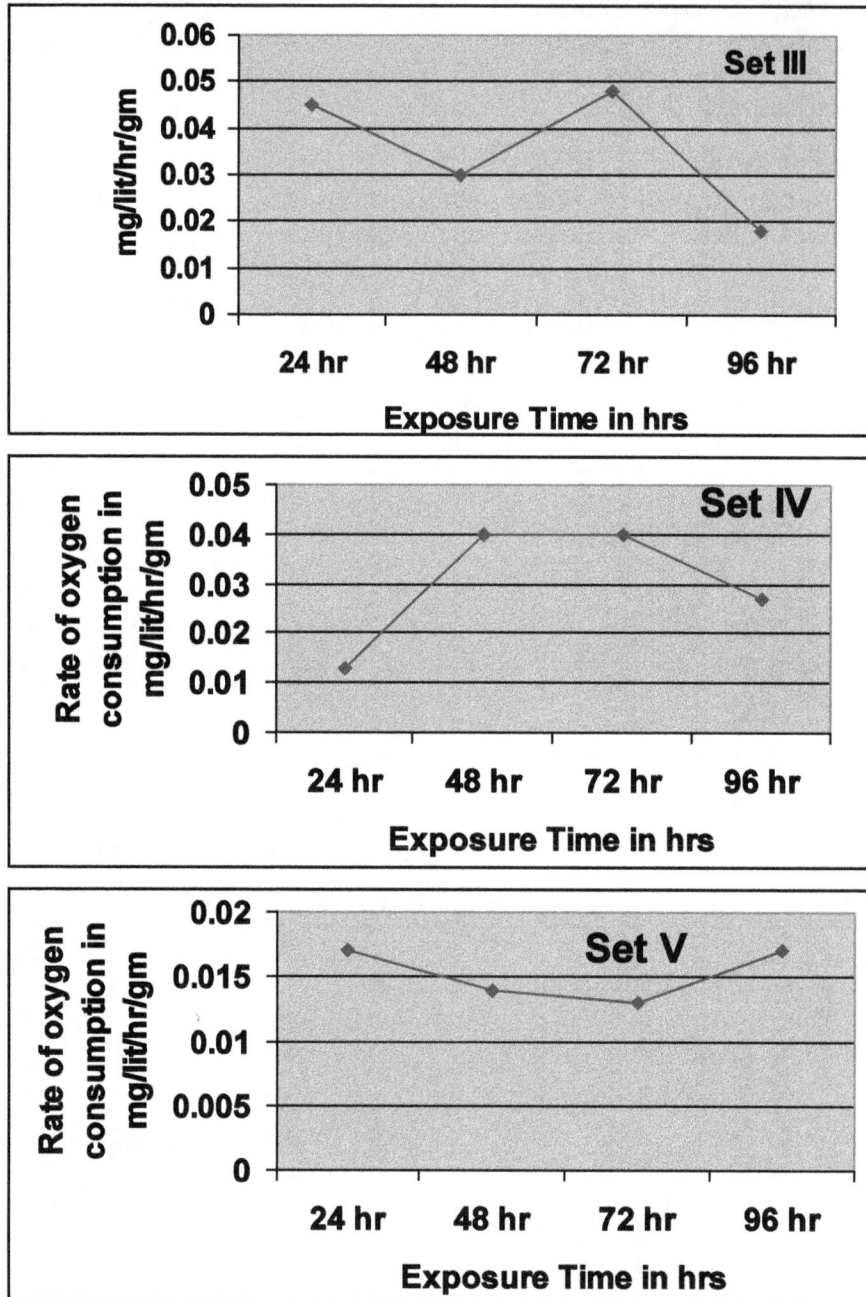

The oxygen consumed by fishes in each set was examined by keeping the fish in respiratory chamber. The weights of animals were noted at each time. The sets were continued for 24 hrs, 48 hrs, 72 hrs and 96 hrs.

Estimation and Measurement of Oxygen Consumption

Oxygen consumption of the fishes in each sets (*i.e.* I, II, III, IV and V) was measured by the method of Winkler as described by Welsch and Smith (1953). The fishes were weighed and placed in Winkler's chamber and care was taken to make it air tight and free from leakage of water. The fish was allowed to stabilize in the chamber for few minutes. After few minutes water was collected into narrow bottle and dissolved oxygen was estimated by Iodometric method. After one hour the next sample was estimated in the same way. The difference between initial and final sample will give the actual oxygen consumed by the animal and expressed as oxygen consumption ml/hr/gm body weight of fish.

Results and Discussion

The rate of respiration of freshwater fish, *Channa punctatus* has been found altered when exposed to different concentrations of polluted water sample. Heavy metals discharged into water resources cause hazardous effect on aquatic life (Kaviraj, 1983 a and b).

Table 39.1: Rate of Oxygen Consumption by Fish Under Different Set of Experimental Conditions

Sl.No.	Aquarium Set No.	Weight of Animal	Exposure Time (hrs)	Burette Reading		Total O_2 Consumed mg/lit	Rate of O_2 Consumed mg/lit/hr/gm
				Initial Sample (mean)	Final Sample (mean)		
1.	I	40.00 gm	24 hrs	4.0 ml	2.0 ml	2.252	0.0592
			48 hrs	7.0 ml	6.5 ml	0.564	0.0148
			72 hrs	8.2 ml	7.5 ml	0.788	0.0207
			96 hrs	8.3 ml	7.5 ml	0.9	0.023
2.	II	40.27 gm	24 hrs	5.0 ml	3.4 ml	1.804	0.044
			48 hrs	7.3 ml	6.3 ml	1.124	0.027
			72 hrs	8.6 ml	8.0 ml	0.676	0.016
			96 hrs	8.0 ml	7.5 ml	0.576	0.014
3.	III	40.02 gm	24 hrs	6.0 ml	4.5 ml	1.688	0.045
			48 hrs	7.5 ml	6.5 ml	1.124	0.03
			72 hrs	8.7 ml	7.0 ml	0.0516	0.048
			96 hrs	8.6 ml	8.0 ml	0.676	0.018
4.	IV	41.21 gm	24 hrs	8.0 ml	7.5 ml	0.564	0.013
			48 hrs	7.5 ml	6.0 ml	1.688	0.04
			72 hrs	8.5 ml	7.0 ml	1.688	0.04
			96 hrs	8.0 ml	7.0 ml	1.124	0.027
5.	V Control	40.12 gm	24 hrs	6.6 ml	6.0 ml	0.672	0.017
			48 hrs	7.0 ml	6.5 ml	0.568	0.014
			72 hrs	8.5 ml	8.0 ml	0.56	0.013
			96 hrs	8.1 ml	7.5 ml	0.676	0.017

*: Number of Animals in each set = 05.

The oxygen uptake measured at every 24 hr till the end of experiment *i.e.* up to 96 hr. The results showed that the decrease in rate of oxygen consumption throughout the experiment when compared with control group of animals in each set.

The rate of respiration by the control group at 24 hr (0.017), 48 hr (0.014), 72 hr (0.013), and 96 hr (0.017). upon exposure to polluted water the rate of oxygen consumption in first set (500 ml pollutant + 1500 ml tap water) was at 24 hr (0.0592), 48 hr (0.0148), 72 hr (0.0207) and 96 hr (0.023). The rate of respiration in set II (1000 ml pollutant + 1000ml tap water) was at 24 hr (0.044), 48 hr (0.027), 72 hr (0.016), and 96 hr (0.014). The rate pf respiration in set III (1500 ml pollutant + 500 ml tap water) at 24 hr (0.045), 48 hr (0.030), 72 hr (0.048), and 96 hr (0.018). The rate of respiration in set IV (2000 ml pollutant) at 24 hr (0.013), 48 hr (0.040), 72 hr (0.040), and 96 hr (0.027).

From the above result it is very clear that the rate of oxygen consumption was found decreased in all first three sets (I, II, III).Dissolved oxygen decrease due to mixing of sewage into river water (Agarwal *et al.*, 2000).In set IV it was observed that rate of oxygen consumption was increased as the time exposure period increase up to 72 hr and decreased at 96 hr.The metal induced changes in a respiration is complicated and vary from metal to metal and from species to species and from one experimental condition to other(Mali *et al.*, 2009).

Thus, from this it was clear that various pollutants affect the fish life directly or indirectly. Freshwaters are highly vulnerable to pollution since they act as immediate sinks for the consequences of human activity always associated with the danger of accidental discharges or criminal negligence (Vutukuru, 2005). The extent of damage depends on the quality and quantity of the pollutants and the species of fish. The decreased rate of oxygen consumption when exposed to pollutant is due to depletion of dissolved oxygen content of water and increase in BOD. The decrement may be due to the respiratory distress as a consequence of the impairment of oxidative metabolism (Prashanth *et al.*, 2003). The pollutants also cause the damage of mucus membrane of the gills which directly affects the rate of respiration in freshwater fishes.

References

Agarwal, T.R., Singh, K.N. and Gupta, A.K., 2000. Impact of sewage containing domestic waste and heavy metal on the chemistry of varun river water. *Poll. Res.*, 19(3): 491–494.

Balaparameswara Rao, M. and Padmavati, V.V., 2004. Effect of oral administration of Doxycycline onthe oxygen uptake of the *Indian mossambica. Ecotoxicol. Environ. Safe.*, 8, 289–293. Major carp, *Catla catla. J. Aqua. Biol.*, 19(1): 173–176.

Janardhan Reddy, V., 1995. Studies on the toxicity and effects of nitrite, nitrogen on the fish *Puntius sophore* (Hamilton) and *Channa punctatus* (Bloch.). *M.Phil. Dissertation,* Nagarjuna University, Andhra Pradesh.

Jadhav, S.M. and Sontakke, V.B., 1997. Studies on respiratory metabolism in the freshwater bivalve, *Corbicule striatella* exposed to carbaryl and cypermethrin. *Poll. Res.*, 16(4): 219–221.

Kaviraj, A. 1983a&b. Effect of mercury on behaviour survival growth and reproduction of fish and on aquatic ecosystem. *Envior. Ecol.*, 1: 4–9.

Leatherland, P. and Woo, T.K., 1998. *Fish Diseases and Disorders: Non-infectious Disorders.* CABI Publishing Oxon, U.K.

Mali, R.P., Jagtap, A.R., Kothole, S.D. and Afsar, Shaikh, 2009. Effect of zinc sulphaton oxygen consumption and heart beat in the freshwater female crab *Barytelphusa guetini*. *J. Aqua. Biol.*, 24 (1): 139–144.

Mali, R.P. and Ambore, N.E., 2003. Impact of copper sulphate on the oxygen consumption in the freshwater female crab, *Barytelphusa guerini*. *J. Comp. Toxicol. Physiol.*, 1(1): 14–18.

More, T.G., Rajput, R.A. and Bandela, N.W., 2003. Impact of industrial effluents on DNA contents in the whole body of freshwater bivalve, *Lamellidans marginalis*. *J. Industrial Pollution Control*, 19(2): 195–202.

Prashant, M.S., David, M. and Kuri, Riveendra C., 2003. Effect of cypermethrin on toxixity and oxygen consumption in the freshwater fish, *Cirrhinus mrigala*. *J. Ecotoxicology. Envron. Monit.*, 13(4): 271–277.

Schreck, C.B. and Brouna, P., 1975. Dissolved oxygen depletion in static bioassay system. *Bull. Environ. Contam. Toxicol.*, 14: 149–152.

Svobodam, 2001. Stress in fish–review. *Bull. Vu RH Vodnany*, 37: 69–194.

Subbaiah, M.B., Usha Rani, K., Geetanjali, K.R., Purushotham, R. and Rama Murthy, R., 1984. Effect of Cupric chloride on oxidative metabolism in the freshwater teleost, *Tilapia mossambica*. *Ecotoxicol. Environ. Safe*, 8: 289–292.

Schaperclaus, W., 1991. *Fish Diseases*, Vol. 1, Oxonian Press Pvt. Ltd., New Delhi.

Singh, S.R. and Singh, B.R., 1979. Change in oxygen consumption of siluroid fish, *Mytus vittatus* put to different concentrations of some heavy metals. *Ind. J. Exp. Biol.*, 17: 274–276.

Saikh, I.S., 1996. Toxic effect of heavy metals on some physiological aspects of crab, *Barytelphus guerini*, *Ph.D. Thesis*, Dr. Babasaheb Ambedkar Marathwada University, Aurangabad.

Sastry, K.V., Gupta, P.K. and Malik, P.V., 1979. A comparative study of effect of acute and chronic treatement of $HgCl_2$ on a teleost fish, *Channa punctatus. Bull. Environ. Contam. and Toxicol.*, 22: 28–34.

Tilak, K.S., Vardhan, K.S. and Kumar, Suman, 2005. The effect of ammonia, nitrite and nitrate on oxygen consumption of fish *Ctenopharyngodon idella* (Valencinnes). *J. Aqua. Biol.*, 20(1): 117–122.

Vutukur, S.S., 2005. Acute effects of hexavalent chromium on survival, oxygen consumption, hematological parameters and some biochemical profiles of the Indian Major Carp, *Labeo rohita*. *Int. J. Environ. Res. Public Health*, 2(3): 456.

Chapter 40

Water Quality of Majalgaon Dam with Special Reference to Zooplankton

☆ *S.B. Ingole, G.A. Kadam, S.R. Naik and G.K. Kulkarni*

Introduction

Limnology is the study of the physical, chemical, geological and biological aspects of all natural freshwater ecology. Whether an animal lives in water or on land its protoplasm hold about 70 per cent to 90 per cent of water. Water is essential to the maintenance of life. It not only forms a major ingredient of living protoplasm but also help several vital processes being carried on in the animals.

Water covers more than 70 per cent of the earth's surface. The high specific heat of fusion,latent heat of evaporation, high surface tension, high density and powerful solvent nature of water plays a significant role in regulation of different activities in organism. It also makes the existence of plankton possible. Being a good solvent water has many chemicals dissolved in it in nature. By utilizing these substances in their various metabolic activities aquatic plants and animals bring about changes in the chemical composition of water. Though freshwater habitats occupy a relatively small portion of the earth surface when compared with other habitats they are extremely important to man as disposal system.

The physical properties of water in any aquatic system are largely regulated by the existing meteorological condition and chemical properties. The effect of physical forces such as light and heat are of great significance as they are solely responsible for certain phenomena, such as thermal stratification, chemical stratification, diurnal and seasonal qualitative and quantitative variation in the plankton, micro and macro organisms and also is the quality of water. The even increasing population and rapid industrial growth in the present era are contributing to a maximum extent in influencing the physico-chemical properties of most water bodies.

Plankton act both as predators and consumers play an important role in transformation of energy from one tropic level to the next highest ultimately leading to fish production which is final product of the aquatic environment. The zooplankton populations in a small water bodies are subjected to extreme fluctuation, the cause of which is not adequately understood even though exhaustive literature on plankton studies are available. The Indian notable contributions to the knowledge of zooplankton are of Arora (1931) and Sewell (1934) who studied on planktonic rotifers. Gouder and Joseph (1961) documented detailed information on copepods. George (1961) has worked on the distribution of zooplankton in pond and lake. Govind (1963) investigated on the relation between copepods and physico-chemical parameters in Tunbgabhadra reservoir. Hospet, Karnataka. Michael (1968) worked on several aspects such as distribution and abundance of zooplankton in different water bodies near Chennai.

At present limnology plays an important role in the decision making processes for problems of dam construction, pollution control, fish enhancement and aquaculture practices. Applied limnology has great scope in healthy existence of natural and man made water bodies and to harvest the natural resources at sustainable level, Goldman and Horn (1983). In order of utilize a freshwater body successfully for fish productions its very important to study a abiotic and biotic factors influencing the biological productivity of the said water body. Such investigation involves physical factors like temperature, rainfall, turbidity, conductivity, total solids, total suspended solids, total dissolved solid etc. chemical factors like pH (Hydrogen ion concentration), dissolved oxygen (DO), Free carbon dioxide (Free CO_2), BOD (Biochemical oxygen demand), COD (Chemical oxygen demand), total hardness, calcium and magnesium. In biological investigation study of micro and macro flora and fauna always provides the clear picture of the ecological relationship existing in the water body.

Hence the present work is an attempt to accumulate information pertaining to various aspect of hydrobiology of standing water bodies from this part of peninsular India. The present investigation has been carried out on 'Majalgaon Dam' located on river Sindphana (Godavari Basin) near 2 Km. U/s from Majalgaon city (Taluka place) of Beed districts in Maharashtra State. Which falls 16° 16⁸ N latitude and longitude 73° 26 E. It is multipurpose type like irrigation and power production (Hydro Electric Project). As a representative of these 'Majalgaon Dam' was selected for the limnology studies. The present study is aimed to investigate some of the important physical and chemical parameters along with the flora and fauna of the reservoir. Similarly by studying the phytoplankton and zooplankton quantitatively to find out what type of exotic fishes can be introduced in the reservoir in future so as to utilize the water body successfully for fish production.

Climatological Condition of Majalgaon Dam Reservoir

The climate of Beed District is on the whole dry except in the south west monsoon season. There are three seasons; Winter (October–January), Summer (February–May), Monsoon (June–September). The cold weather commences towards the end of November when air temperature beginning to fall up to 12 °C in an average and it rises from March and goes up to 39 to 42 °C. May is the hottest month.

Air Temperature

The air temperature values during the two years (June 2001 to May 2002 and June 2002 to May 2003) were given in Table 40.1.

In the year June 2001-May 2002 the lowest air temperature was found to be 21°C in the month of August and December, while highest air temperature (42°C) was observed in the month of May.

In the year June 2002-May 2003 the minimum air temperature was recorded in the month of August, which is found to be 20°C and highest air temperature was recorded in the month of May, which is found to be 41°C.

Rainfall

At Majalgaon Dam the record of the rainfall is kept by the authorities of the rain gauge station at the reservoir. They provide the detailed record of rainfall up to date. The rainfall in the two years *i.e.,* June 2001-May 2002 and June 2002-May 2003 is given in Table 40.2.

In the year June 2001-May 2002 the total rainfall was observed to be 714 mm. The maximum rainfall (263 mm) was observed in the month of August and minimum rainfall in the month of June as 31 mm.

In the year 2002-2003, the total rainfall was found to be 680 mm. The highest rainfall was observed as 262 mm in the month of June, while lowest rainfall was observed in the month of October as 32 mm.

Table 40.1: Monthly Air Temperature at Majalgaon Dam (2001-02, 2002-03)

Months	Temperature (°C)			
	2001-02		2002-03	
	Mini	Maxi	Mini	Maxi
June	28	39	26	37
July	25	30	24	29
Aug	21	28	20	27
Sept	22	29	21	27
Oct	24	27	21	28
Nov	24	30	23	29
Dec	21	30	22	30
Jan	23	28	22	29
Feb	25	34	23	32
Mar	27	36	29	33
Apr	34	40	33	39
May	36	42	37	41

Table 40.2: Monthly Average rainfall (mm) at Majalgaon Dam (2001-02, 2002-03)

Average rainfall (mm)	
2001-02	2002-03
134	262
31	42
263	190
90	154
196	32
–	–
–	–
–	–
–	–
–	–
–	–
–	–

Sampling Sites of Reservoir

The investigations on physical, chemical and biological parameters were carried out from June 2001 to May 2003. The water samples were collected from four selected sites (S_1, S_2, S_3, S_4) from the reservoir. The following aspects were considered while selecting the sampling sites–the shape and size of the reservoir, human activities at different sites and the vegetation and position of inlets and outlets of the reservoir.

The samples were collected from four different sites from the reservoir every month using thoroughly prewashed clean bottles for the analysis. All collections and observations were made between 10: 00 AM and 2: 00 PM throughout the period of the study.

Four sampling sites are as follows:

S_1: Kesapuri camp (near)

S_2: Main Door of Dam

S_3: L. out let canal (near)

S_4: Near Dhorgaon (rehabilitated)

For studying the hydro biological aspects of the reservoir, I have selected four different sites. S_1 site is located on the north side of the dam, near Kesapuri Camp. S_2 site is located on east side 1 km from S_1 site beside the main door of the dam. S_3 site is located on east side near left side of the L. outlet canal which is 1.5 km from S_2 site. S_4 site is located on south side near Dhorgaon, which is rehabilitated and is 1.2 km away from S_3 site.

Plankton Studies on Majalgaon Dam

In any body of water, the amount of animal life is directly proportional to the amount of plant life in it. All major divisions of plants and animals are well represented in aquatic communities. Aquatic habitats fall into three large categories namely freshwater, marine water and estuarine. Method of investigation for each of these habitats are basically similar. Aquatic organisms may be classified according to their form or life habit region or sub habitat or according to their position in the food chain, freshwater habitats may be divided into two categories namely standing water or lentic habitat and running water or lotic without injuring the three tree in tee process. Freshwater habitats occupy only a small percentage of the earth surface. Both plants and animals are well represented in aquatic communities. Algae are the most important producers. Mollusca, Insects, Crustacea and fish are the major consumers. Both bacteria and fungi of the freshwater are of equal importance as decomposers.

The freshwater plankton particularly in lakes nearly all the number of plankton belong to the littoral area or the bottom. These species which have pushed out into the free pelagical zone from the shore and bottom communities are essentially those which are able to cope with the pelagic conditions in virtue of their physiology and body structure.

Since a freshwater habitat is limited in size and enclosed by land it is subjected to a number of diurnal seasonal changes such as chemical and physical stratification depending on the time of the day or year and the depth of the water. Many of these changes are reflected in the vertical distribution of the organism. A study of this distribution of the pelagic organisms will shed light on the biotic and mechanical factors operating in a freshwater habitat. Number of individual of a particular species may very depending on the rate of production level and the rate of depletion, surface tension is responsible for the occurrence of a number extremely microscopic organism above and below the surface of the calm water of a freshwater habitat.

It is difficult to be sure of the factor which determine the vertical distribution of species population. Mechanical factors are more easily obvious than physiological factors. Specific gravity can cause stratification of living and non living bodies generally in the lake or reservoir. Nearly 40 per cent of the plankton communities live at a depth of 1-5 meters. The population being densest at a depth of 7-8 meters.

An ideal method to study plankton will be one which will not only estimate the quantity of living organisms but also the concentration of different species of planktonic forms as well. No single method has yet been devised to satisfy these requirement. It will therefore be best to choose the method most suited to the particular problem that is to be investigation taking into consideration. The area in which the investigation is to be carried out.

A theoretical knowledge of the patchiness of plankton distribution is very important in both qualitative and quantitative studies of plankton. The phytoplankton consisting essentially of algae is found only down to such depth which has sufficient light penetration for photosynthesis. The zooplankton is found at all depth of water since they have power of movement. Which though feeble help them to move up and down. In many of the zooplankton species such vertical movement are regular and rhythmic and occur daily. There migratory forms lives at certain depth during the day and rise to the surface towards the evening sinking back to their normal depth in the morning. Horizontal variation in the distribution of freshwater plankton is caused by the action of local current and wind. The plankton are floating organism whose movement are more or less dependent on currents, while some of the zooplankton exhibit active swimming movement that aid in maintaining vertical positions. Plankton as a whole is unable to more against appreciable current.

Classification of Plankton

Plankton individuals differ greatly in size. Generally animal plankton (zooplankton) are larger, while the plant plankton (phytoplankton) are smaller.

Material and Methods

For the study of limnological to the biological aspect flora and fauna or plankton studies to selected four site at Majalgaon Dam reservoir on the river Sindphana. Samples were collected from different four sites and investigated in the laboratory with the help of latest technical instruments and observe and finally the result are recorded as the every month in the year June 2001-May 2002 and June 2002-May 2003.

The water samples for plankton studies were collected at 10.00 AM to 2.00 PM on first Sunday of every month of two years. Sample collected and preserved in 4 per cent solution of formalin. The quantitative and qualitative analysis were carried out by taking 20 ml of concentrate obtained by siphoning the supernatant liquid. The genera of phytoplankton were identified and quantitative determination was carried out referring Needhan and work of Edmondson. Phytoplanktons were counted by drop count method and the results were converted to organisms per ml of water. The sample was collected from the surface water by filtering 100 liter of water with a planktonic net having a mesh size of 30 micron. While taking the sample, care was taken that water is not disturbed and the samples were transferred into wide mouth bottle and preserved in 4 per cent formalin. For zooplankton analysis, sample of zooplankton were taken into Sedgwick Rafter Cell and identification of zooplankton was carried out and the counting was done following the work of Edmondson (1965), APHA, AWWA and WPCF (1985), Trivedy and Goel (1984), Tonapi (1980), Standard key and other literature were used for identification of different species and the identified species were expressed in no. per liter.

Biodiversity and Fluctuation of Zooplankton

The largest group of zooplankton is the Crustacea. In a zooplankton community the smaller organism tends to be herbivorous where as the largest species tends to be carnivorous. Freshwater zooplankton is represented by fewer phyla and tends to be smaller in size than marine zooplankton. Rotifera constitute a significant proportion of the freshwater zooplankton in many habitats though the crustaceans (especially the Cladocera, Copepoda and Ostracoda) dominate zooplankton distribution changes in abundance both spatially and temporally. Many zooplankton species migrate vertically concentrating near the surface of water at night but at the bottom during the day. The growth rate of zooplankton is slower in colder water and faster in warmer water.

Seasonal qualitative and quantitative fluctuation occurs in the plankton of a pond. In freshwater pond of Uttar Pradesh peak of plankton are reported to occur during spring and monsoon.While slack period is observed during summer months. During peak production, the phytoplanktons were the dominant types, while zooplanktons were predominating during the period of minimum production. A Bio-model pattern of plankton production has been reported in various ponds, the monsoon peak consisting mostly of phytoplankton, while the winter peak consists mainly of zooplankton. There marked seasonal fluctuation in the occurrence and abundance of plankton are due to difference in the physico-chemical condition of the pond.

Zooplankton is an important component of secondary production in an aquatic system. They are an important food source of higher organisms including fishes. The quantitative analysis of zooplankton does not only reflect energy transfer in the system but it also certainly provides the information about the happening in the food cycle.

The zooplankton community of the Majalgaon dam reservoir consisted of four groups. The monthly recorded zooplankton values on the four different sites are shown in the Tables 40.3–40.6 and Figures 40.1–40.4.The groups rotifera, copepoda, cladocera and ostracoda. These groups comprising about 29 species are recorded *i.e.* Rotifera–11, Copepoda–09, Cladocera–07 and Ostracoda–02.

Zooplankton

Rotifers

1. *Brachionus angularis*
2. *Brachionus forticualta*
3. *Brachionus fulcatus*
4. *Keratella troipa*
5. *Filinia longiseta*
6. *Brachionus caudatus*
7. *Kelliocottia* sp.
8. *Notholcaocum minota*
9. *Filinia opoliensis*
10. *Rotaria* sp.
11. *Keratella tourocephala*

Copepoda

1. *Nauplius* sp.
2. *Mesocyclops* sp.
3. *Heliodiaptomus* sp.
4. *Cyclopoid* sp.
5. *Rhinediaptomus indicus*
6. *Neodiaptomus diaphorus*
7. *Senecella calconoides*
8. *Heliodiaptomoas vidunus*
9. *Cyclops bisuspidatus thomasi*

Cladocera

1. *Daphnea pulex*
2. *Ceriodaphina parthenogenetic*
3. *Moina micrura*
4. *Adona longicanda*
5. *Macrothrix* sp.
6. *Diaphnosoma brachyarum*
7. *Bosminopsis deitersi*

Ostracod

1. *Cypris* sp.
2. *Stenocypris* sp.

Table 40.3: Monthly Variation of Zooplankton per liter during Year 2001–2002

Months	Rotifera					Copepoda				
	Sampling Sites					Sampling Sites				
	S_1	S_2	S_3	S_4	Total	S_1	S_2	S_3	S_4	Total
Jun	4	1	3	10	18	3	1	0	10	14
Jul	0	2	0	5	7	2	3	9	4	18
Aug	1	1	7	0	9	1	4	10	1	16
Sept	5	2	1	0	8	6	9	3	3	21
Oct	8	3	2	7	20	10	9	12	7	38
Nov	11	5	7	5	28	8	15	11	14	48
Dec	12	6	1	13	32	20	17	12	8	57
Jan	6	12	0	9	27	13	15	9	12	49
Feb	18	20	10	21	69	11	14	7	3	35
Mar	6	21	17	5	49	10	13	18	16	57
Apr	18	20	10	9	57	20	17	25	11	73
May	15	12	7	21	55	9	5	18	0	32
Total	104	105	65	105	379	113	122	134	89	458
Percentage					27.765					33.553

Table 40.4: Monthly Variation of Zooplankton per liter during Year 2001–2002

Months	Cladocera					Ostracoda				
	Sampling Sites					Sampling Sites				
	S_1	S_2	S_3	S_4	Total	S_1	S_2	S_3	S_4	Total
June	4	0	2	10	16	0	1	3	0	4
July	15	4	6	7	32	5	2	7	0	14
August	1	10	8	2	21	4	1	2	4	11
September	7	12	6	11	36	2	2	0	1	5
October	5	0	3	8	16	3	2	0	3	8
November	8	0	5	10	23	4	5	8	4	21
December	12	9	1	1	23	9	4	11	1	25
January	10	12	16	8	46	10	1	2	10	23
February	9	13	11	6	39	4	2	3	0	9
March	12	20	15	0	47	0	5	1	10	16
April	9	1	9	0	20	12	8	3	7	30
May	5	6	8	10	27	3	7	5	1	16
Total	97	87	88	74	346	56	40	45	41	182
Percentage					25.348					13.33

Table 40.5: Monthly Variation of Zooplankton per liter during Year 2002–2003

Months	Rotifera					Copepoda				
	Sampling Sites					Sampling sites				
	S_1	S_2	S_3	S_4	Total	S_1	S_2	S_3	S_4	Total
June	5	0	7	11	23	1	0	4	1	6
July	12	6	2	1	21	4	0	6	9	19
August	6	3	2	4	15	12	7	6	2	27
September	0	1	5	12	18	3	5	4	0	12
October	1	6	9	7	23	1	9	6	3	19
November	5	0	0	3	8	1	12	14	0	27
December	3	12	1	6	22	5	8	9	5	27
January	5	3	10	12	30	12	7	9	0	28
February	15	12	9	17	53	3	5	12	17	37
March	3	8	13	9	33	10	13	10	7	40
April	6	13	15	16	50	9	12	14	10	45
May	12	3	4	6	25	10	12	17	20	59
Total	73	67	77	104	321	71	90	111	74	346
Percentage					23.143					24.946

Table 40.6: Monthly Variation of Zooplankton per liter during Year 2002–2003

Months	Cladocera					Ostracoda				
	Sampling Sites					Sampling sites				
	S_1	S_2	S_3	S_4	Total	S_1	S_2	S_3	S_4	Total
June	5	6	0	1	12	5	0	1	3	9
July	2	5	7	10	24	0	1	6	1	8
August	12	4	0	15	31	2	5	1	6	14
September	17	18	4	7	46	0	9	12	0	21
October	5	10	8	7	30	10	6	7	7	30
November	6	9	12	0	27	0	3	9	2	14
December	0	15	13	10	38	10	15	6	1	32
January	10	12	7	5	34	19	13	8	3	43
February	1	8	19	10	38	4	8	0	12	24
March	12	0	16	20	48	13	19	10	15	57
April	6	8	15	5	34	0	17	20	11	48
May	3	7	10	5	25	9	6	12	6	33
Total	79	102	111	95	387	72	102	92	67	333
Percentage					27.90					24.00

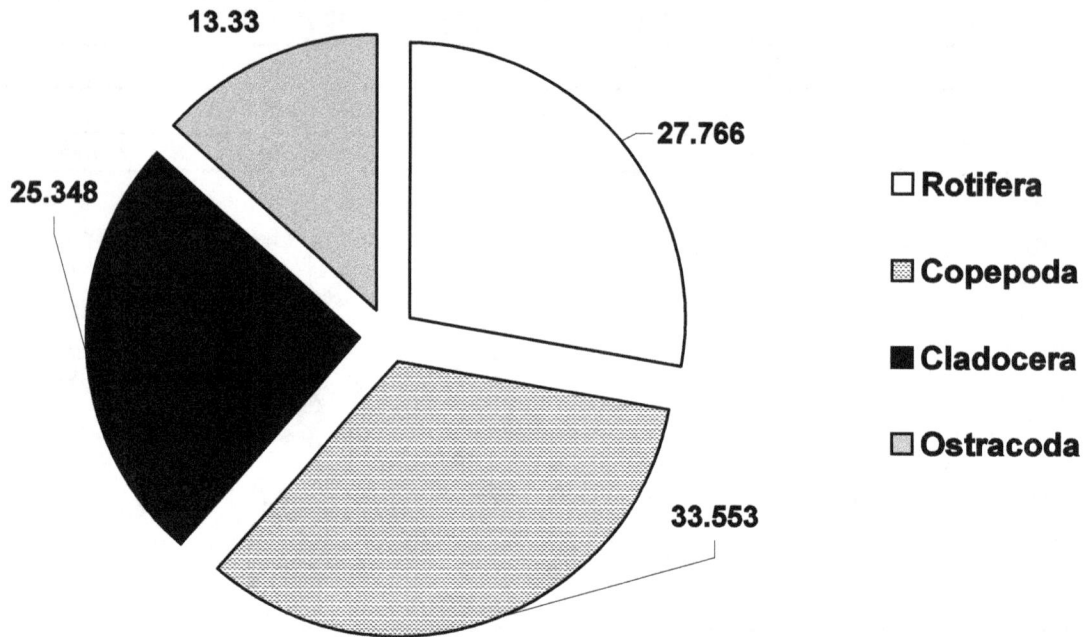

Figure 40.1: Yearly Variation in Zooplankton Percentage
during June 2001–May 2002

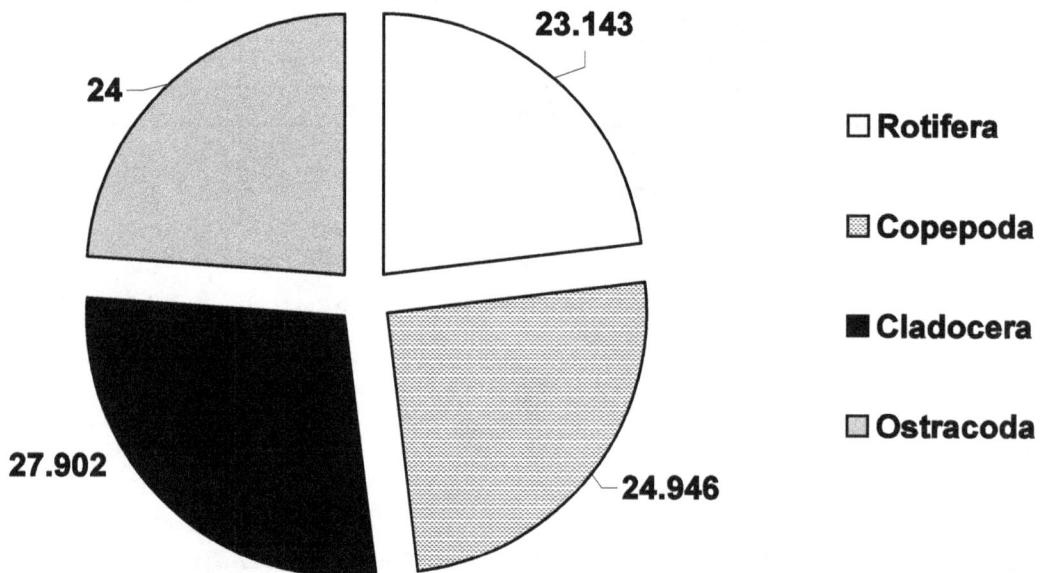

Figure 40.2: Yearly Variation in Zooplankton Percentage
during June 2002–May 2003

Figure 40.3: Monthly Variation in Zooplankton/liter during 2002–2003

□ S1 □ S2 □ S3 □ S4

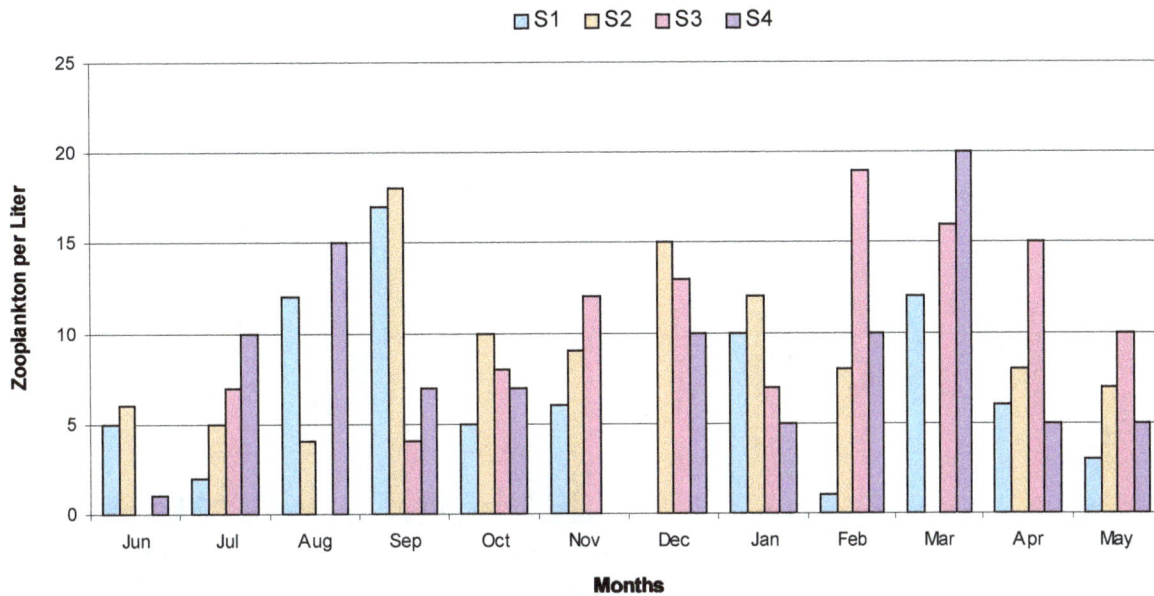

Rotifera

□ S1 □ S2 □ S3 □ S4

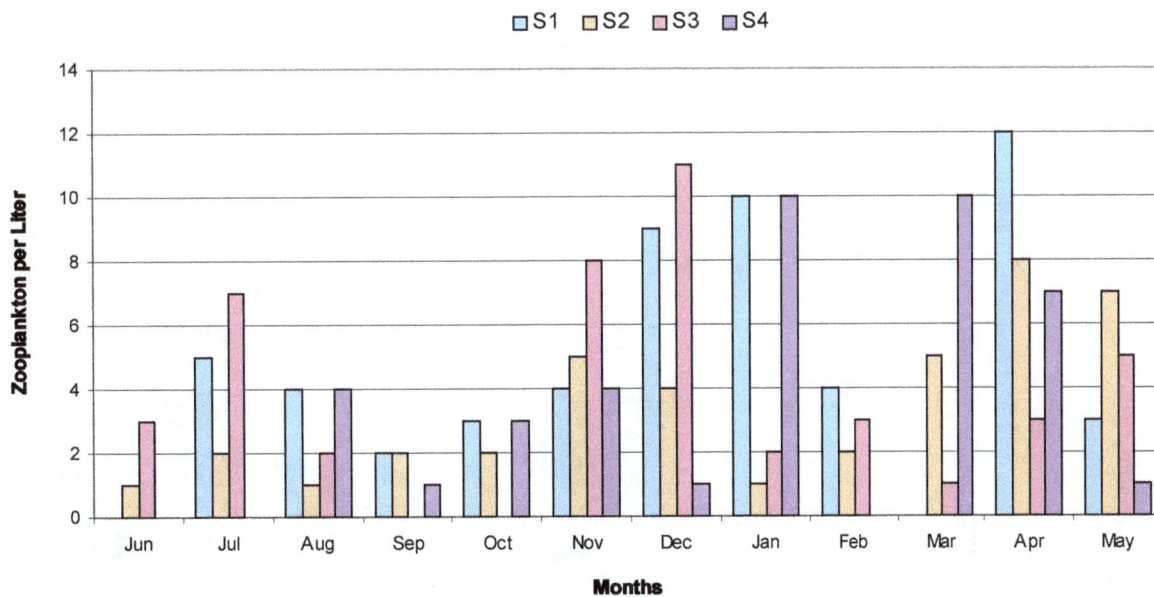

Copepoda

Contd...

Figure 40.3–Contd...

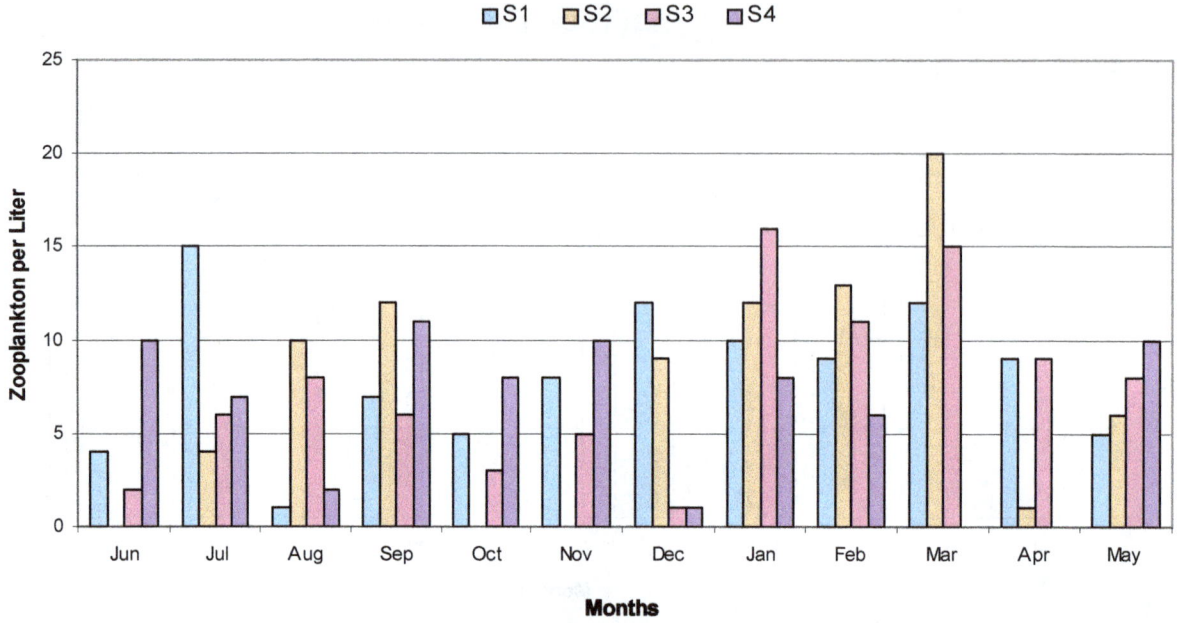

Cladocera

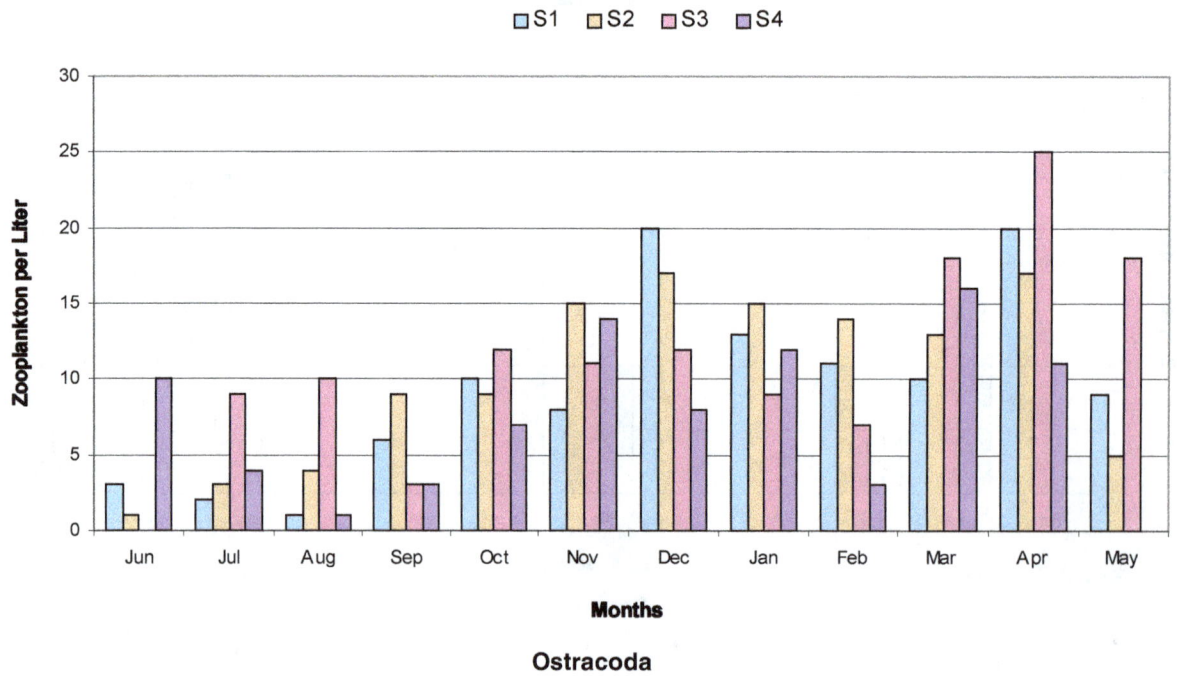

Ostracoda

Figure 40.4: Monthly Variation in Zooplankton/liter during 2001–2002

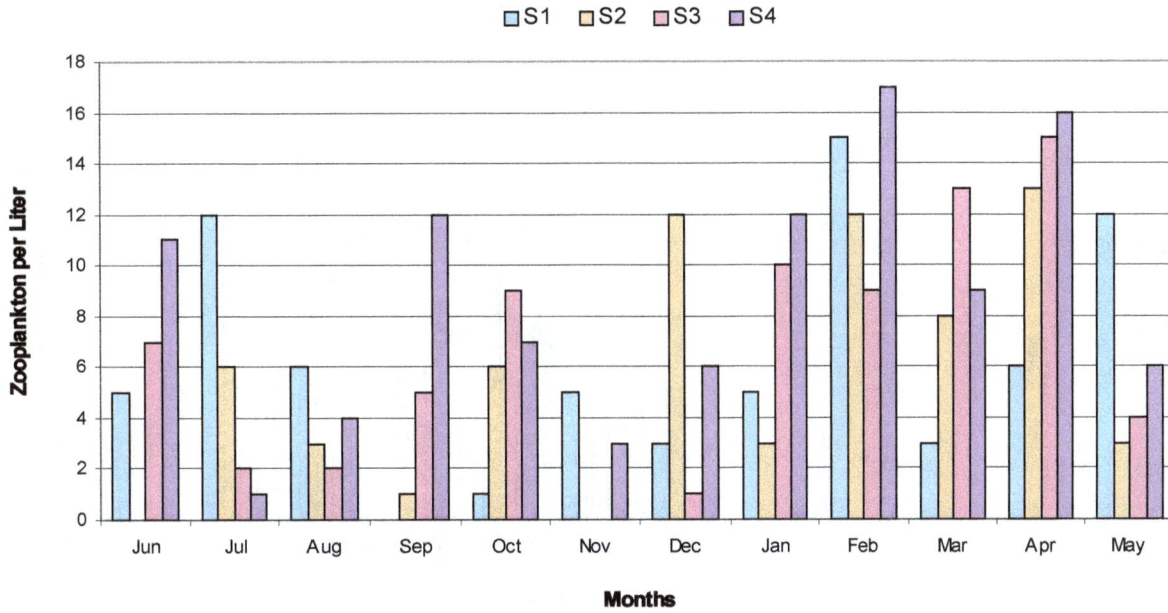

Rotifera

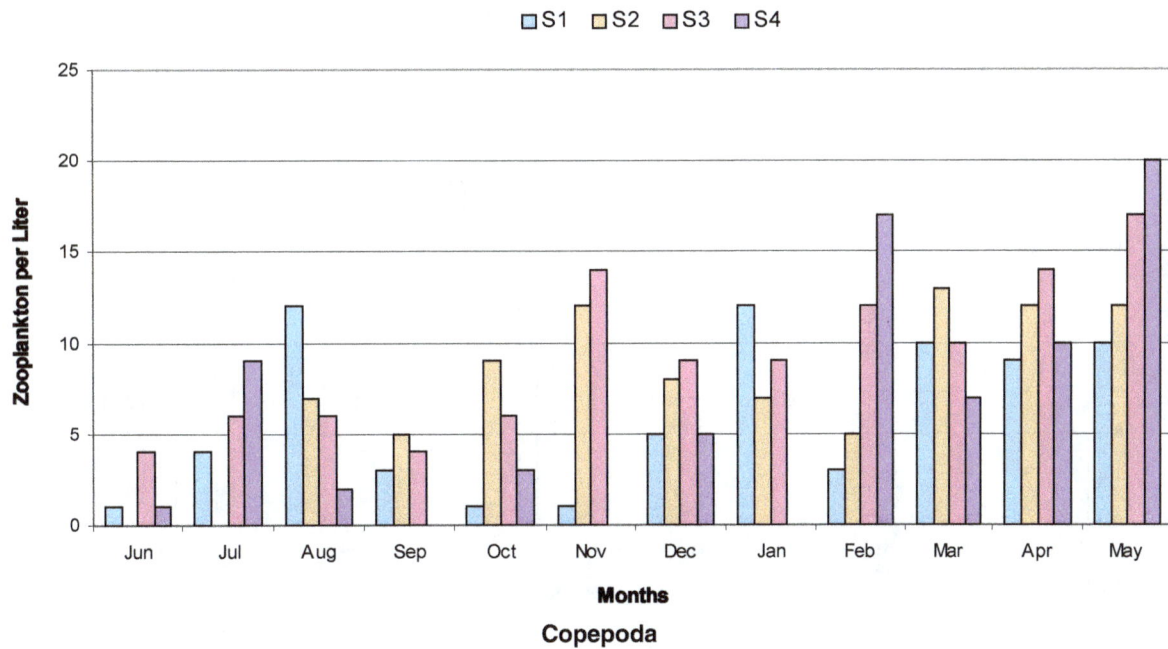

Copepoda

Contd...

Figure 40.4–Contd...

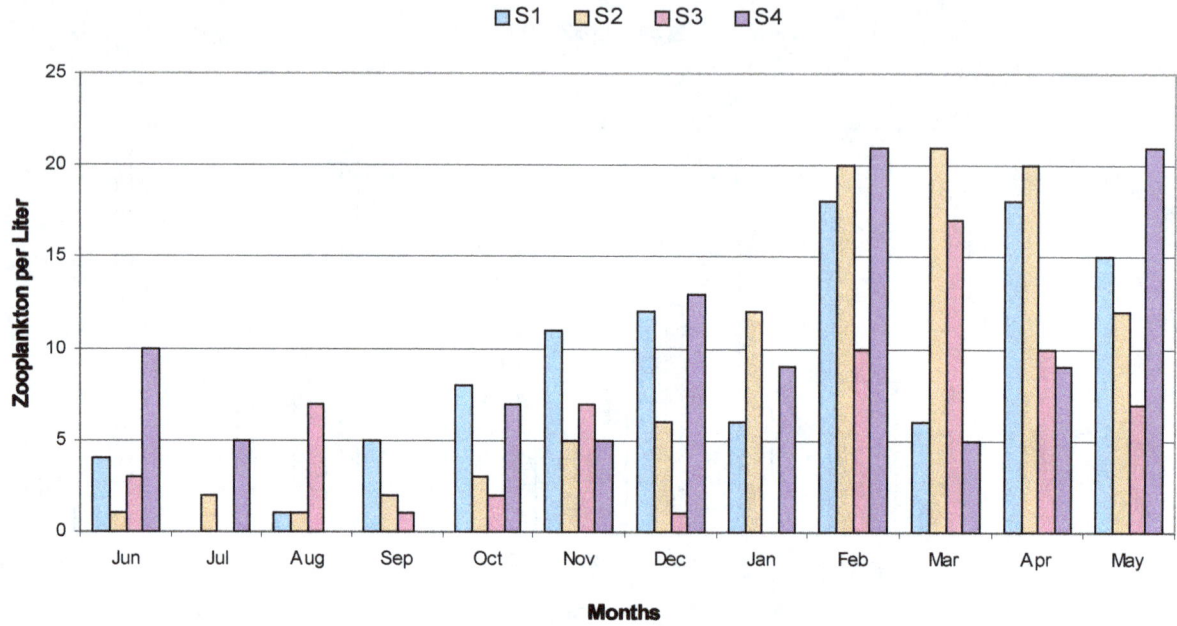

Cladocera

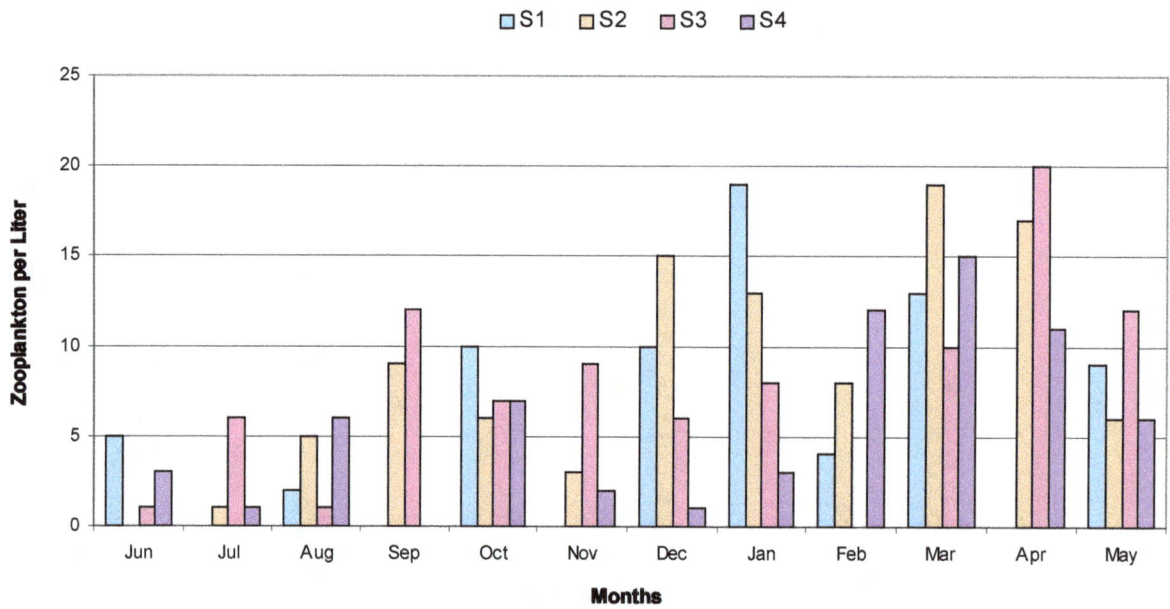

Ostracoda

Rotifers

Rotifer population is very useful in indicating the water quality, particularly in pollution studies. The difference in periodicity and population density of different rotifers species can be analyzed by considering the nutritional ecology and biotic interacters.

Rotifers play an important role as grazes suspension feeder and predators within the zooplankton community. Nanoplankton biomass exerts direct effect on rotifers fertility, while toxicity inhibits their fecundity.

Copepoda

Water temperature and availability of food organism affect the copepoda population. Interrelationship between high populations of the rotifers and cladocera and low population of copepoda during winter may be due to the feeding pressure of stocked fish on the latter and if copepoda are removed, then there is sudden increase in the population of rotifera and cladocera. During winter its biotic interaction operating through feeding pressure rather than water quality seen to affect the zooplankton diversity, particularly the stocked fish species play an important role in harvesting species on copepoda and cladocera there by reducing their predatory pressure on other groups of animals.

Cladocera

The cladocera of the most of the species are primary consumer and feed on microscopic algae and the fine particulate matter in the detritus.Thus influencing cycling of matter and energy in benthos. The cladocera component of zooplankton plays an important role in the benetic throphodynamics.

Quadri and Yousuf (1980) investigated the influence of some physico-chemical factors on the seasonal variations of cladocera. The maximum population of cladocera in winter may be attributed to favorable temperature and availability of abundant food in the form of bacteria, nanoplankton and suspended detritus. Edmonson (1965) reported that during summer and monsoon the parameter like water temperature, dissolved oxygen, turbidity, transparency and conductivity plays an important role in controlling the diversity and density of cladocera. In India therefore the limnetic zooplankton communities are invariably dominated by the species of cladocera.

Ostracoda

Ostracoda species are like the small seed like shape are bivalve. It is coloured species, green coloured species live with algae while gray coloured species live with ooze. The abundance of it provides good food for aquatic organisms. The water temperature and availability of food for organisms may affect the ostracoda population.

Tonapi (1980) reported that the higher population of ostracoda during monsoon might be due to the abundance of fine detritus to which omnivorous organisms food over during monsoon from there natural benthic habit and bacteria mould and algae as food. The decrease in the population of ostracodas during winter and summer may be due to feeding pressure of stocked fishes. Ostracodas population was found to be maximum while other zooplankton components were minimum during monsoon may be due to the dilution effect that population of ostracoda in the month of November.

Discussion

The importance of plankton in fisheries is well established. It has been clearly demonstrated that the zooplankton constitute the only food for the fish fry and the adult fish not only eat them, but also select them as a delectable item. Thus zooplankton have a direct bearing in the fish industry. In India, several studies were conducted in reservoirs elucidating the characteristics of zooplankton (Khan *et al.*, 1990; Adholia and Vyas, 1992 and Sugunan, 1995).

The zooplankton peak was found during summer followed by winter and rainy season. Dominance of Rotifera in seasonal data were reported in the findings of Michael (1968), Saha *et al.* (1971), Pandey *et al.* (1992) and Sarawar and Praveen (1995). The peak value of zooplankton during summer might be due to optimal thermal and nutritional condition and higher concentration of oxygen (Singh, 1991). The lowest zooplankton recorded during winter may be related to low temperature. Marshall and Orr (1972) also observed minimum zooplankton population at lower temperature.

In the Majalgaon Dam reservoir, Rotifera, Cladocera, Copepoda and Ostracoda recorded maximum density and species richness during the pre-monsoon and post-monsoon period respectively. When condition were relatively stable. They recorded low diversity and low species richness during the monsoon period. This concurs with Hawkes (1979) suggestion that low diversity is a reflection of environmental stresses. Margalef (1968) recorded that higher diversity is a clear indication of longer food chains. Evenness index of rotifer species were higher during the pre-monsoon period in the reservoir while that Cladocera species were higher during the post-monsoon period.

References

Adholia, U.N. and Vyas, A., 1992. Correlation between copepods and limnochemistry of Mansarovar reservoir Bhopal. *J. Env. Biol.*, 13(4): 281–290.

Arora, G.L., 1931. Fauna of Lahore II–Entomonstraca water fleas of Lahore. *Bull. Dept. of Zoology, Punjab Uni.*, 1(10): 62–100.

APHA, AWWA and WPCF, 1985. *Standard Methods for the Examination of Water and Wastewater*, 16th Edn. American Public Health Association, Washington D.C.

Edmondon, W.T., 1965. *Freshwater Biology*. John Wiley and Sons Inc., New York.

Goudar, B.Y.M. and Joseph, K.J., 1961. On the correlation between the natural population of freshwater zooplankton, cladocera, copepoda and rotifera and some ecological factors. *J. Karnataka Univ. Sci.*, 6: 89–96.

Govind, B.V., 1963. Preliminary studies on plankton of the Tungabhadra reservoir. *Indian J. Fish.*, 10(1): 148–158.

Goldman, C.R. and Horna, A.J., 1983. *Limnology*. International Student Edition, McGraw Hill International Book Company, Tokyo, Japan.

George, M.G., 1961. Observation on the rotifers from shallow ponds in Delhi. *Current Sci.*, 30(7): 268–269.

Hawkes, H.A., 1979. *Invertebrates as Indicator of River Water Quality*. A. James and Lilian Evison (eds) John Willey and Sons, Great Britain, p. 16–30.

Khan, M.A., Srivasthav, K.P., Dwivedi, R.K., Singh, D.N., Tyagi, R.K. and Melhotra, S.N., 1990. Significance of ecological parameters in fisheries management of a newly impounded reservoir-Bachhara reservoir in contribution to the fisheries of inland open water system in India. Part I. A.G. Jhingran, V.K. Unnithan and Amitabh Ghosh. Inland Fisheries Society of India, Barrackopre, p. 100–108.

Margalef, R., 1968. *Perspectives in Ecological Theory*. University of Chicago Press, Chicago, p. 112.

Marshall, S.M. and Orr, A.P., 1972. The charophytes of the Dal lake Kashmir. *Proc. 21ˢᵗ. Ind. Sci. Congr.,* 295 (Abst).

Michael, R.C., 1968. Studies on the zooplankton of a tropical fish pond, India. *Hydrobiologia,* 32(1–2): 47–68.

Pandy, B.N., Lal, R.N., Mishra, P.K. and Jha, A.K., 1992. Seasonal rhythms in the physico-chemical properties of river Mahananda (Katihar), Bihar. *Env. and Eco.,* 10(2): 354–357.

Qadri, M.Y. and Yousuf, A.R., 1980. Limnological studies on lake Malpur. *Geobios,* 7: 117–119.

Sarwar, S.G. and Praveen, A., 1995. Community structure and population density of zooplankton in two interconnected lakes of Srinagar, Kashmir. *Poll. Res.,* 15(1): 53–58.

Sewell, R.B.S., 1934. Fauna of Chilka lake, Copepoda. *Mem. Ind. Mus.,* 5: 771–857.

Singh, R., 1991. Rotifer fauna of Jamalpur, Munger, Bihar. *J. Curr. Biol. Sci.,* 8: 45–48.

Sugunan, V.V., 1995. *Reservoir Fisheries of India.* FAO Fisheries Technological Paper No. 345. FAO, Rome, p. 423.

Tonapi, G.T., 1980. *Freshwater Animal of India: An Ecological Approach.* Oxford and IBH Publishing Comp., New Delhi, pp. 341.

Trivedy, R.K. and Goel, P.K., 1984. *Chemical and Biological Methods for Water Pollution Studies.* Environmental Publication, Karad (India).

Index